AF535079

Concepts and Techniques in Oil and Gas Exploration

Editors

Kamal C. Jain
Venex Corporation
Houston, TX

and

Rui J. P. deFigueiredo
Rice University
Houston, TX

SOCIETY OF EXPLORATION GEOPHYSICISTS

ISBN 0-931830-22-2

Library of Congress Catalog Number: 82-50079

Society of Exploration Geophysicists
P.O. Box 3098
Tulsa, Oklahoma 74101

Published 1982
Printed in the United States of America

CONTENTS

CONTRIBUTORS

Chapter 1 Synergism in Exploration

M. Ray Thomasson
Spectrum Oil and Gas Company
Houston, Texas

Chapter 2 Exploration Strategy

S. Duff Kerr, Jr.
Kerr, Jain & Associates
Denver, Colorado

Chapter 3 Imaging the Subsurface

Larry C. Wood
America Resource Consultants, Inc.
Houston, Texas

Chapter 4 Modeling — The Forward Method

Ken R. Kelly, Richard M. Alford, and
N. D. Whitmore
Amoco Production Research Company
Tulsa, Oklahoma

Chapter 5 Migration — The Inverse Method

Jeffrey D. Johnson
Amoco Production Company
New Orleans, Louisiana

William S. French
Tensor Geophysical Service Corp.
Metaïrie, Louisiana

Chapter 6 Physical Basis of Well Logging

O. Gomez-Rivero
Petroleos Mexicanos
Mexico City, Mexico

Chapter 7 Well Logs in Exploration

Kamal C. Jain
Venex Corporation
Houston, Texas

Chapter 8 Graphic Data Bases

Ray C. Dillahunty
Digicon Geophysical Corp.
Houston, Texas

Donald W. Townsend
Geosource, Inc.
Houston, Texas

Chapter 9 Technological Impact of Exploration

Norman S. Neidell
Zenith Exploration Corp.
Houston, Texas

Chapter 10 Pattern Recognition Approach to Exploration

Rui J. P. deFigueiredo
Rice University
Houston, Texas

PREFACE

The quest for petroleum is like looking for a needle in five haystacks. To find a prospect, one needs to cull through and synthesize a massive amount of data using developments from such diverse disciplines as geology, geophysics, petrophysics, micropaleontology, geochemistry, and signal estimation and detection. Obtaining the mineral rights for drilling the prospect adds to the difficulty. To top it all, one has only a twenty percent or so probability of discovering a new field in spite of tremendous technical advancements in recent years. The drilling costs are high, ranging from several hundred thousand to millions of dollars per well depending upon its depth and location. If the well is dry, the loss is total. However, when a new field is discovered the rewards are tremendous. Such high stakes have motivated the petroleum industry to pursue innovative ideas vigorously to improve the odds for finding oil and gas.

This book presents an overview of the modern exploration practices. Because of the many disciplines involved, we have not attempted to cover the subject matter comprehensively. Instead, with the help of selected topics, the book provides insight into such major exploration tools as seismic and well log prospecting. Emphasis is generally on the concepts, although several techniques are presented to make the reader familiar with the degree of sophistication and/or ignorance in exploration practices. The book also looks into areas of new development and the importance of unsolved problems in the context of the exploration effort as a whole.

This publication is an outgrowth of a one-day colloquium on "Large Data Systems in Oil and Gas Exploration," which was cosponsored by the IEEE, SEG, and AAPG societies as a part of the 1980 IEEE International Symposium on Circuits and Systems. Several participants in the colloquium, which was a popular interdisciplinary event attended by over 300 people, indicated a need for a general book providing an overview of some of the modern practices in exploration without becoming highly specialized in any one area of exploration. Several speakers at the colloquium accepted the challenge and have made written contributions to this book where emphasis is changed from the "large data systems" in the colloquium to "concepts and techniques" in the book.

The book has been organized for an overall continuity.

The first two chapters cover the geologic basis for the habitats of hydrocarbons and the key parameters characterizing them. The parameters also provide a logical rationale for the use of seismic data and well logs. These have become major exploration tools in recent years following the developments in computer and electronic technologies.

Chapters 3–5 describe the seismic approach. The chapter "Imaging the Subsurface" provides a general background on the reflection seismic methods emphasizing data acquisition and processing techniques. The next two chapters enlarge on two basic elements of interpretation: modeling and migration. The image of the subsurface is seismically constructed by a combination of the forward approach, which is modeling, and the inverse approach where the actual seismic data, generally after migration, are processed to generate a detailed subsurface profile.

Chapters 6 and 7 present the physical basis for well logs and some of their applications in exploration. Well logs can provide a detailed description of the sediments in and around the borehole. The explorationists try to understand the ancient geologic processes including the depositional model by combining and interpolating descriptions from many wells in the general proximity. Such geologic models, when combined with a detailed subsurface profile derived from seismic data, lead to prospects. The actual procedure for locating a prospect is very tedious and complex. An understanding of the information available from well logs and proper use of such information can make prospecting less risky. These chapters provide a conceptual sense of the usefulness of well log data.

The last section in the book, consisting of three chapters, is concerned with the computer and technology issues pertaining to exploration. Chapter 8 discusses computerized data management systems for decision making where data bases from many disciplines are integrated. The next chapter dwells on the impact that the technological innovations in geophysics have had on the science and technology in general. Finally, the last chapter presents some of the basic concepts in the field of pattern recognition and their likely impact on exploration practices.

In preparing this book, we received help from numerous sources which we gratefully acknowledge. Ken Larner (Western Geophysical) orchestrated the publication of the book by SEG as a member of the SEG Executive Committee. Manus Foster (Mobil) and Norman Domenico (Amoco) provided technical liaison with the SEG Publication Staff as chairmen of the SEG Publication Committee. Peter Rose (Telegraph Exploration), Ken Larner (Western Geophysical), and Norman Neidell (Zenith Exploration) provided valuable guidance in initial planning.

The major chore of reviewing the written manuscripts was shared by the contributors to the book. Harry Stephanou (Exxon), Charles Rabe (Raven Banner), and Norman Domenico (Amoco) also reviewed portions of the book in their areas of specialty.

Marie Bone (Venex Corporation) managed the preparation of the book from the idea stage through the final printing by undertaking a variety of organizational tasks over the two-year period. Belynda Bland and Jerry Henry of the SEG Publication Staff did the editing and preparation of the book for final printing.

The Editors

THEME

Chapter 1

SYNERGISM IN EXPLORATION

Introduction

Exploration for oil and gas has witnessed dramatic changes in its nearly 120 year history. Initially, prospects were located by surface shows or seeps; random drilling was predominant. In the early twentieth century, the anticlinal theory became a dominant element in locating traps. Direct mapping of structures by using magnetic, gravity, and seismic data began in the mid 1920s.

In the last 10 years a true revolution has occurred in the use of seismic data in exploration. To a minor degree, the first use of seismic data to locate reefs and carbonate buildups took place in about 1950, but the main era of more quantitative stratigraphic trap detection began in the late 1960s when direct hydrocarbon detection by the so-called "bright spot" concept was first used in the Gulf Coast Cenozoic offshore. In a very short time since then, an increasingly sophisticated seismic mapping approach has swept through the exploration industry. It is now possible to map seismically subsurface stratigraphy, model stratigraphic analogs, and make comparisons of the recorded data with known analogs in order to "read the subsurface."

We are now on the verge of an era of synergism in revolution where exploration techniques are being integrated with reservoir delineation and production engineering methods. Synergism, according to Webster, is "the joint action of agents . . ., which when taken together increase each other's effectiveness." Recognizing this need as a key to success, several companies have begun to integrate the know-how from geology, petrophysics, and reservoir engineering in developing plays such as the Ozona-Sonora gas play in West Texas.

Play Concept

Predicting accurately all the variables in the subsurface requires the solution of an extremely complex equation. This is because there are so many parameters which cannot be scientifically measured adequately ahead of (or, for that matter, after) the drilling bit. Thus, our exploration efforts are designed to reduce the risk of being wrong in our solution of the "subsurface equation."

Risk reduction in exploration can be greatly facilitated by bringing all the necessary technical expertise to bear on the problem. The exploration hexagon in Figure 1 illustrates the interrelation among six broad technologies that can be used to minimize the risk of drilling a dry hole. Frequently, there are insufficient data available

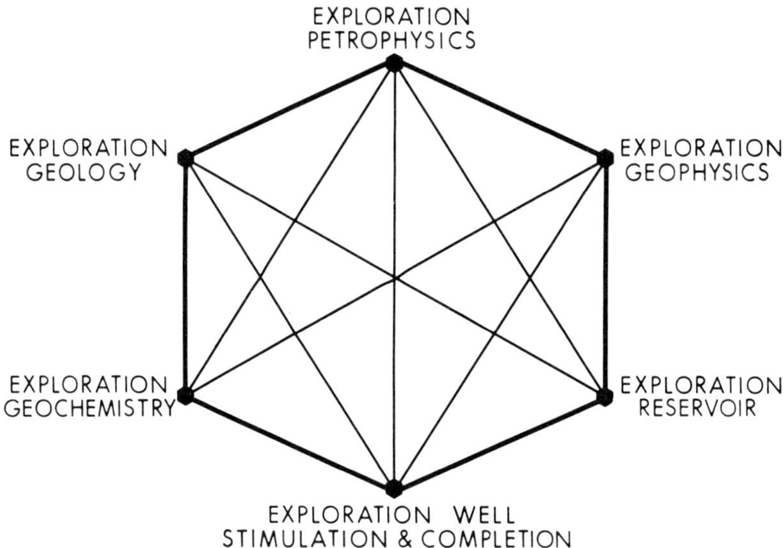

FIG. 1. Map of a play area with several prospects.

to make use of all disciplines and often only one or two create a play.

A play is an exploration activity involving a geographically designated and geologically definable volume of rock in which one or more targets for hydrocarbons can be described (see Figure 2). The geologic strata within a certain geographic limit may include the entire sedimentary section from the surface to Precambrian basement or only a limited stratigraphic horizon or horizons. A prospect involves a relatively small area with possible hydrocarbon traps which are usually stacked vertically and which support drilling in one or more locations. The distinction between play and prospect is one of scale. Normally several prospects can be developed within a play.

In making a play, a company budgets large sums on people, leases, and seismic and subsurface data. The technical staff is usually composed of geologists and geophysicists. A ratio of two interpretive geologists to one interpretive geophysicist is fairly common, although a one-to-one situation may exist in some plays. Whenever necessary, geophysical data processors and seismic crews provide support for the interpretation group. It is also common to call on specialists in petrophysics.

The amount of available subsurface data varies from play to play, but in many areas very large data packages can be assembled and analyzed using advanced interpretation techniques. Many plays involve extensive seismic work for which substantial funds are needed. Very large plays, particularly those offshore, can require seismic budgets of tens of millions of dollars.

Plays extend over relatively large areas. In the early phases of a play, it is not possible to delineate all the prospects. Therefore, it is a common practice to lease blocks of land in the play trend without fully developing prospects. This leasing program is usually directed by leads developed in early stages of the play. A lead

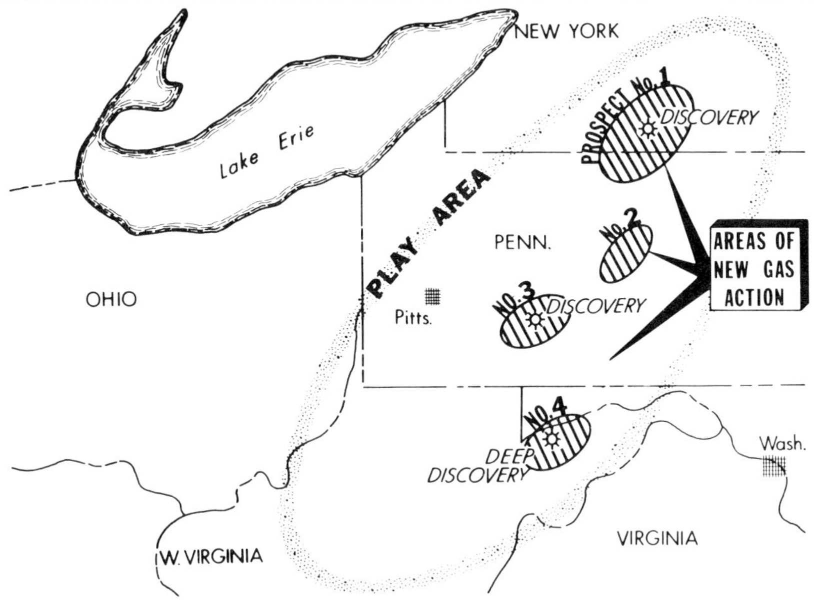

FIG. 2. Undiscovered recoverable resources of crude oil and natural gas for the United States. Reported as a range of values at 95–5 percent probability in billions of barrels for oil and trillions of cubic feet for gas.

is an area for which there are sufficient data to indicate the possibility of a prospect but not enough data to recommend a drill site. It is often considered better to lease without completing all the technical work if the land is sufficiently inexpensive than to risk not having it later in the right places. Thus, very large expenditures are sometimes made for land in the early phases of making a play.

In the plays, besides the requirement of large budgets, there is need for a large variety of specialized technical expertise. Very few companies routinely use petrophysics in play generation, and yet it is a powerful hydrocarbon-finding tool in exploration. Similarly, only a few companies have the resources of a highly technically trained staff capable of using concepts of stratigraphic geophysics to generate subtle prospects and to delineate reservoirs once the hydrocarbon pool has been discovered.

Case Histories

A play may start from any discipline or a combination of disciplines. Examples of recent plays, the technologies used to initiate them, and the technologies that eventually turned out to be the most important are presented in Table 1. The following generalizations may be made from a review of the data in Table 1.

Several plays have been attempted many times; these attempts either failed or the plays were imperfectly executed prior to the blending of necessary disciplines. For instance, the Michigan basin Niagaran reef play had been attempted previously, but it was not until geophysical data processing was able to resolve surface velocity

Table 1. Plays and fields resulting from synergistic exploration

Play area	Play name	Field examples	Main technical expertise used to initiate or reinitiate play	Technical expertise needed to consummate the play
Rocky Mountains				
Eastern Powder River basin	Minnelusa stratigraphic traps	Wolff	Stratigraphy and geophysics	Stratigraphy, structural geology, petrophysics, geophysics
Williston basin	Red River	Mondak	Geophysics	Geophysics, carbonate sedimentation, and stratigraphy
Central Williston basin	Madison (Mission Canyon)	Little Knife	Geophysics	Geophysics, petrophysics, carbonate stratigraphy, structural geology, and well stimulation and completion technology
Western Wyoming, Northwest Utah, Eastern Idaho, and Western Montana	Overthrust belt	Pineview, Whitney Canyon	Geophysical acquisition and processing	Geophysics, petrophysics, structural geology, geochemistry, stratigraphy, well stimulation and completion
Central Alberta basin	Cretaceous sand	Elmworth	Petrophysics and well stimulation	Clastic stratigraphy, petrophysics, well stimulation and completion technology
West Texas				
Eastern Midland basin	Fusselman-updip pinchout	Coahoma	Carbonate stratigraphy and geophysics	Geophysics, petrophysics
Gulf Coast				
Southern and central Texas Gulf Coast	Edwards reef trend	Pawnee, Stuart City	Geophysical data processing and well stimulation	Geophysics, petrophysics, carbonate stratigraphy, well stimulation and completion
Southcentral Mississippi and southern Alabama	Smackover and Norphlet deep sour gas	Thomasville, Piney Woods	Regional stratigraphy and geophysics	Geophysics, petrophysics, stratigraphy, well completion technology
Offshore Tertiary	Bright spots in Miocene, Phiocene, and Pleistocene	Eugene Island, Block 130, etc.	Clastic stratigraphy, petrophysics, geophysical processing	Clastic stratigraphy, petrophysics, geophysical acquisition and processing, structural geology

Table 1. (continued)

Play area	Play name	Field examples	Main technical expertise used to initiate or reinitiate play	Technical expertise needed to consummate the play
Eastern U.S.				
Appalachian basin of West Virginia and Pennsylvania	Shallow sands	Athens	Reservoir engineering, well stimulation and completion technology	Stratigraphy, well stimulation and completion, petrophysics, and reservoir engineering
Basin and Range				
Nevada	Eagle Springs	Trap Springs	Geochemistry, geomorphology, stratigraphy, geophysics data	Stratigraphy, geophysics, igneous petrology
Midcontinent				
Western Missouri	Pennsylvanian heavy oil sands	None designated	Well stimulation and completion technology	Stratigraphy, structural geology, well stimulation and completion technology
Michigan basin	Niagaran reefs	Mayfield, Paradise	Carbonate stratigraphy, geochemistry, geophysical data processing	Geophysical data acquisition and processing, petrophysics, carbonate stratigraphy
Anadarko basin	Springer-Morrow sands in deep basin	West Reydon	Regional stratigraphy and geophysics	Geophysics, stratigraphy, geochemistry, well completion technology
West Coast				
Southern San Juan basin	Cretaceous sands	Trico	Geochemistry, regional stratigraphy, and geophysics	Stratigraphic geophysics, regional stratigraphy, and reservoir fluid dynamics

problems that a combination of carbonate stratigraphy and geophysics could be used to locate Niagaran reefs. Similarly, the Western Overthrust belt play was made several times, but the geophysical data were too poor for a structural resolution of the prospects. Wave-equation migration techniques also helped this significant exploration opportunity.

The Edwards reef trend in Texas is a good example of selecting drilling locations on the basis of stratigraphic geophysics. In this play, favorable carbonate porosity can be defined geophysically within a very long stratigraphic trap. Of equal importance are the advances made in well stimulation techniques.

The Trap Springs field discovery in Nevada is an example of synergism using geochemical understanding to define the generation and migration of hydrocarbons, geomorphologic analysis to locate optimum trapping conditions, and geophysical response to map the field. It is a unique play in that the reservoir in this case is a porous ignimbrite with both matrix and fracture porosity sealed by clastics in a valley fill sequence.

Bright spots have been used in offshore exploration for some 10 years. The techniques of modeling and direct hydrocarbon detection perfected in the offshore Cenozoic of the Gulf Coast have now been used in several onshore plays. For instance, in the Sacramento basin shallow gas sand plays, it is possible to detect high-amplitude gas sands at approximately 5000 ft depth with considerable accuracy.

The Cretaceous sand play in the Central Alberta basin is an example of the use of petrophysics as the dominant technique to make the play possible. Many wells had been drilled through the objective horizon (8000–10,000 ft depth) and shows had been encountered. A petrophysical analysis of these wells allowed the identification of gas fields of major proportion in relatively tight sands. Exploratory drilling has now extended these fields, and many trillions of cubic feet of gas will be produced from this play. Better techniques for stimulation and completion in these tight gas sands made the play successful.

The Smackover-Norphlet sour gas play in Mississippi was based on stratigraphy; individual prospects were defined seismically. But it has required new well drilling completion technology to handle the high geopressures, high temperatures, and highly corrosive (sulfur compounds) conditions encountered at 20,000 ft and greater depths.

The Williston basin has gone through several play cycles. Abundant oil in fractured Mississippian carbonates with some matrix porosity was found in Little Knife field where over 100 million bbl ultimate is thought possible. This field was discovered on the basis of regional seismic maps which defined a regional uplift in the deep mature part of the basin. The Ordovician Red River play, which resulted in Mondak and several other fields, is a classic case of economics coming to the rescue of the geophysical play. The prospect had been on the books of a major oil company for many years. Red River prospects are small in areal size and most require detailed quarter-mile spaced shooting. An average anomaly produces on the order of 1 million bbl. This was dormant until price increases for oil made it a realistic economic target at 11,000 ft. It has turned out to be one of the hottest plays in the Williston basin.

Even though some of the plays listed in Table 1 were initiated by a particular expertise, in every instance a combined effort from several disciplines was necessary to consummate the play. Certainly the more sophisticated plays, like the Basin and Range (Trap Springs), the Overthrust belt, and the Minnelusa, could only be made with the interplay of many disciplines. Also, the expertise involved in initiating plays is quite varied. Thus, an advantage of having a complete range of expertise is that any one of the disciplines can originate a play if it can rely on the "technical core" for backup in the execution phase of the play.

Future Plays

Figure 3 shows the range of undiscovered reserves for all the major provinces within the United States as estimated from the data given in USGS circular 725. Thus, a range of 7 to 19 billion bbl with a mean of 11 billion bbl of oil and a range of 24 to 72 trillion cu ft of gas with a mean of 43 trillion cu ft remain to be discovered in the Rocky Mountains. In the Overthrust belt, the estimated range of 0 to 200 million bbl and 0 to 1.1 trillion cu ft of gas may eventually be exceeded by 50 fold or more.

Figure 4 shows some of the more recent discoveries in the Rocky Mountains and possible volume ranges given by explorationists who found the fields. The Overthrust belt, Trap Springs in Nevada, and Little Knife and Mondak are all plays made successfully in the last 10 years. All these plays involved some surprises for the geologic community at large, and all involved the use of a synergistic approach to exploration.

Statistical projections of estimates of "to be discovered resources," no matter how convincing they may appear, can easily be limited by a lack of creative thinking. The latest, more highly technical, synergistic era may lead to hydrocarbons at almost every location in many basins, and many of these may eventually become economically productive. It would not be surprising if synclines were found to be productive and if stratigraphic traps totally different from any we currently have on our drawing boards were discovered. Tools such as stratigraphic geophysics should lead us to larger volumes of oil and gas, both in more mature provinces and in lightly explored, less mature provinces—certainly to a much greater degree than we currently foresee.

The future exploration opportunity can be diagrammed as a resource triangle, as shown in Figure 5. Most natural resources are distributed as in a triangle; as the triangle expands, the richness of the deposit decreases but the total volume increases. Geologists are familiar with the concept of large volumes of low-grade to small volumes of high-grade ore deposits. However, they are less accustomed to thinking about oil and gas in these terms.

Petroleum reserves can be described by a triangular distribution similar to ore deposits. As reservoir quality deteriorates only slightly, larger and larger volumes of oil and gas are trapped in lower porosity sands. Many very large fields fall into this category, such as Altamont in Utah with 80–100 million bbl ultimate reserves and the San Juan field of New Mexico with 25 trillion cu ft of gas.

Economics will in the future, as it has in the past, play a key role in allowing this type of development. For example, the Milk River gas field, now the largest in Canada with approximately 9 trillion cu ft of recoverable gas, was actually discovered by a Canadian Pacific Railroad well drilled for water in 1883 but was not economically producible until 1973.

We believe that with new exploration and exploitation capabilities, stimulated by price increases, the explorationist will be encouraged to look for lower-grade oil and gas deposits and that there will be an increasing volume of hydrocarbons available for exploitation.

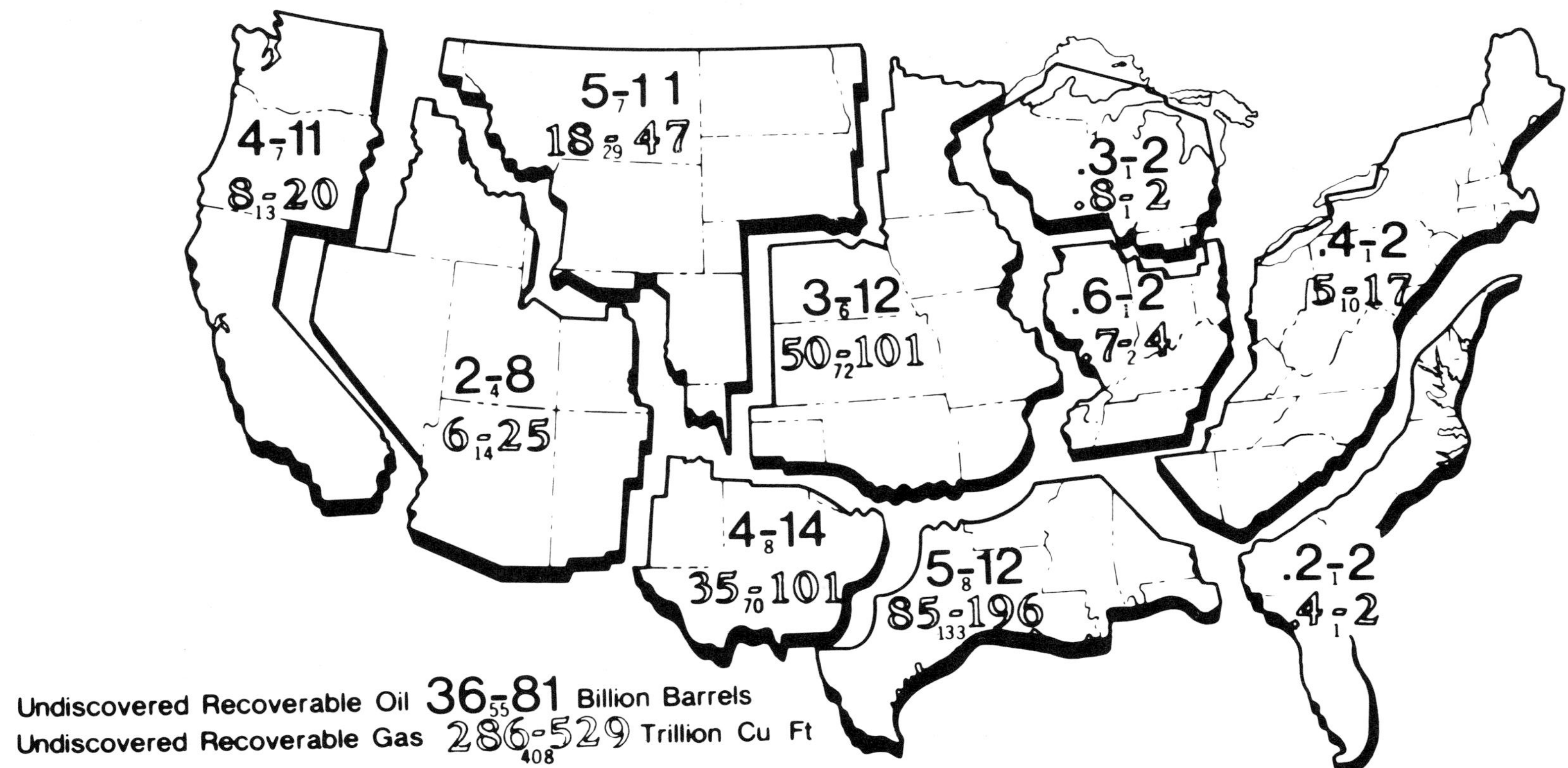

FIG. 3. Estimated range and statistical mean of undiscovered recoverable resources—crude oil and natural gas.

Many plays and prospects can be developed from our current data base by going directly to the hydrocarbons. Thus, manipulation of the massive amount of available petrophysical data synergistically linked to a better understanding of reservoir conditions and stimulation techniques will create new opportunities.

Training for Synergism

Because the exploration of subtle hydrocarbon habitats is becoming an increasingly difficult and complex proposition, synergistic team effort involving diverse skills

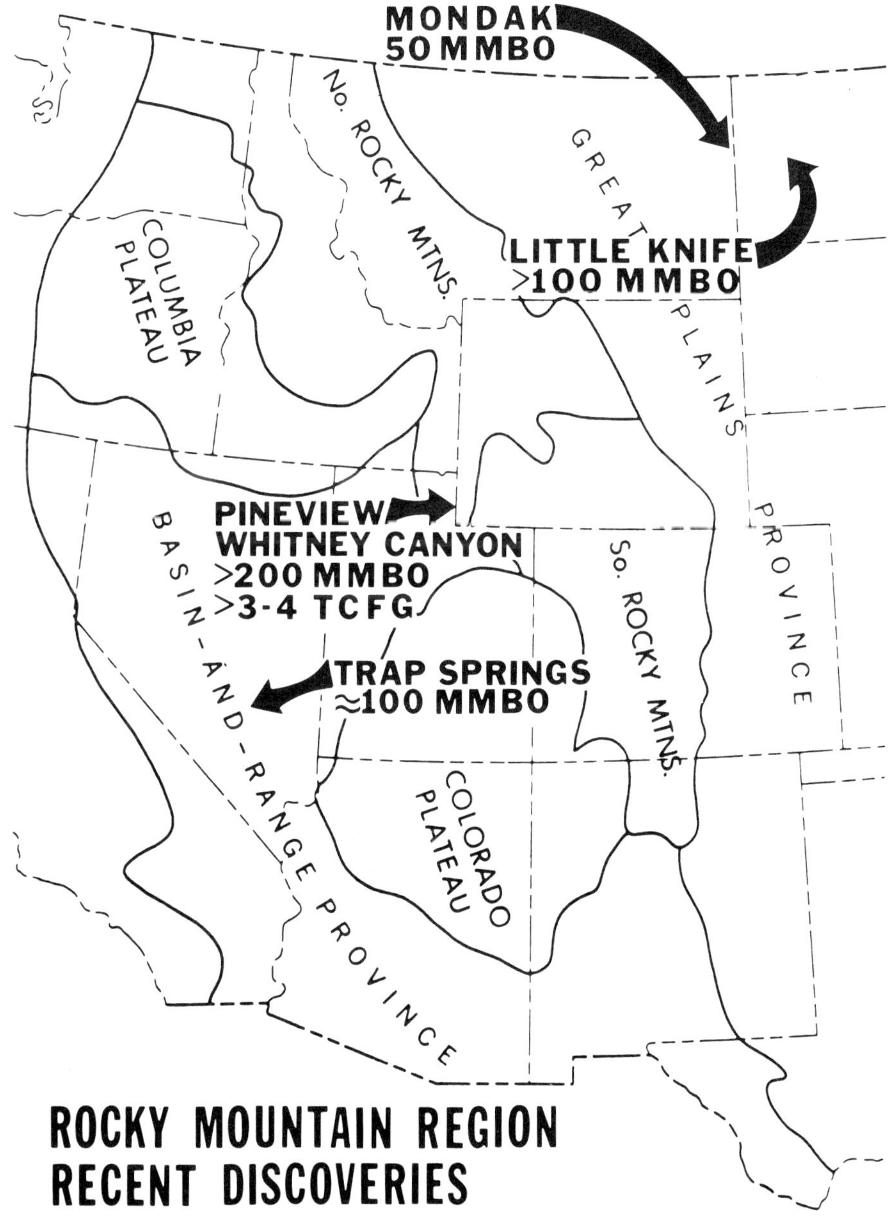

FIG. 4. Recent major discoveries in the Rocky Mountains.

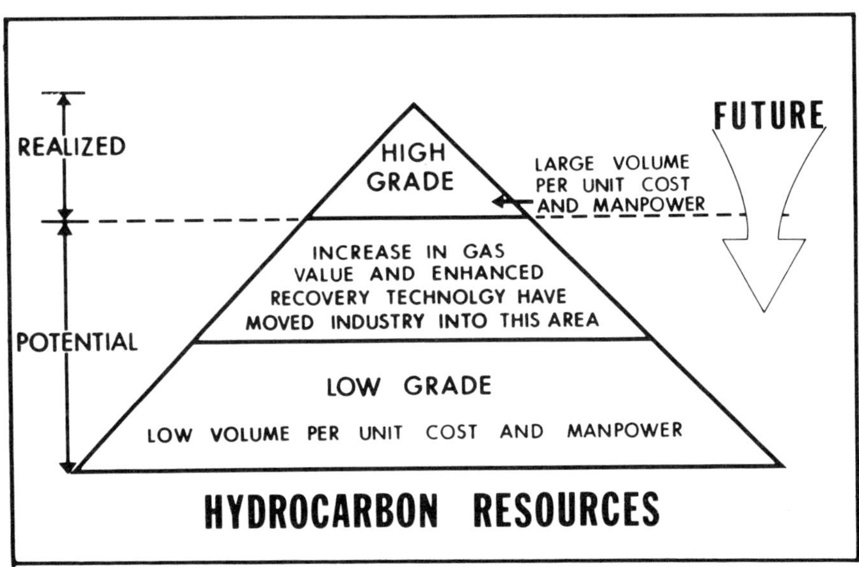

FIG. 5. Resource triangle (cf., Masters, J. A., 1979, Deep basin gas trap, Western Canada: AAPG Bull. v. 63, p. 152–181).

and disciplines is necessary for its successful planning and implementation. To this end, a carefully conceived educational program for the professional explorationist can play a significant role. Such a program should begin within the framework of university education and should be carried on all through the professional years.

The team concept so important to synergistic work is not encouraged in most college curricula; rather, it is often discouraged inadvertently by an emphasis on independent work. At the university level, the education of professional explorationists should be broad based. In addition to the conventional basic subjects, such an educational program ought to provide optional training in related pertinent disciplines such as organic geochemistry, petrophysics, and stratigraphic geophysics. Some exposure of trainees to computer science, signal processing, statistics, and systems theory also appears desirable. One objective of the educational experience should be to develop sufficient overlap in training among specialists so that they can communicate effectively with one another in the context of general exploration objectives. This will promote the synergism discussed here.

Within exploration organizations, recognition of the potential for working in a synergistic mode must take the form of continuing education. For instance, it is critical that the geologist be able to converse with the petrophysicist and geophysicist in a joint effort at making stratigraphic-geophysical plays. In working together, each professional needs to understand how members of the other disciplines think and work. Transfer to the synergistic viewpoint will take time, but eventually a more complete species of explorationist will evolve. This will undoubtedly lead to a manifold increase in success for finding hydrocarbons.

Chapter 2

EXPLORATION STRATEGY

Introduction

Hydrocarbons have been sought in North America for over 120 years. After such a long period of exploration, one might expect that little remains to stimulate further search. The fact is that surprisingly few areas have been totally depleted and abandoned; old areas are being continually reworked and new fields are being discovered in the established provinces. The thrust of new exploration is maintained by the presence of large areas underlain by prospective rocks which have received little or no attention in the past and by the search for new pay horizons, usually deeper, in established producing areas. Thus, the areal extent of oil and gas production, as well as the number of producing wells, is continually increasing. It is important to point out, however, that the average production per well has been declining over a long period because of a decrease in the size of the fields being discovered.

The intensity of hydrocarbon exploration has been cyclic. Prosperous times of increased demand and availability of capital have seen high levels of exploration activity, whereas periods of economic downturn have seen a corresponding decline. With each new cycle, economic incentives coupled with the demand for more petroleum have caused corresponding increases in technical capability for exploration.

Technical improvements have motivated and provided for an extension of exploration programs within both established and newer areas. Today, with the tremendous increase in pressure to meet demands for oil and gas, we are seeing advances in technology beyond those of any previous time in the history of hydrocarbon exploration. These advances also apply to the recovery of discovered hydrocarbons.

Fundamental Requirements for Exploration

There are three basic elements for hydrocarbon accumulation in the subsurface: reservoir, trap, and source. A *reservoir* is a body of rock with pore space capable of containing hydrocarbons. A *trap* is a configuration of rocks which provides isolation of a reservoir in such a manner that escape of hydrocarbons is shut off by the sealing property of overlying and adjacent rocks. Finally, a *source* consists of organic-rich rocks which, under appropriate conditions, generate hydrocarbons that migrate into reservoirs and are preserved there if a trap is developed.

Reservoir

A reservoir is a body of rock containing pore space of proper size which permits the flow of fluids. Sandstones, limestones, and dolomites are the most commonly encountered reservoirs.

Capillary properties of the pore spaces are critical to the character of the rock as a reservoir. That is, the size and shape of pores and the nature of their interconnection affect the capability for transmission of fluids. Pores of a wide variety of sizes exist in rocks; they range from submicron size in chalk to cavern size in carbonate reefs. The critical aspect of size is the interconnecting passages between individual pores through the body of rock. For instance, many sedimentary rocks are composed of sand-sized particles (.06 to 2 mm diameter) deposited by river or ocean currents. Such material in natural packing contains pore spaces between individual particles of very uniform size, and these pores are connected by narrower passages of smaller but uniform size, generally very favorable for passage of fluids (Figure 1). Any variation in the size of rock constituents will change both the distribution of pore sizes and the size of connecting channels, thus profoundly changing the character of the rock as a reservoir (i.e., "transition" in Figure 1).

Trap

The delineation of a trap is the essence of the challenge of exploration. Exploration strategy is to understand the framework of structure and sedimentary strata within a basin well enough to conceive of all the potential trapping situations and to design methods of detecting them.

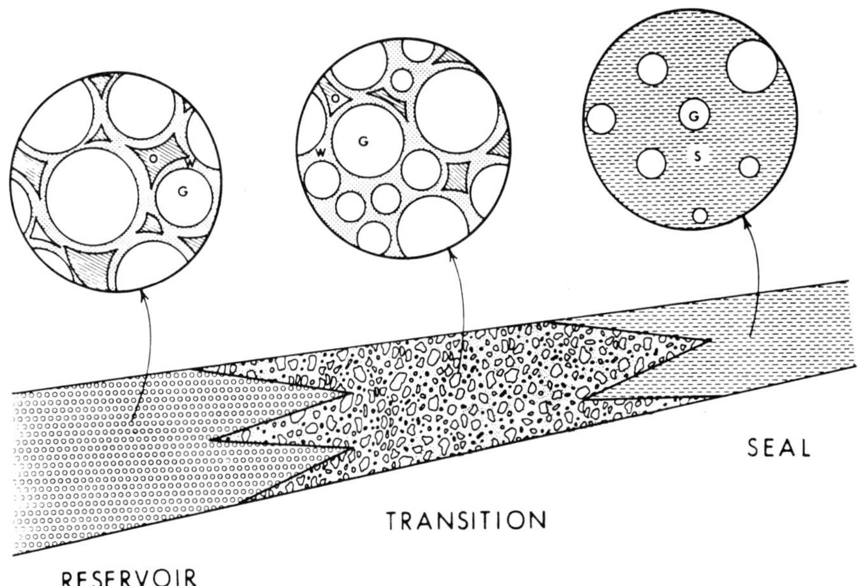

FIG. 1. Reservoir and seal. Uniform pore sizes in reservoir rock allow a uniform distribution of oil throughout the reservoir. Nonuniform pore sizes in transition rock limit oil to larger pores. Pores in seal are all too small to be entered by oil. (G = grain, S = shale matrix, O = oil, W = water).

A trap is formed by discontinuities in the rock framework which juxtapose a reservoir rock against a seal, so that hydrocarbons cannot escape from the reservoir. A seal, in contrast to a reservoir, is a rock in which pore spaces and especially the interconnections are either absent or so small that hydrocarbons cannot pass. Seals are highly variable (Figure 2). They may range from a sediment with some porosity having capillary properties (which would not allow passage of relatively viscous oil, but which might be permeable to water and gas) to a totally nonporous rock such as bedded salt or anhydrite or uniform shale. The latter are so-called perfect seals in that essentially no gases or liquids may pass.

Traps may occur in a great variety of configurations, but they are generally classified as structural or stratigraphic (Figure 3). Stratigraphic traps are formed by changes within a layer or rock resulting from deposition of lenticular bodies of sediment of reservoir character, enclosed by sealing rocks. Structural traps are created by folding or faulting of layered rocks, breaking the normal continuity by juxtaposing a reservoir against a seal. Many traps result from a combination of both structural and stratigraphic configurations.

SEAL EFFECTIVENESS

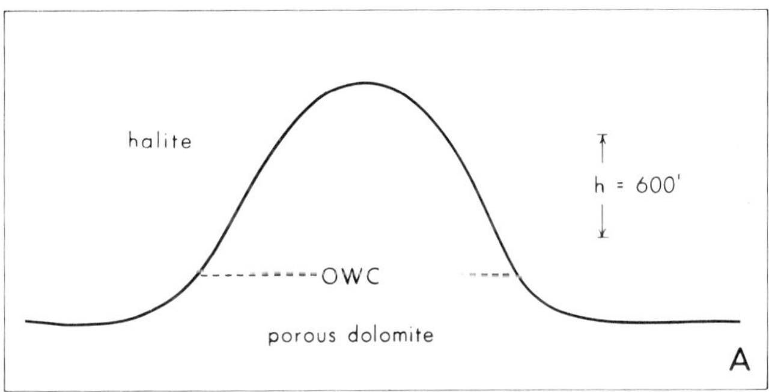

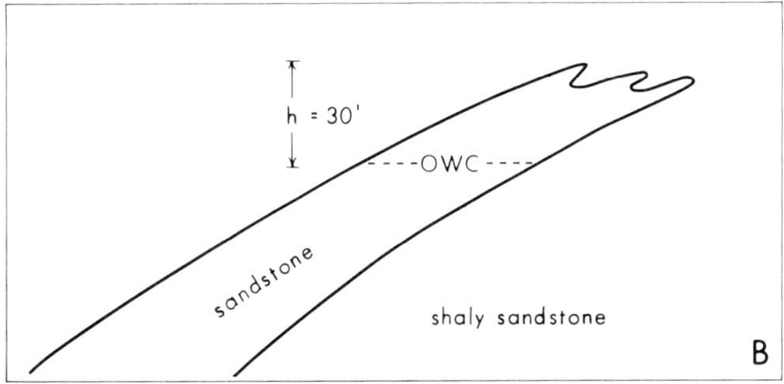

FIG. 2. Seal effectiveness. Impermeable halite will not pass fluids under any condition and thus may trap a very thick oil column, such as in a pinnacle reef. Shaly sandstone may have reduced permeability to oil so that capacity to trap a buoyant column of oil is severely limited.

Source

The question of the ultimate source for hydrocarbons has been the subject of continuous research throughout the history of hydrocarbon exploration. It is only within the past few decades that a better understanding of geochemistry and physics has allowed significant synthesis of the patterns of hydrocarbon generation and migration. It is now generally accepted that hydrocarbons are generated from certain organic-rich rocks in which once-living organic material buried with fine-grained sediments has been preserved from oxidation. An organic-rich rock may then be buried to sufficient depth over an adequate period of time to allow alteration of the organic materials, resulting in hydrocarbon generation (Figure 4). The process of organic metamorphism is largely thermally controlled. The source rock first goes through a sequence of oil generation which is followed at further elevated temperatures, usually associated with increasing depth of burial, by gas generation. In the deepest sediments only gas (methane) is likely to be generated or preserved. The type of organic matter also controls the type of hydrocarbon generated. Only lipid organic materials generate a full range of heavy and light hydrocarbons; humic materials, such as coal, generally produce only light hydrocarbons, principally methane.

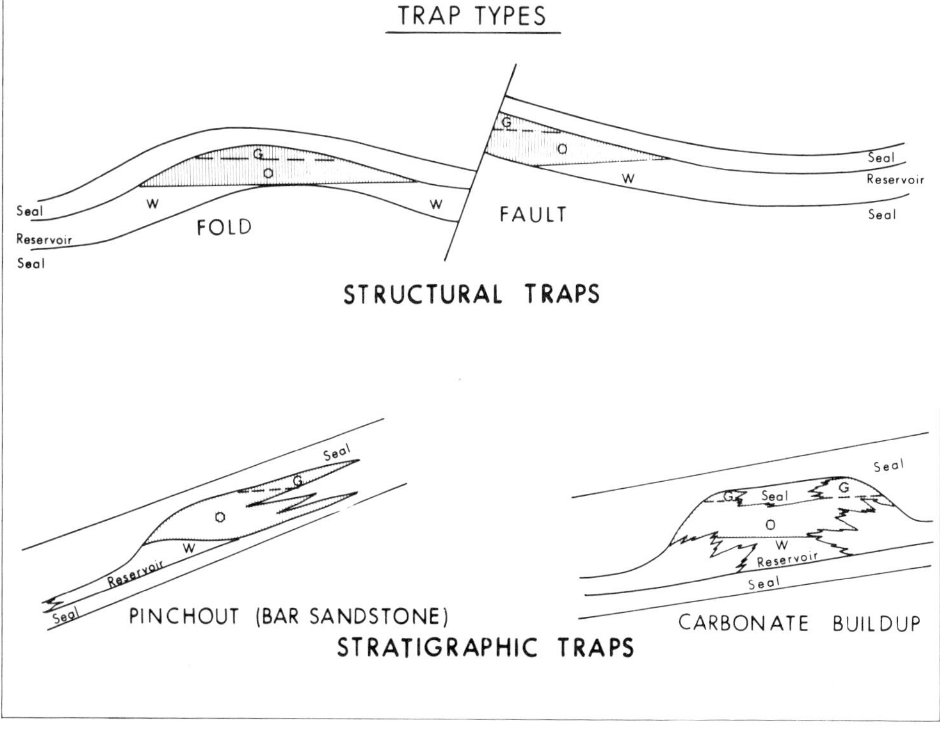

FIG. 3. Trap types. Traps may exist in a nearly infinite variety of associations. Key factors involve a reservoir bounded laterally and vertically by seals in such configuration that buoyant hydrocarbons cannot migrate beyond. Traps are classified as stratigraphic if the configuration is produced by sediment depositional processes, or as structural if produced by later folding or faulting.

An unsolved question, the subject of intense research at present, concerns the precise mechanisms by which hydrocarbons are expelled from source beds and how they migrate to and accumulate in traps.

Exploration

Perhaps the greatest challenge in exploration is that reservoirs, seals, traps, and source rocks cannot be directly observed in the subsurface. Exploration requires indirect means of both detection and measurement of the properties of rocks within the subsurface. Two avenues of approach have been used most often in exploration techniques. The first of these is the development of geologic analogs, that is, understanding of the processes and principles of geology pertaining to the rocks in which hydrocarbons may occur and the framework of basins in which these rocks may exist so that a model or pattern may be developed. Such a model must be based

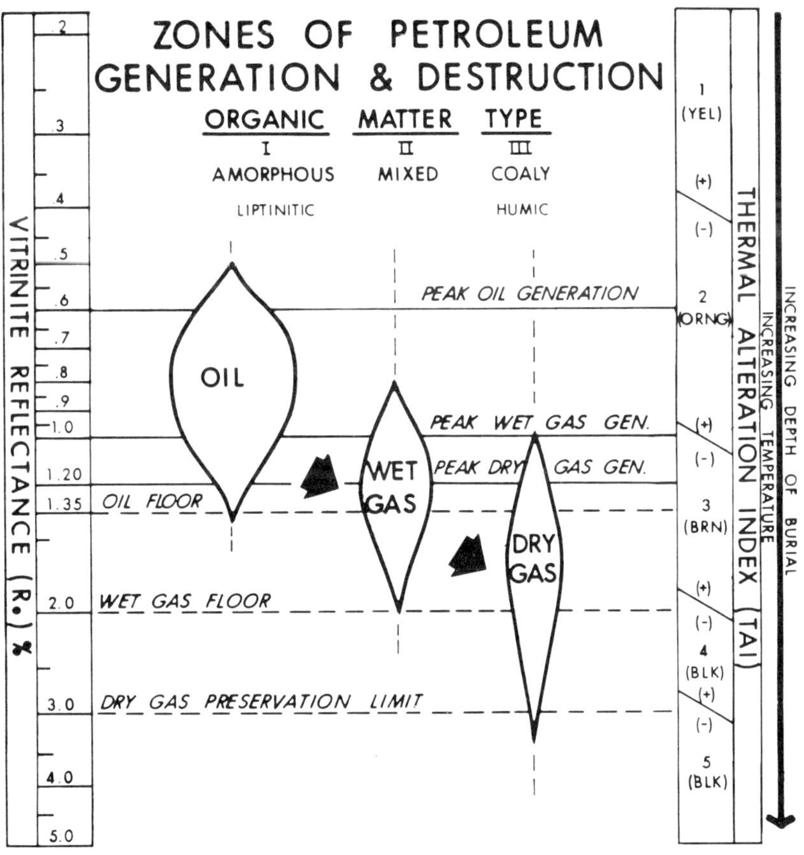

FIG. 4. Hydrocarbon generation occurs when organic-rich sediments are subjected to increasing temperature over a period of time. (Modified from J. M. Hunt, unpublished AAPG Continuing Education lecture series).

on widely scattered points of information. The other thrust has been to develop various means of measuring the properties of rocks within the subsurface; thus the application of geophysical techniques has evolved.

Logic and practice have determined that an interplay between the above two approaches is an essential element in successful exploration problem solving. Traditionally, geologists have utilized some sort of a geologic concept or model to predict traps located between limited available control points. Geophysicists pursued indirect measurement techniques to develop pictures of the subsurface. Giving both geologists and geophysicists access to the other's results and combining these separate means of investigation have allowed substantial gains in the practice of exploration during the last decade.

The process of indirect measurement is the subject of several other chapters; therefore, in this chapter emphasis will be on the concepts of geologic models and the approach to exploration strategy as developed through geologic research.

Stratigraphic Exploration

Of the many subjects of exploration research during the past two decades, possibly the most fruitful and productive in increasing understanding has been on the deposition of sediments and their alteration (diagenesis) as they are converted into rocks. About 25 years ago in many industrial geologic research centers, a concerted effort was given to the study of modern sedimentary environments. The goal was to gain enough knowledge to predict accurately the distribution of potential reservoir rocks in the subsurface.

Sedimentary rocks generally exist as layers, and these layers can be traced from place to place in the earth. In outcrops the layers may be recognized by various characteristics such as color, hardness, weathering, composition, etc. In the subsurface, well logs may be used to define the various layers and trace them between wells by correlating similar physical properties.

Through time, the layers are deposited one upon the other, forming the stratigraphic sequence in which reservoirs, seals, and source rocks occur. Each stratigraphic layer is composed of a variety of materials deposited by different sedimentary processes. In order to explore intelligently, it is desirable to identify layers and geographic areas in which reservoir, seal, and source rocks are most probable. Therefore, stratigraphic research programs were directed toward understanding the processes by which various sediments were deposited.

Sedimentary rocks are generally divided into three categories: clastic, carbonate, and chemical sediments. *Clastic* sediments, such as sands and clays, consist of particles derived from weathering and breakdown of older rocks, which are transported by currents from their place of origin to a (usually distant) site of deposition. These sediments are generally composed of resistant minerals such as quartz or clays. *Carbonate* sediments are composed originally of calcium carbonate and are mostly derived from the hard parts of a wide variety of organisms. *Chemical* sediments include evaporite minerals, usually composed of sodium chloride (rock salt) or calcium sulphate (anhydrite or gypsum), which may be directly precipitated from water concentrated by evaporation.

Clastic Sedimentary Studies

Sandstones are recognized as one of the most prolific and widespread reservoir rocks, based upon data from existing oil and gas fields. Most sandstones are dis-

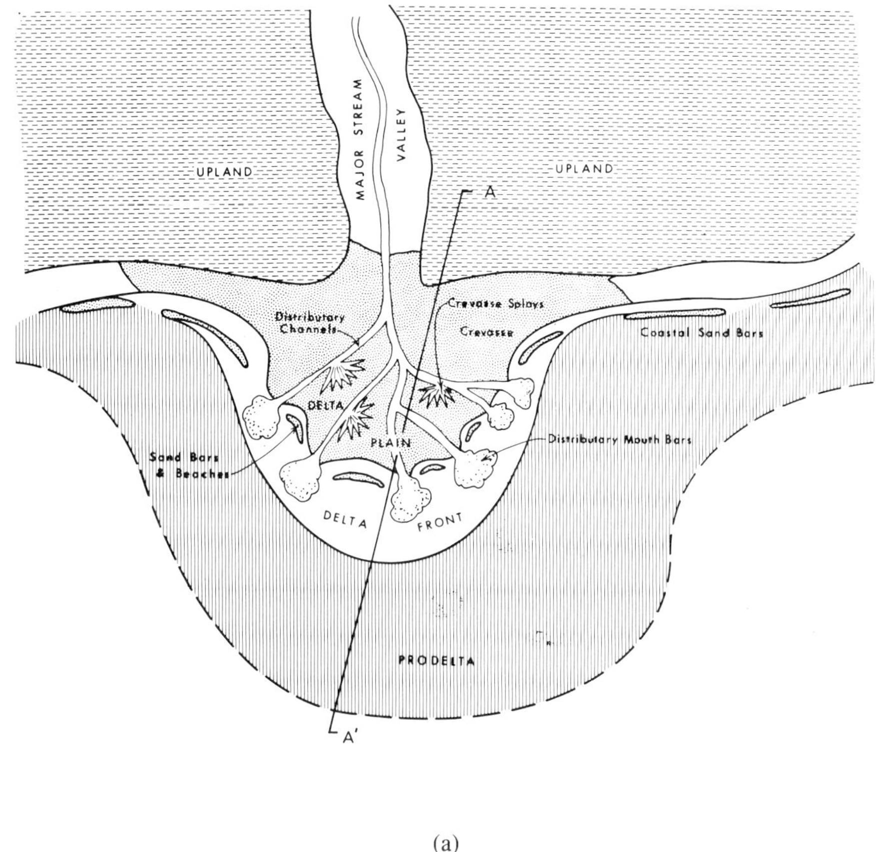

(a)

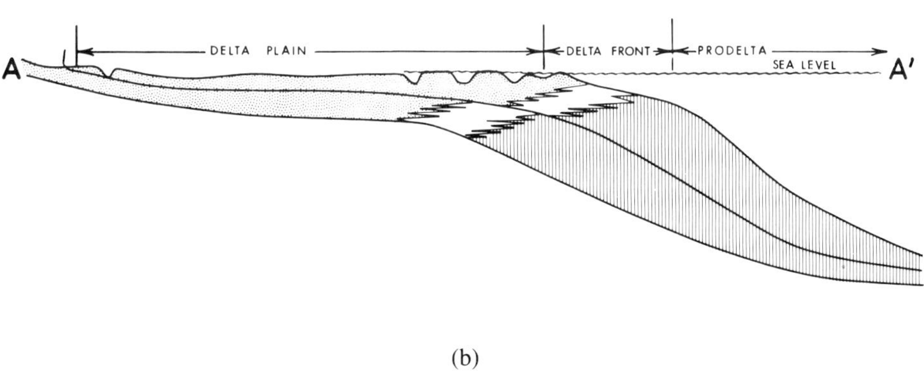

(b)

FIG. 5. Deltaic sedimentary model. (a) Potentially permeable reservoir quality sandstones are present in distributary channels and in bars and beaches in the marine fringe. (b) In section (A–A′) succeeding layers develop as clastic material is deposited over a period of geologic time.

tributed in lenticular bodies, that is, showing an oval or elongated outline in plan view. Some of these patterns resemble those of modern stream deposits, whereas others resemble shoreline sand accumulations. Major river systems collect detritus from land areas and transport it to the margin of lakes and seas, where it is deposited in the form of deltas (Figure 5). If one could understand the distribution of clean, well-sorted sand bodies within modern deltaic environments, in contrast with the clays and finer grained sediments that exist within the same depositional areas, models could be developed which might be useful in predicting the position of reservoirs and seals between subsurface control points. Initial extensive studies were made in the delta of the Mississippi River along the northern Gulf of Mexico. These studies developed the deltaic model in considerable detail and allowed a number of generalizations which could be extrapolated to many other sedimentary settings.

Similar studies followed, covering other types of sandy accumulations. It was recognized that many sandstone bodies occur in linear trends along shorelines in the form of barrier islands which contain associations of shoreface, dune, and lagoonal sediments (Figure 6). Identification of particular kinds of sandstones deposited in these environments and their recognition in the subsurface would allow application of a sedimentary model as a guide in exploration. At the same time, research was carried out on sands deposited in other environments, such as deep ocean basins, where sediments may be transported by density currents. Some of these currents are generated by catastrophic events such as earthquakes, while others result from the continuous flow of sediment into deeper water from shallow coastal environments such as river mouths on rapidly prograding deltas.

In each of these examples, the process of analysis involved a careful study and description of one or a small group of modern sediment units identified by their surficial distinctiveness. A three-dimensional description was possible by studying shallow borings and trenches. After sufficient study and description, generalizations of the patterns of these modern sedimentary bodies were possible. Next, an attempt was made to find an ancient example, preferably the reservoir in a producing oil or gas field or a well-exposed surface outcrop. Careful study of all the data available from such an oil field or outcrop and its comparison with a modern sedimentary model allowed the development of a set of criteria for recognizing the sedimentary environments in ancient rocks and, in particular, the use of subsurface tools in making these identifications. Once these modern and ancient analogs or models were fully developed, it became possible to take widely scattered subsurface well control and reconstruct patterns based on one or more of these depositional models. Identification of a particular depositional model in a subsurface borehole and its position with respect to other models of sandstone deposition provides a means of extrapolating between the limited number of borehole control points generally available in exploration areas. Understanding a wide variety of sedimentary environments makes a regional stratigraphic analysis possible, delineating beds which have some hydrocarbon trapping potential.

Carbonate Stratigraphic Studies

Initial successes in studying clastic sedimentary environments and recognition of such studies' usefulness in exploration for stratigraphic traps led to more aggressive research programs. One of the first extensions was to investigate the other major group of potential reservoir rocks, that is, the carbonate realm of limestones and dolomites.

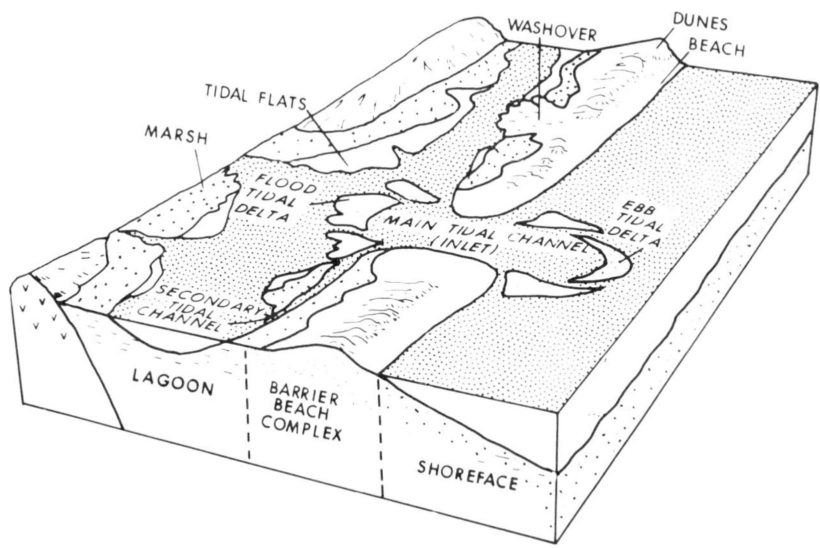

Block diagram illustrating the various subenvironments, in a barrier–island system. (from Reinson, 1979)

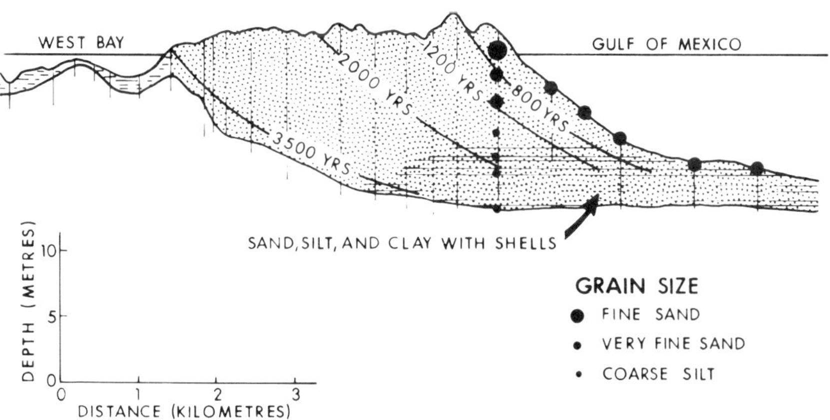

Cross section of the prograding Galveston barrier island, (from Bernard et al., 1962).

FIG. 6. Barrier island sedimentary model. A narrow, linear body of sandstone with potential for continuity over great distances parallel to the shoreline.

It was soon recognized that there are significant differences in the basic controls of deposition of clastic and carbonate sediments. The principal difference is that clastic sediments are generated by erosional processes, removed from their source, and transported by some process to a site where they are deposited. Carbonate sediments, on the other hand, are derived from the life processes of various organisms, and they accumulate generally at or near places where the organisms lived and died. The sediments may be molded into particular shapes and local patterns by means of current action, but production and accumulation are generally processes of the life and death of organisms. Therefore, the study of carbonate sedimentary environments involves the recognition of places where carbonate particles have been produced and related places of preferred accumulation.

The basic approach of stratigraphic research to modern carbonate sediments was quite similar to that for clastic sediments. A distinctive body of modern sediments was examined and described, and an attempt was made to generalize the basic features of the deposit. One of the reasons that carbonate research lagged so greatly in its initial phase was the difficulty in recognizing and delimiting particular areas of highest productivity of carbonate sediments. It was long recognized that great volumes of carbonate sediments are produced in features such as reefs in modern seas; the growth of coral and related organisms in building great masses of carbonates was evident. However, concurrent studies of ancient carbonate rocks based upon both surface outcrops and subsurface borehole data indicated that massive coral reefs were not common in the geologic record. Other organisms, various ones at different geologic periods, filled the ecological niche presently dominated by corals. It was eventually recognized that carbonate sediments evolved through time in response to the evolution of organisms themselves; paleobiology became an important adjunct to much of this early research.

Sufficient data have been accumulated from stratigraphic studies to allow generalization on a fairly broad level in carbonate sediments similar to clastic sediments. Two major depositional patterns (Figure 7) are recognized in carbonates. One is that of the ramp where the sea bottom extends from the shoreline into deeper water with a more or less regular continuity of increasing depth of water. The other is the platform wherein shallow water covers an extensive, relatively low-relief surface upon which carbonate sediments accumulate. The edge of this platform facing the open sea is characterized by sharp relief across which water depths increase rapidly and from which there is a rapid descent of the depositional surface into deeper water. Since carbonate sediments accumulate most rapidly where there is the greatest current activity, each of these two models has a preferential area of rapid sedimentation. In the ramp model, this preferred area of most rapid sedimentation occurs near the shoreline and results in sedimentary patterns that are curvilinear, parallel to the shoreline. In the platform model, the preferred area for rapid sedimentation is at the break in slope which occurs at the edge of the platform facing relatively deeper water. This is the area in which organic reefs may develop and where accumulation rates are greatest. This preferential accumulation of sediment may construct mound-like features with distinct relief of considerable thickness. This ability to create sedimentary packages with a distinctive vertical geometry is unique to carbonate rocks and it plays an important role in exploration programs.

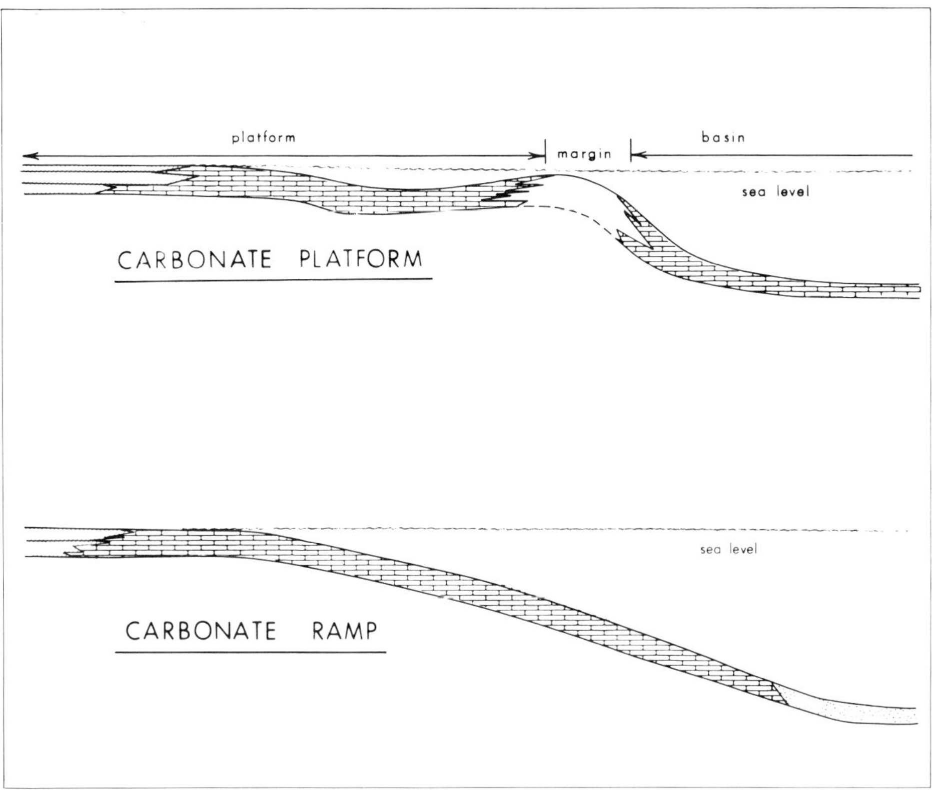

FIG. 7. Carbonate depositional patterns. In the platform model, reservoir development may occur along the margin trend and also near the inner limit of marine sediments. Either part of the model may have great linear extent. In the ramp model, reservoirs are concentrated in a relatively narrow belt near the shoreline.

Source Rocks Studies

The major thrust of research on source rocks and the generation and migration of hydrocarbons has been through the field of organic geochemistry. Widespread application of geochemical principles in petroleum exploration has only recently been utilized.

Tissot and Welte (1978) presented a very thorough review of the geochemistry of hydrocarbon source rocks and some of the geologic consequences and controls on hydrocarbons in the earth. There are several stages in the history of organic accumulations:

(1) Generation and accumulation of organic material in the sedimentary basin,
(2) preservation of the organic material prior to burial by succeeding sediment layers,
(3) a pattern of normally increasing temperatures during the burial history in a basin which causes the alteration of organic material into oil or gas (maturation), and
(4) expulsion of oil or gas from a mature source rock and migration into a trap.

Certain environments are more favorable for the accumulation of organic material, because it must be protected against destruction by oxidation or other degradation prior to burial. Therefore, localities having high rates of circulation and turbulence in the depositional process are unfavorable for preservation of organic material, but they are favorable locations for deposition of coarse-grained sediments which form potential reservoir rocks. Sites favorable for preservation of organic material include the deep portions of lakes or seas and isolated lagoons or swamps on their margins, where depositional energy is low and fine-grained sediments may accumulate.

Different types of organic materials have varying potential for generation of gas on one hand or oil on the other. These were divided into three classes by Tissot and Welte (1978) reflecting varying content of lipid or humic materials (Figure 4). Generally, lipid materials will generate oil and humic materials will generate gas. Class I is dominantly lipid in content and reflects organic material generated in special lacustrine or marine environments principally from algal sources. This includes such rocks as the Green River oil shale. Class II reflects many marine environments where a wide variety of sources for organic material are present. Class III is dominantly humic and is typified by coaly material generated in marginal to nonmarine environments. Important coal-bearing sequences, such as the Pennsylvanian and Upper Cretaceous of North America, should be included here.

The thermal history of a given organic sediment is critical to assessing the stage of maturation of hydrocarbons at the present time (Figure 4). Maturation is a function of rates of burial from time of deposition to the present and the geothermal gradients through which the sediment has passed. Several techniques of measurement, both quantitative (vitrinite reflectance) and qualitative (spore, pollen, and conodont coloration), have been developed to measure the present state of maturation.

Geologic assessments of the hydrocarbon potential of a basin must include determination of (1) stratigraphic and geographic location of organic-rich sediments and the type of material present, (2) the degree of their maturation, and (3) possible paths of migration from source to reservoir. With such data, predictions may be made as to volumes and kinds of hydrocarbons likely to be present and the most probable areas in which traps may be charged. Current research includes the development of predictive models with computer simulation of basin history using these and other factors (Welte and Yukler, 1981).

Basin Analysis

Application of the understanding of stratigraphic relationships derived from geologic research is very important to the design of an exploration program. Modern exploration strategy frequently utilizes the results of this knowledge to predict stratigraphic patterns within a basin as part of the evaluation of its exploration potential; this technique is called *basin analysis*. By gathering together all related data from subsurface and surface control points within a basin, sedimentary units may be subdivided. This process consists of either erecting a framework of time correlation based upon paleontologic data, simply recognizing major rock units, or preferably combining both methods. Once the stratigraphic sequence has been divided into major layers of a workable nature, then each of these units may be studied to provide an assessment of the relationship of the various rock facies pertinent to reservoir, trap, and source within each of these layers.

Each basin has its own unique characteristics, and the amount of geologic information available also varies from basin to basin. Many of the basins in North America have been extensively explored, and hence large quantities of data available within these basins must be reduced to a manageable level. In other parts of the world and in some frontier areas in North America, few or no subsurface data are available. Each of these cases presents a different problem. In the first case, the sheer magnitude of data may inhibit a thorough basin analysis and thereby inhibit future exploration. Some basins have well penetrations numbering in the tens of thousands, with each well containing information on strata from a few thousand to more than 20,000 ft deep. Within these drilled intervals there generally is information available, such as (1) rock cuttings derived from the drilling process for each 10 to 30 ft interval; (2) from 1 to as many as 7 or 8 wire line well-log curves over the entire interval; (3) other well data such as drill stem tests, cores, and other samples or tests taken within the well; and (4) geologic information such as formation tops that may involve from several to as many as 2 or 3 dozen entries. The handling and integration of all of these data in large-scale well history systems is still in a relatively primitive state, and their availability and retrievability on a widespread basis is extremely limited.

Basin analysis, where well control is available, is carried out in a number of stages. These stages are designed to provide information on the distribution of rock facies which may control hydrocarbon traps, such that the distribution may be used in future exploration strategy decisions. The four stages are as follows:

(1) Construction of regional cross-sections covering the entire area of interest to allow subdivision of major stratigraphic units;
(2) if production has been established, a review of the existing fields to define (a) type of reservoir, (b) type of seal, (c) geometry and character of trap, and (d) nature of the hydrocarbon charge;
(3) analysis of the sedimentary facies progressions shown in cross-sections and tied to producing examples, if any, to isolate preferred areas or zones for reservoirs, seals, and source rocks; and
(4) construction of maps to allow geographic integration of key stratigraphic relationships listed above.

The final group of maps produced at this stage of stratigraphic analysis should portray the arrangement of potential reservoirs and seals in different layers that have been subdivided. Most frequently, analysis and exploration strategy questions will deal with one layer at a time, but they will involve successive layers as overall geologic patterns and availability of data dictate. If the basin has previously been extensively developed, analysis may be limited to a single horizon in which additional opportunities may be expected. In undeveloped basins, each layer must be considered. However, because of the lack of data there may be only a limited number of subdivisions made at this particular stage.

In addition to locating known or probable reservoirs and seals, distribution of source rock facies and the overall history of burial must be assessed to determine the time and locus for generation of hydrocarbons. The rate of burial and the thermal history together determine the point (Figure 4) at which hydrocarbons are first generated and, in a deep or hot basin, when thermal alteration may occur. Such timing of generation and migration may be critical in relation to the time of formation of structural traps and essential in predicting the phase of hydrocarbon present. In young or shallow basins, the time of initiation of hydrocarbon generation may limit the area of potential hydrocarbon charge in either structural or stratigraphic traps to the most deeply buried portion of the basin.

Stratigraphic basin analysis is essential to determination of the distribution of elements critical to hydrocarbon trapping. It is, however, only the beginning phase in the rationale of exploration strategy within a particular basin. Two cases illustrating a limited approach follow.

Cretaceous of the Rocky Mountain Area

As a first example, let us examine the distribution of rock types within the Cretaceous system in the central and southern portion of the Rocky Mountains (Figure 8). These rocks have produced large volumes of oil and gas over the past several decades. The rocks in question were deposited in a very large basin that covered much of the west-central portion of North America during the Cretaceous period (Figure 8b). This large basin was then broken up into a number of individual basins by tectonic events that occurred in the late Cretaceous and early part of the Tertiary (Figure 8a). These later tectonic basins are superimposed on the stratigraphic units which were developed in the earlier large basin.

Figure 9 is a regional cross-section of stratigraphic units present in a large portion of the Cretaceous depositional basin (Weimer, 1960). In this cross-section, we can observe that the overall Cretaceous system consists first of a basal group of sandstones (Lower Cretaceous) which are quite widespread and which represent the initial transgression. This is followed by a major deepening of the basin which initially deposited the Mancos shale over most of these basal sandstones. Also during this time a very long, regressive sequence began in which sandstones were deposited in progressively wider and wider areas in a step-wise fashion, eventually filling the basin. Since the sandstone units represent potential reservoirs and the shale units represent potential seals as well as source rocks, relationships between shale and sandstone may provide attractive exploration opportunities. We can expect particularly attractive opportunities to exist where the lower sandstone is capped by shale and within each of the later sequences where shale overlaps and separates sandstones. In areas landward from the tongues of shale which interrupt the sandstones, there is only a limited trapping opportunity because of the abundance of sandstones and the absence of obvious shale seals. This is not to say that opportunities cannot exist within those sequences. They certainly do, but to exploit these opportunities would require a more complete understanding of the stratigraphy than is likely to be achieved in a first round of basin analysis.

The basinal shale facies (Figure 9) includes most of the potential source rock in this sequence, and the burial history of these shales in the later tectonic basins (Figure 8a) is critical to the generation of hydrocarbons. Therefore, in addition to tracing potential reservoir and seal facies based upon data from the earlier depositional basins, each tectonic basin must be analyzed for evidence of hydrocarbon generation before exploration opportunities can be evaluated. Since the burial history may vary from basin to basin, some areas seemingly attractive because of reservoir-seal relationships may be condemned or reduced in priority for exploration because burial was too shallow to reach a favorable stage in the generation cycle (Figure 4).

Figures 10 and 11 show the relationships in two selected producing fields within the overall Cretaceous interval. These fields represent two different styles of traps in which different sedimentary facies are involved. The first (Ecker, 1971; Land and

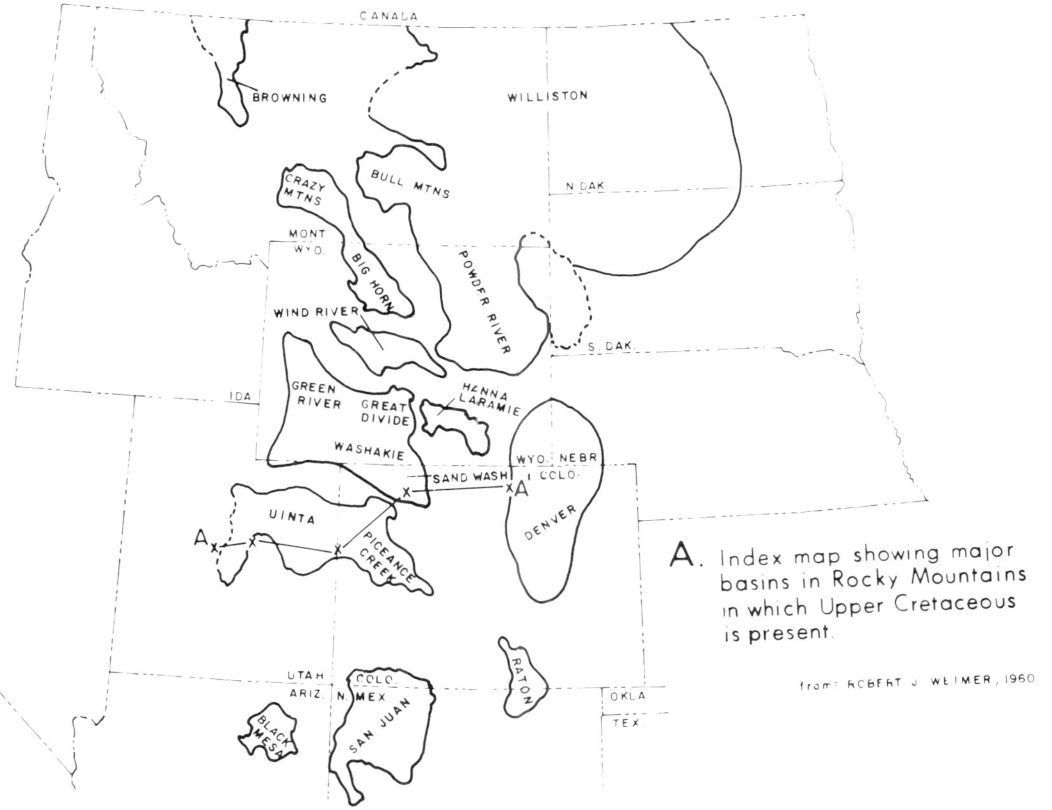

FIG. 8. The many basins now present through the Rocky Mountains area were formed after deposition of Upper Cretaceous sediments. Thus, to reconstruct original depositional patterns, data from these basins must be assembled to document the single, widespread depositional basin of the Upper Cretaceous. Paleogeography may be constructed by tracing positions of maximum and minimum extent of the sea from cross-sections such as A–A′ (Figure 9).

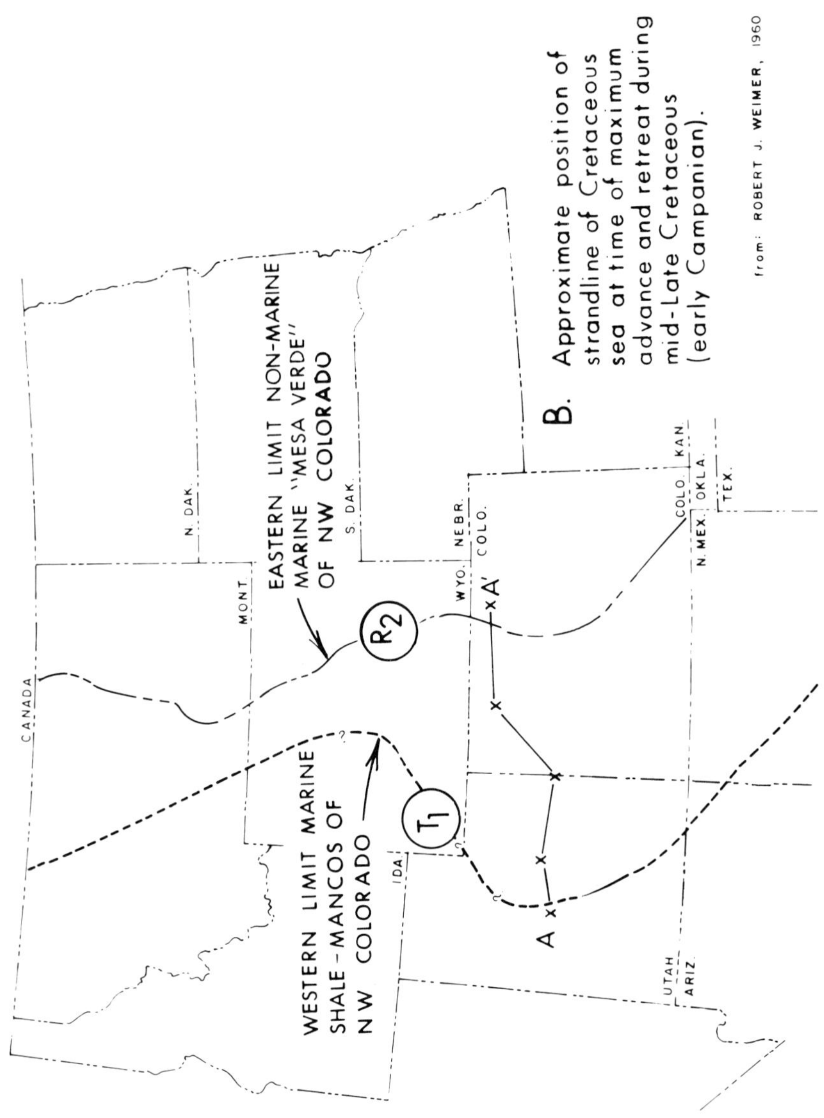

B. Approximate position of strandline of Cretaceous sea at time of maximum advance and retreat during mid-Late Cretaceous (early Campanian).

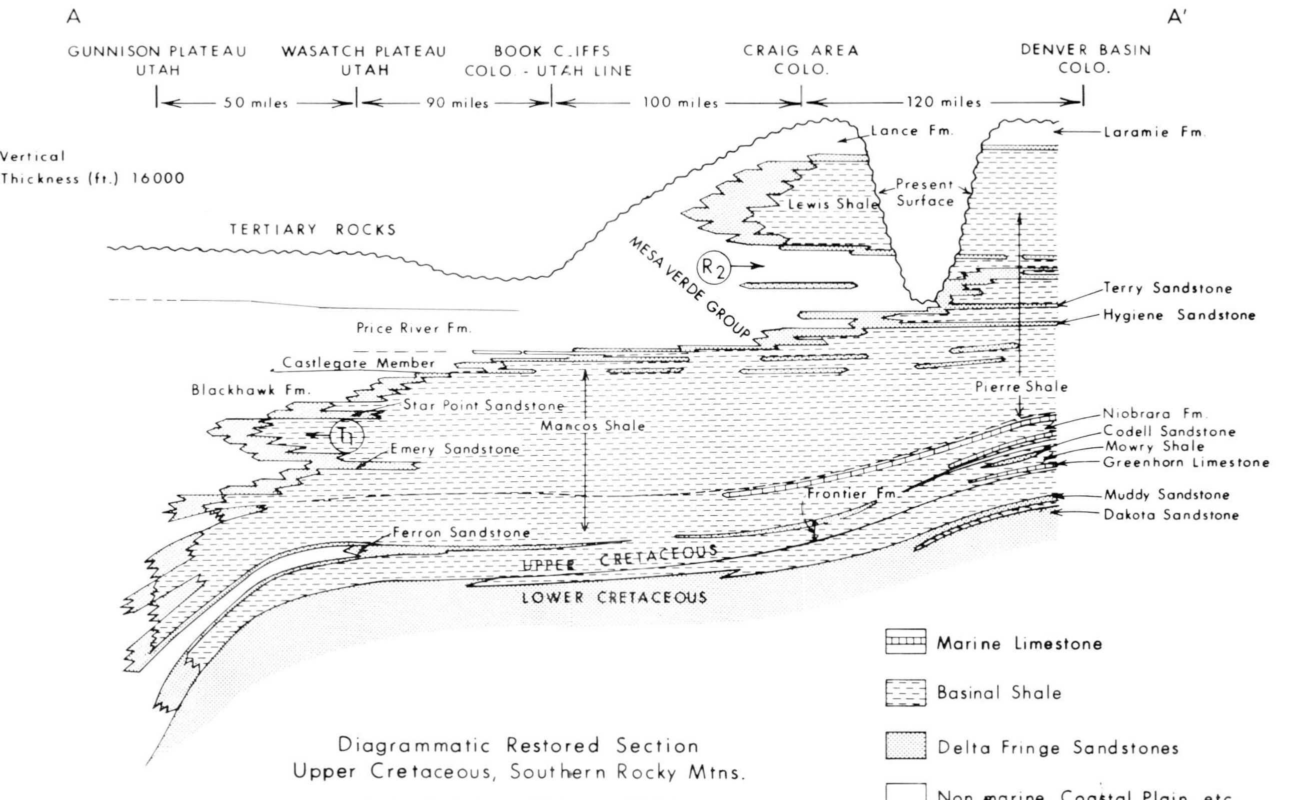

FIG. 9. A restored section of Upper Cretaceous sediments in the southern Rocky Mountains (A–A′ in Figure 8) allows reconstruction of Cretaceous paleogeography. Positions of shoreline may be established, such as T_1 and R_2, and correlated by means of similar sections from place to place in the region. By this means, trends in Figure 8 may be established. Once positions of potential reservoirs are established, their distribution may be predicted from basin to basin by this technique.

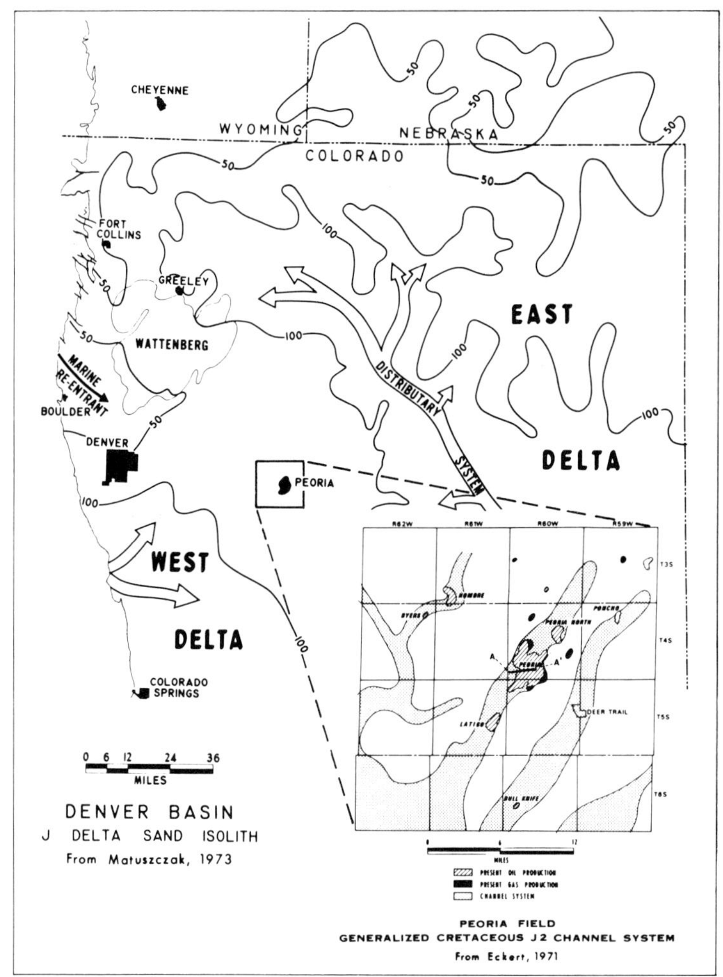

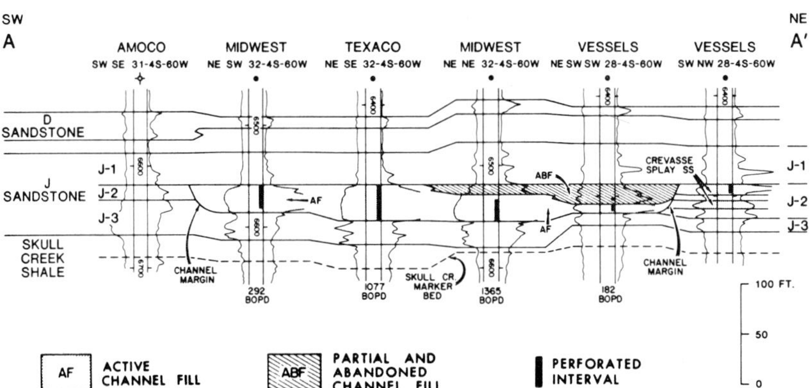

FIG. 10. *J* sandstones in the Denver basin are part of a widespread deltaic complex. Production, such as in Peoria field, is from lenticular sandstones formed in individual distributary channels. Cross-section A–A′ shows relation of the individual channel to the overall *J*-2 layer. (b) Electric log cross-section. (From Land and Weimer, 1978).

Weimer, 1979; Matuszczak, 1973) illustrates the style of trap found in a deltaic sequence. In this sequence distributary channel sands provide a reservoir, and the irregular distribution of these sands and resulting lenticular shape provide a discontinuous reservoir with an inherent trapping capacity where it is surrounded by shaly sediments (Figure 10b). Figure 10a is a map at the stratigraphic level which includes the Peoria field; it is used to illustrate the distribution of this overall facies which provides a potential trap. This layer is characterized by the building of deltaic sequences over very broad areas from both the east and west sides of the basin. This map can be used to assess the potential for distribution of reservoir beds outside the presently developed and productive areas. Because of the later superposition of tectonic basins within the overall depositional basin, data are required from each of these later basins to provide a continuity for extrapolation between basins.

A second example, shown in Figure 11, is from a younger portion of the sequence. It illustrates a different stratigraphic style, that is, a barrier island or shoreline sandstone body which also provides inherent trapping conditions because of its highly lenticular nature. This sandstone body tends to have elongated distribution parallel to the shoreline. It is important to establish the position of the shoreline, in order to extrapolate it from tectonic basin to tectonic basin. Figure 11 includes a map showing the distribution of the shoreline at the stratigraphic interval. By displaying the original depositional pattern along the shoreline and its relationship to the various later tectonic basins, potential reservoir sandstones may be traced and areas between established production may thereby be assessed.

San Andres-Grayburg of the Permian Basin

For a second example, we turn to the Permian basin of West Texas and southeast New Mexico. In this area, great thicknesses of principally carbonate facies with associated evaporites and some sandstones were deposited during the middle portion of the Permian period. This, too, is a very highly developed basin with large numbers of control points. In order to develop new prospects, exploration strategy must involve subdivision of the sequence into very small stratigraphic units and careful study of each unit. Figures 13 and 14 illustrate two kinds of trapping conditions in these carbonate-evaporite sequences which are productive.

This basin provides an excellent illustration of the carbonate platform depositional model (Figure 7). Widespread carbonate platforms surrounded basins filled by relatively deep water during deposition (Figure 12). A stratigraphic cross-section illustrates the relations of the various rock layers between the platform and basin (Figure 13). Development of a distinctive accumulation of particulate sediments marks the platform edge, as predicted from the Recent model (Figure 7). The platform edge deposit is a preferred locus for good reservoir rock, and associated evaporite facies in the adjacent platform provides a seal. This relationship is demonstrated in the McElroy field (Figure 13) and may be followed for great distances around the ancient platform edge, thus providing a basis for extensive exploration programs.

Trapping conditions may also exist on the platform, well removed from the platform edge. Widespread fine-grained carbonate layers may develop porosity and become reservoirs, owing to chemical changes (diagenesis) which occur after deposition. Associated sedimentary facies include evaporites in more shoreward, or restricted, areas (Figure 14). The stratigraphic sequence results when these units step

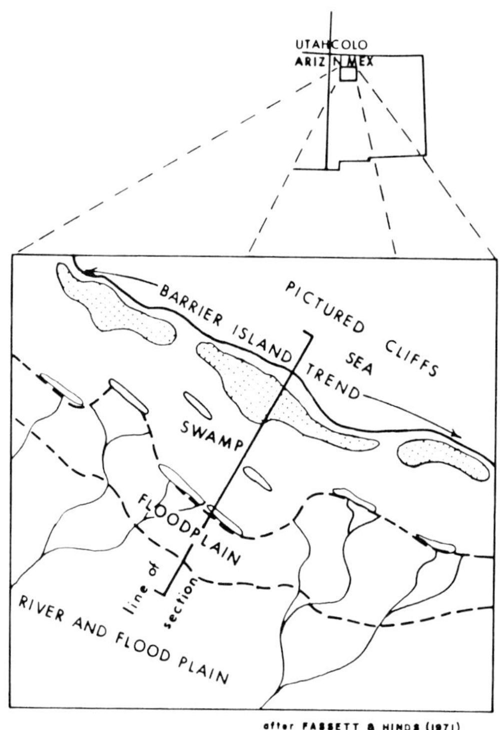

Diagrammatic Map and Section of
Pictured Cliffs Barrier Island Sandstone Trends

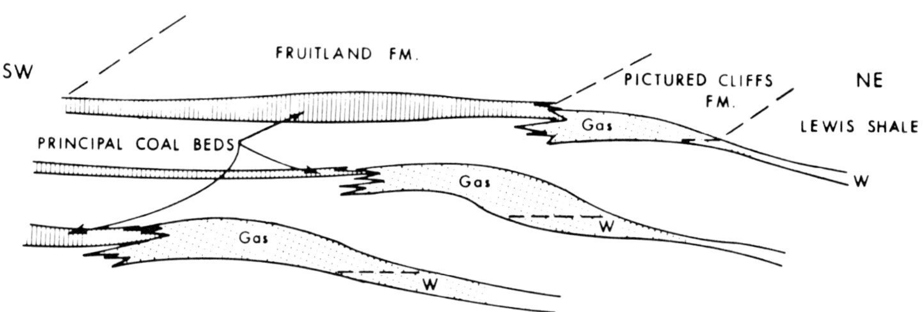

FIG. 11. Barrier island sandstones of the Pictured Cliffs formation form an important part of the San Juan basin (see also Figure 16). Associated coal beds are major energy sources for the Four-Corner electrical generation complex and are also the source for methane of the gas field.

over each other toward the basin, and the laterally and vertically adjacent anhydrite seals the dolomite reservoir. These changes occur near the ultimate shoreline and may be traced laterally in a similar manner to the platform margin.

Undrilled Exploratory Basins

One of the most challenging problems for exploration is that of a totally unexplored basin. Basin analysis must be carried out, but with techniques quite different from those where abundant well control is available. Two avenues of information may be used. The first and most direct is from geophysical measurements within the basin. In the early phase of exploration, seismic is generally the easily available data source, and it is the interpretation of these data that is critical. Velocity analysis may provide information as to relative patterns of rock types within the basin-fill sequence. The various patterns of reflections may provide information about the geometry of various stratigraphic units within the basin. A second tool to be employed here is the analog model based on studies in other basins and results of research on modern sediment patterns. A thorough understanding of the principles of sedimentation and the application of these principles within many basins have provided general models which may be extrapolated into new basins.

Extensive studies of seismic stratigraphy based on examples from many parts of the world were summarized by Vail et al (1977). They present an interpretation of

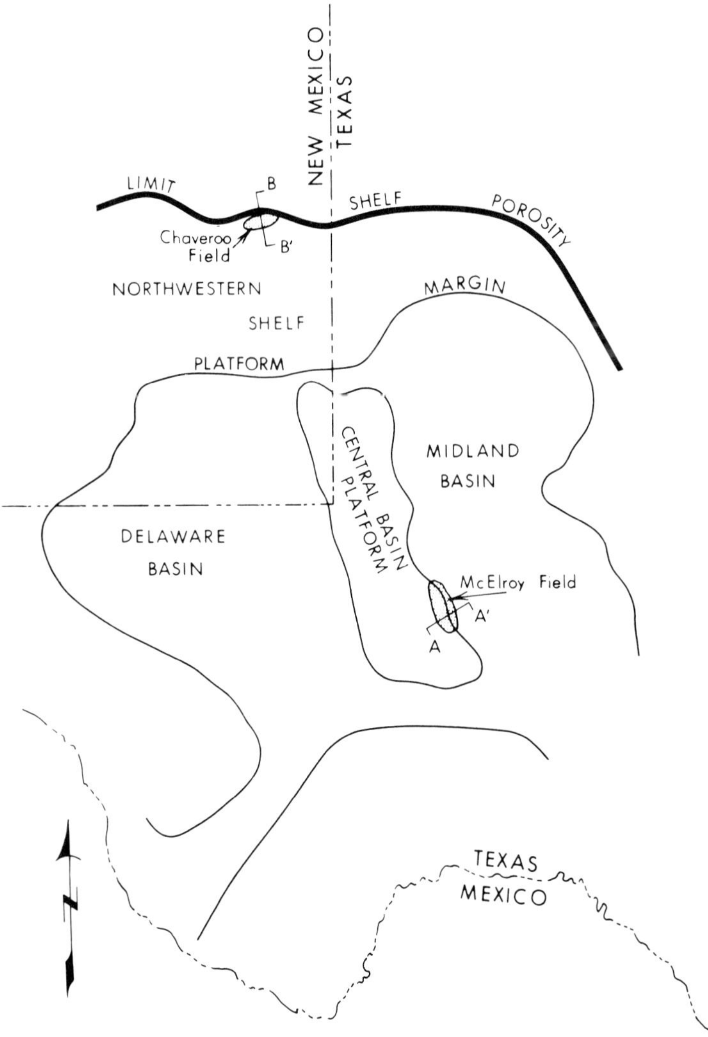

FIG. 12. Index map of Permian basin showing the alignment of platform margins (McElroy field) and limit shelf porosity (Chaveroo field).

the stratigraphic sequences in totally undrilled or sparsely drilled shelf-slope areas facing major ocean basins. This technique uses both an understanding of the overall stratigraphic record developed in areas in which control does exist and an extrapolation to undrilled areas.

At the present time, one of the greatest challenges of this type in North America is the exploration of the Atlantic Coast margin of the North American continent. Until a few years ago, this was a totally unknown stratigraphic sequence, with the exception of a few scattered wells in onshore positions which represent situations far landward from the positions of prime interest. Limited data existed from early geophysical measurements conducted by academic and government research institutions. Stratigraphic interpretation in this area had to be based upon the projection of models from developed basins in other parts of the Atlantic and Gulf of Mexico. Figure 15 shows some of the extrapolations which were possible from seismic information along the East Coast (King, 1977). One of the exploration targets appears to be a Mesozoic carbonate shelf margin comparable to a similar carbonate shelf margin of Cretaceous age known to extend around the Gulf of Mexico and which has been found to be highly productive in the Mexico portion of the Gulf.

Assessment of source rock potential and the burial history of the basin is especially critical in a lightly explored basin. By its very nature, a producing basin indicates that source rocks are present and are mature at least in a part of the basin. A non-producing basin does not provide that insight, and the potential for hydrocarbon presence must be inferred from whatever data are available. In the case of a totally undrilled basin, the presence of source rocks and the burial history must be inferred from geophysical measurements, usually seismic surveys, or from surface outcrops around the periphery.

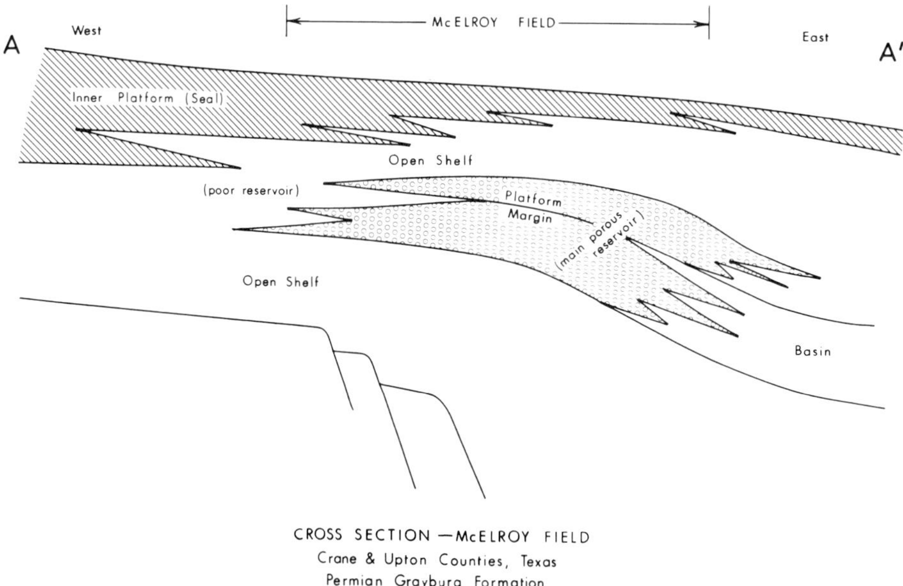

FIG. 13. Major reservoir facies of the platform margin of the Permian Grayburg formation is a high-energy oolite facies. Main seal consists of evaporitic anhydrite facies. This may be separated from the main reservoir in a situation similar to the principle shown in Figure 1.

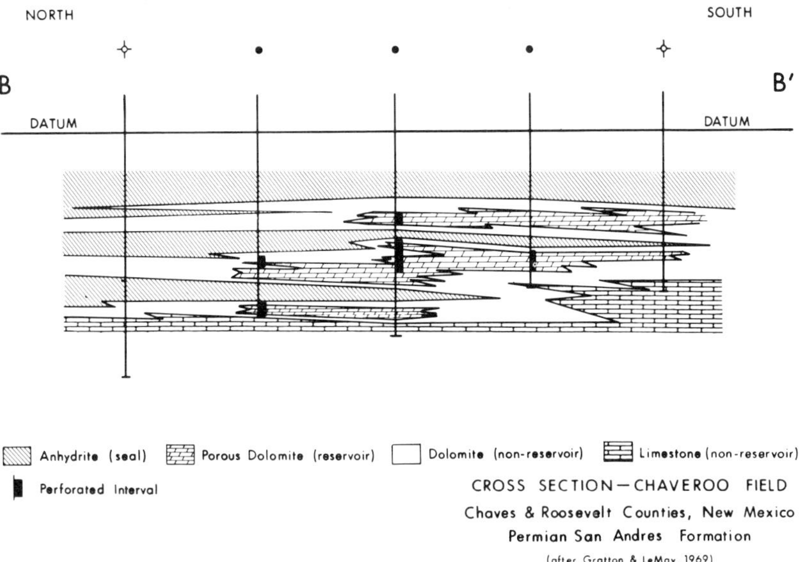

FIG. 14. Chaveroo field is an example of termination of porosity formed in widespread, tabular shelf layers. Mineralogy of reservoir and nonreservoir and the seal are all related to depositional processes in shallow seas in arid climates.

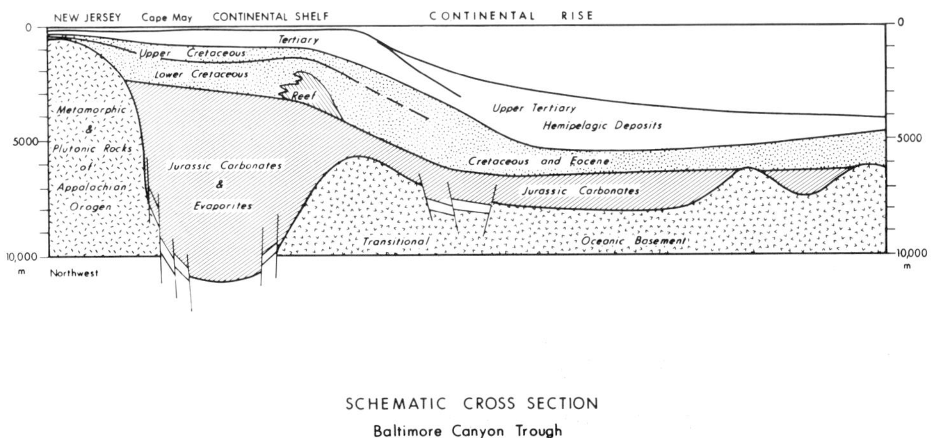

FIG. 15. Framework of the Atlantic Coast. Very few boreholes have actually penetrated deep sediments of the Atlantic Coast. However, modern seismic techniques together with other geophysical tools and application of geologic models based on related areas allow this reconstruction of the geologic framework.

Developing Exploration Strategy

Exploration strategy today still relies heavily on the anticlinal theory, that is, the fact that hydrocarbons will generally occur at the structurally highest elevation within a given reservoir bed. An exploration cycle in a given basin usually consists of three phases. These phases may be extended over a long period of time or may be temporally compressed through a high level of activity. In some basins, various stages of these three phases may be repeated with respect to different drilling depths or different parts of the basin.

The first phase consists of structural exploration where the target is a closed anticlinal or fault trap. The presence of a particular reservoir bed across the crest of a structure in the presence of hydrocarbons provides the simplest target. A structural mapping program utilizing surface geology, remote sensing, or seismic surveys or any combination thereof provides information on the structural attitude of the beds. The exploration for structures may be highly difficult and may require sophisticated techniques in structurally complex areas.

The second phase of exploration usually begins when the apparent structures within a given basin have been tested. At this stage, assessment of the stratigraphic trapping potential is made, that is, the exact distribution of reservoir beds within the basin is mapped and evaluated. There are many situations wherein anticlinal closures are of relatively low importance to the actual trapping of hydrocarbons. This phase of exploration generally consists of finding the larger scale stratigraphic variations whereby the termination of porous units are juxtaposed against seals. Figure 1 shows an example of updip termination of porosity. In many cases, such stratigraphic traps have resulted in some of the largest accumulations. In North America many of the larger traps are dependent upon some sort of a stratigraphic termination of a reservoir bed. Prime examples are (1) the East Texas field, where the Woodbine sandstone terminates against an unconformity and is sealed by younger shales; (2) the Prudhoe Bay accumulation on the North Slope in Alaska where Permian Sadlerochit sandstone terminates against an unconformity; and (3) the Pennsylvanian Scurry atoll in West Texas where a carbonate buildup nearly 1000 ft thick is encased in sealing shales.

The third phase of exploration involves a search for obscure trapping conditions where hydrocarbons accumulate under conditions that do not fit conventional stratigraphic or structural trap descriptions. Many accumulations of this type have been found more or less fortuitously during exploration for other targets. One example of this kind of trap is the basin-centered gas accumulation (Figure 16) in the San Juan basin of New Mexico where the upper Cretaceous sandstones are charged with gas over a very thick interval, but the termination around the periphery of the accumulation cannot be conventionally described. Another is the giant Hugoton field (Figure 17) in Permian carbonates of the Western Anadarko basin where gas has accumulated in a layered sequence of porous Permian carbonates and the gas terminates against water-bearing carbonates of the same unit in an updip direction. Again, trapping conditions in the termination of the gas accumulation out to the basin center are obscure and poorly understood. The most recent example of this class of trap is the Elmworth field of northwest Alberta and northeast British Columbia (Masters, 1979). Here, Cretaceous sandstones are charged with gas which apparently terminates in water-bearing sandstones of the same units in updip direction in a situation very similar to that in the San Juan basin. These traps are in apparent contradiction of the buoyant property of gas, and presumably they result

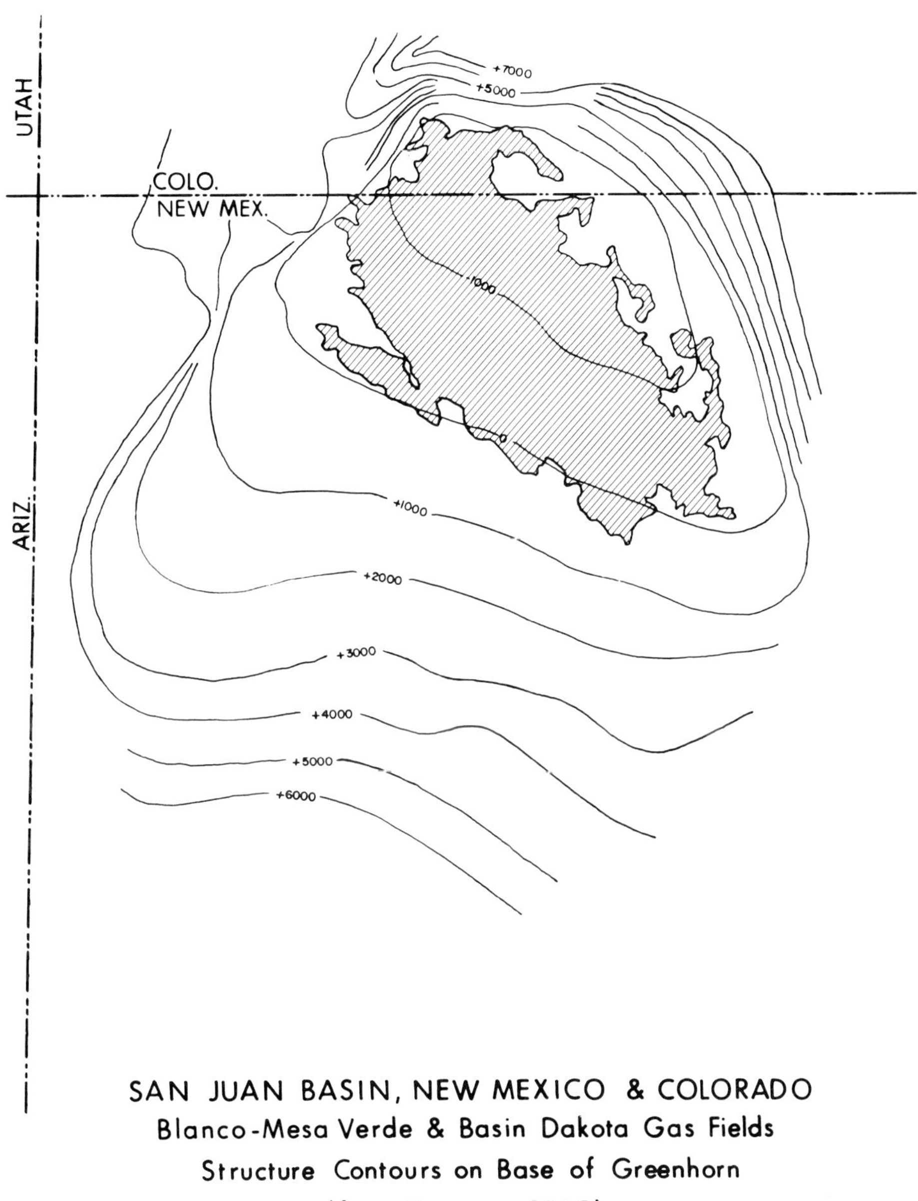

FIG. 16. The Blanco-Mesa Verde and Basin Dakota gas fields lie together in the deep center of San Juan basin. One of the largest gas accumulations in North America is trapped here, apparently contrary to the principle of buoyant hydrocarbons accumulating in structurally high positions.

from obscure changes in permeability possibly coupled with variations in fluid dynamics. There are a number of other kinds of traps that fit this class. The other obscure trap most frequently encountered is formed by fracturing of an otherwise impermeable rock. The reservoir does not exist except by virtue of the fractures, so that the limit of the fracturing is the limit of the reservoir and constitutes the trap.

Ideally, in a modern exploration program some element of all three of these phases should be going on at once. In a new basin, such as the Atlantic Coast offshore, the elements of stratigraphy must be assessed at the same time as the elements of structure. Today, the capability of gathering seismic information and mapping structures in a new basin is largely a matter of the time, effort, and expense of acquiring sufficient data over a sufficient area to cover all of the potential structural features.

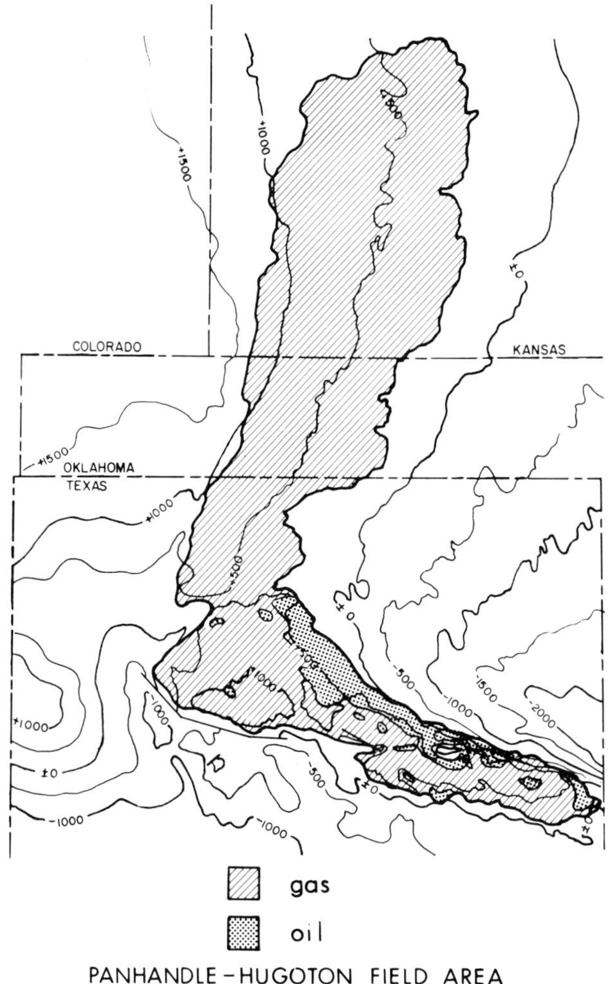

FIG. 17. Panhandle-Hugoton fields are a second example of an obscure trap. The southern area, Panhandle field, includes areas of conventional anticlinal trapping. However, the northern extension, Hugoton field, is controlled by complex fluid dynamics, rather than by reservoir termination.

There are two major questions that must be assessed at the earliest state: (1) Are reservoirs present which are regionally extensive, and (2) is hydrocarbon charge present within the basin and does it actually charge those reservoir beds? In the case of an undrilled basin, the assessment of hydrocarbon charge may require an initial drilling strategy.

For the above reasons, a modern exploration strategy includes assessment of the stratigraphic relationships within a basin from the very outset. In an undrilled basin, every effort is made to collect data from peripheral outcrops and from the limited wells that may have penetrated the periphery of the sedimentary basin. Modern seismic analytical techniques allow a certain level of assessment of stratigraphy because of the known relationships of acoustic parameters of the various sedimentary rocks and their contained fluids. Stratigraphic interpretation may be developed using a system of repeated comparisons of various known sedimentary models with acquired data until a reasonable fit is found. Another technique used during this early phase of exploration is drilling a limited number of stratigraphic tests at key positions within the basin to provide real samples from the stratigraphic sequence, for directly measuring its potential for both reservoir and source. During the early phases of exploration and particularly during the initial drilling phase, all the data must be continuously reviewed in order to assess all possible interpretations. If the initial ventures should prove unsuccessful, reassessment must be aimed at the basic understanding of relationships between reservoirs and seals and the formation of a trap, as well as the potential for generation of hydrocarbons from organic-rich beds and their migration into the reservoir and trap.

Exploration strategy in partially developed basins is a somewhat different problem. If early phases of exploration have concentrated on the more obvious and larger scale stratigraphic or structural traps, then a certain number of data have been gathered through those development programs. These data can provide input for the next phases of exploration. Many opportunities have been bypassed in past exploration programs because they were conducted at a time of lesser degree of sophistication; exploration tools, the understanding of stratigraphic and sedimentologic principles in the description of reservoir and trap, and the evaluation of source and generation have all developed significantly. Another control on the exploitation of developed basins is that of economics. Because of a ten-fold increase in crude prices within the past 7 years, yesterday's marginal or poor discovery is today's commercial well. Many plays have been reopened and reassessed both in the light of present economics and in the light of modern exploration and production technology.

The Williston basin (Figure 18) is a good example of a basin which has suddenly bloomed because of a change in the economics of hydrocarbon pricing, as well as an increase in technological aspects of exploration. The basin contains several examples of the reopening or increasing interest in an exploration play that had been well established but had fallen into inactivity, in part because of economic considerations, but also because of a lack of sufficient technology to carry out the play.

During the course of exploration in the Williston basin, oil was discovered in structural traps in Richland County, Montana, producing from the Ordovician Red River formation. During the middle 1960s and thereafter, limited exploration activity continued for a number of years with some success. Most of the discoveries were fairly small in areal extent, and their exploration constituted a high risk based on the seismic techniques of the period. Economic analysis of several discoveries caused a decrease in interest until the price of oil began to increase in the middle and late

1970s. Concurrently with this price rise, increased reliability of seismic mapping techniques and a better understanding of the reservoir allowed more precise definition of the factors which control the accumulations. A realization of the technical mastery of the mappability problem and the increase in product prices which made the discovered fields highly profitable caused a resurgence of activity within the area. That resurgence set the stage for the discovery of two additional trends: a combination structural-stratigraphic trapping situation within the Mississippian Madison carbonates, and an obscure trapping condition governed apparently by localized fracturing

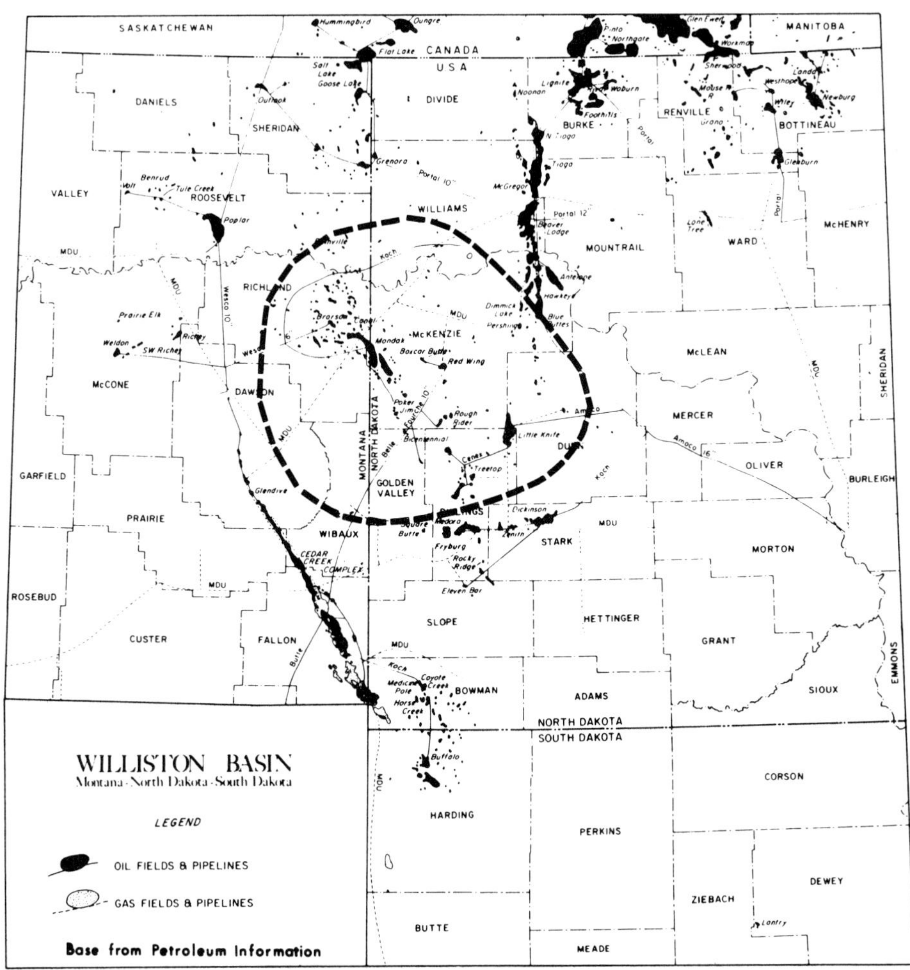

WILLISTON BASIN

WITH CENTER OF ACTIVITY, 1971–1981

FIG. 18. Williston basin first produced oil in 1951, and about 10 years of intensive activity resulted in development of fields along major structural trends and around the periphery of the basin. After a period of dormancy of nearly 10 years, activity was renewed in the early 1970s based on technological advances, principally in seismic exploration. The increase in economic incentives brought about by the OPEC cartel, coupled with technology available, caused an explosion in activity in this basin which many explorationists had nearly written off.

patterns within the same Madison carbonates. Extensive seismic mapping led to the drilling of structures with very limited anticlinal closure but in which trap closure was determined by stratigraphic changes. In this case, porous carbonate strata terminate updip against sealing anhydrite beds. These same stratigraphic units have produced in nearly identical trapping situations in other parts of the basin for more than 20 years. A combination of lack of economic incentive and limited understanding of stratigraphy throughout the basin caused a hiatus in the exploration program within the basin. These same opportunities, brought about by an interaction of economic and technological factors, have been repeated in many of the older producing basins in North America.

Incentives brought about by today's increased product prices have provided the basis for a strong resurgence of technological reassessment throughout the entire exploration industry. Exploration strategy must include a systematic evaluation of every basin and every layer within each basin, utilizing the best principles of stratigraphic geology, a careful assessment of the underlying structural mechanics, as well as application of the most advanced geophysical technology. In addition to geologic and geophysical principles, an exploration program must include a thorough understanding of the principles of fluid flow mechanics and reservoir engineering. The hydrocarbon accumulations remaining to be discovered are unlikely to occur in simple traps; in order to delineate complex traps, it is frequently necessary to quantify the fluid-pore relationships of the rocks with high resolution.

Future exploration successes will require interaction between multiple technologies to ensure accurate data and adequate interpretation to define the best targets. Recent major successes in the Wyoming Overthrust belt, the Deep Tuscaloosa trend, and the Williston basin demonstrate that significant opportunities still remain for the alert explorationist.

REFERENCES

Bernard, H. A., LeBlanc, R. J., and Major, C. F., 1962, Recent and Pleistocene geology of southeast Texas, *in* Geology of the Gulf Coast and Central Texas and guidebook of excursions: Presented at the GSA Annual Meeting in Houston, p. 175–205.

Ecker, G. D., 1971, Peoria field, Arapahoe County, Colorado: Mountain Geol., v. 8, p. 141.

Fassett, J. E., and Hinds, J. S., 1971, Geology and fuel resources of the Fruitland formation and Kirtland shale of the San Juan basin, New Mexico and Colorado: U.S.G.S. prof. paper 676, 76 p.

Gratton, P. J. F., and LeMay, W. J., 1969, San Andres oil east of the Pecos, *in* The San Andres limestone, a reservoir for oil and water in New Mexico: W. K. Summers and F. E. Kottlowski, Eds., New Mexico Geol. Soc. spec. publ. 3, p. 37–43.

King, P. B., 1977, The evolution of North America: Princeton, Princeton Univ. Press, rev. ed., 197 p.

Land, C. B., and Weimer, R. J., 1979, Peoria field, Denver basin, Colorado—J sandstone distributary channel reservoir, *in* Energy resources of the Denver basin: J. D. Puitt and P. E. Coffin, Eds., Rocky Mtn. Assoc. Geol. Symp., p. 81–104.

Masters, J. A., 1979, Deep basin gas trap, Western Canada: AAPG Bull., v. 62, p. 152.

Matuszczak, R. A., 1973, Wattenburg field, Denver basin, Colorado: Mountain Geol., v. 10, p. 99.

Peterson, J. A., Loleit, A. J., Spencer, C. W., and Ullrich, R. A., 1965, Sedimentary history and economic geology of San Juan basin: AAPG Bull., v. 49, p. 2076.

Pippin, L., 1970, Panhandle-Hugoton field, Texas-Oklahoma-Kansas—The first fifty years, *in* Geology of giant petroleum fields: M. T. Halbouty, Ed., AAPG memoir 14, p. 204–222.

Reinson, G. E., 1979, Barrier Island systems, *in* Facies models: R. G. Walker, Ed., Geosci. Can. reprint series 1, p. 57–74.

Tissot, B. P., and Welte, D. H., 1978, Petroleum formation and occurrence: Berlin, Springer-Verlag, 538 p.

Vail, P. R., Mitchum, R. M., Jr., Todd, R. G., Widmier, J. M., Thompson, S., III, Sangree, J. B., Bubb, J. N., and Hatlelid, W. G., 1977, Seismic stratigraphy and global changes of sea level, *in* Seismic stratigraphy—Applications to hydrocarbon exploration: C. E. Payton, Ed., AAPG memoir 26, p. 49–212.

Weimer, R. J., 1960, Upper Cretaceous stratigraphy, Rocky Mountain area: AAPG Bull., v. 44, p. 1.

THE SEISMIC APPROACH

Chapter 3

IMAGING THE SUBSURFACE

Introduction

Exploration geophysics is a young, hybrid science developed during the last 60 years to aid the geologist. Geophysical methods include measurements of the earth's magnetic, gravitational, and electrical fields; however, seismic recording of propagating elastic waves constitutes the principal method for mapping hydrocarbon deposits, accounting for more than 94 percent of all expenditures in geophysical activity (cf., Oil and Gas J., September 14, 1981). Modern exploration seismology uses advanced technology from many diverse fields such as communications engineering, electronics, information retrieval, telemetry, and optics.

Seismic profiles provide a cross-section of the subsurface revealing structural details such as faults, folds, anticlines, and synclines. In some cases compositional information such as porosity, permeability, water saturation, and hydrocarbon content can also be predicted. Presently, seismic data provide the most reliable means of acquiring subsurface structural and stratigraphic information without drilling wells.

The combined use of geology and seismology greatly reduces the risks involved in oil and gas exploration. The success rate of new-field wildcats in the United States was much higher for the years 1979–1980 than ever before, as shown in Figure 1. These statistics correlate directly with major advances in seismic technology (Crook, 1979), underlining the importance of improved geophysical methods.

The seismic crew count has historically been a leading indicator of oil and gas activity. The crews record and process massive amounts of seismic data on a routine basis for the oil industry. Seismic surveys are carried out on a surface grid in order to build a three-dimensional (3-D) picture of the subsurface. Each survey mile contains nearly 100 million bits of information, making modern data processing impossible without high-speed digital computers. Figure 2 shows a history of seismic exploration activity in the United States from 1932 through March, 1982. The largest number of crews (approximately 670) were in the field during 1952, while August of 1981 represented an all-time record with 720 seismic crews of which 670 were land and 50 marine (Figure 2). This comparison is somewhat inappropriate, since a modern crew is equivalent to several crews of 1952 vintage, due to size and operating efficiency.

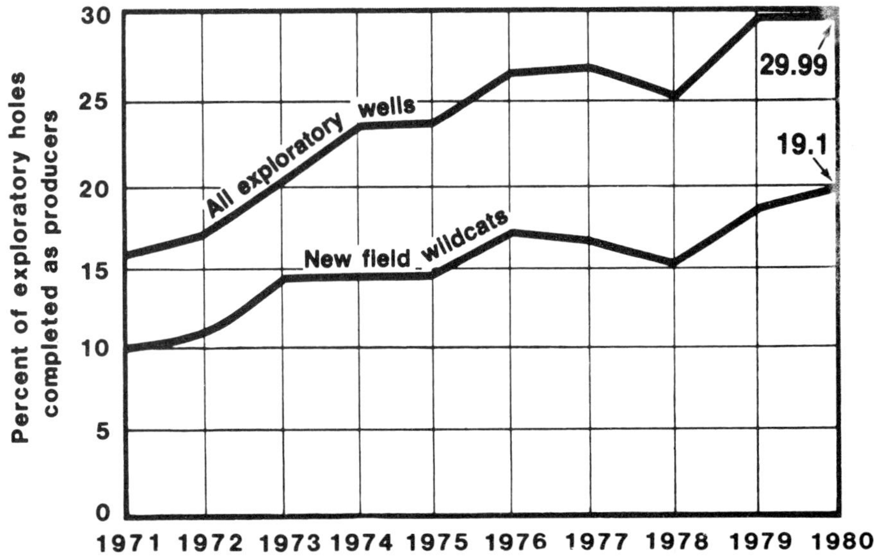

FIG. 1. The domestic exploration success rate. (Courtesy AAPG)

During 1980, about 1.1 million miles of seismic data were acquired at a total cost of approximately 2.91 billion dollars. Petroleum geophysical spending for the years 1970–1980 is shown in Figure 3. In the United States, typical drilling costs on land in 1980 were $30/ft for wells drilled to a depth of 5000 ft, $80/ft for depths between 5000 and 15,000 ft, and $185/ft for wells exceeding 15,000 ft. Offshore wells drilled in hostile environments cost as much as $1000/ft for 15,000-ft tests. The above information on exploration success rate versus costs points out that using seismic in exploration can significantly lower exploration costs by reducing the number of expensive dry holes.

The seismic data are used in exploration, among other things, for mapping the structural and stratigraphic traps, direct detection of hydrocarbon accumulations, and estimation of total hydrocarbon volumes in a field. Time differences between reflected seismic signals map structural deformation, whereas amplitude changes of reflected signals indicate changes in rocks and/or the presence of hydrocarbons.

The first part of this chapter describes some acquisition considerations and key concepts related to time and amplitude adjustments, whereas the latter part deals with advanced processing techniques, including digital filters and interpretation.

Data Acquisition

Readers are referred to Anstey (1970) for field procedures, Kramer et al (1968) for energy sources, and Evenden and Stone (1971) for recording instruments. The procedures currently employed in common-depth-point (CDP) surveying, first conceived by Mayne (1962), are reviewed in a later section. Refer to Nettleton (1940), Heiland (1940), and Jakosky (1950) for historical background of exploration geophysics.

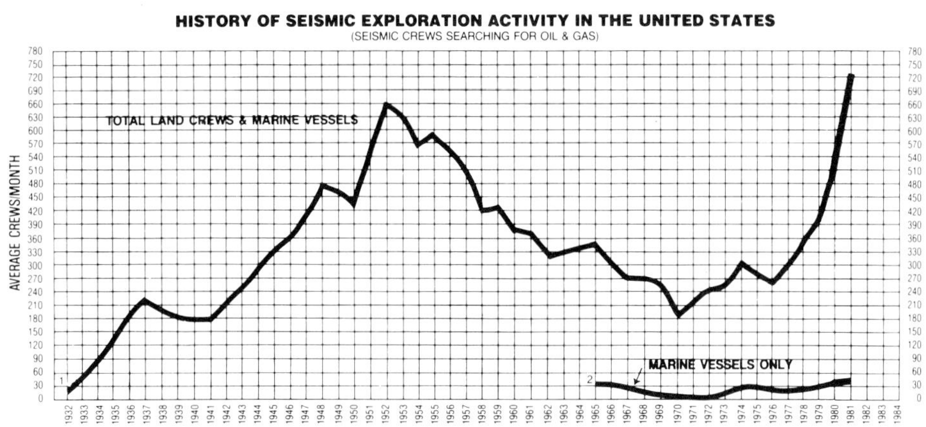

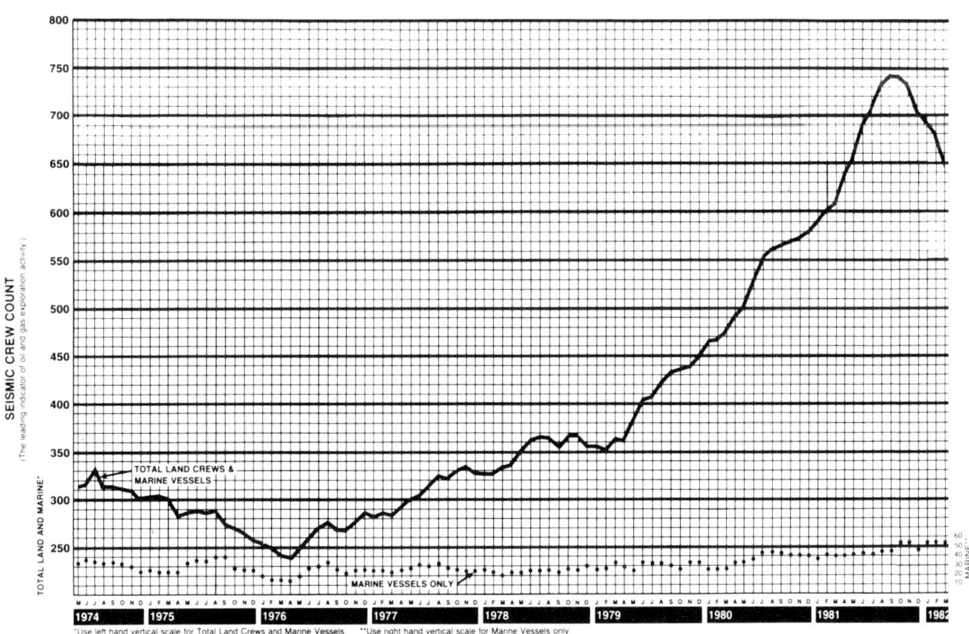

FIG. 2. Seismic exploration activity in the United States. (Both figures from GEOPHYSICS, v. 47, no. 3)

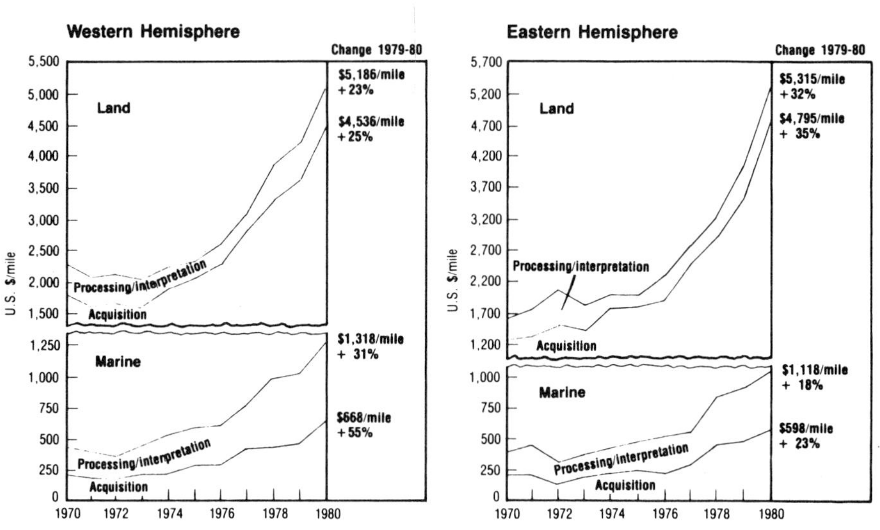

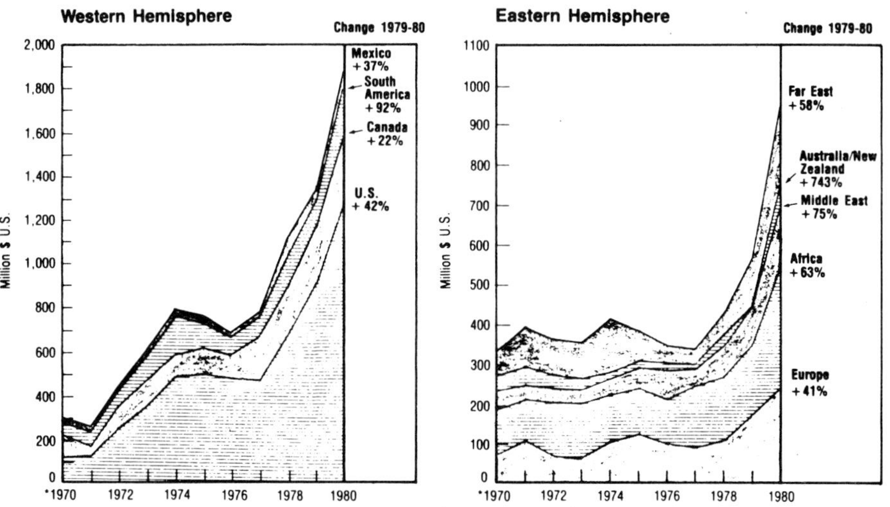

FIG. 3. Petroleum geophysical expenditures. (Source: Society of Exploration Geophysicists)

Figure 4 shows a seismic data acquisition system. Elastic waves created by seismic energy sources propagate through the earth, where interfaces between geologic strata reflect spreading wavefronts. Arrival times at surface receivers of single-bounce echoes (primary reflections) permit the determination of depths and dips of reflectors when subsurface velocities are known. The receivers are combined in arrays (groups) of transducers (seismometers) for attenuation of surface and near-surface generated noise waves. Each group may have up to 50 individual geophones laid out in various linear or spatial patterns, with group intervals ranging from 25 to 400 ft. For each energy source impulse, 24, 48, 96, or a larger number of group stations (traces) are recorded on digital tape. Seismometers are basically damped harmonic oscillators whose natural frequency varies from 6–50 Hz. As much as surface terrain permits, the seismic surveys are conducted along straight lines extending from 5–100 miles or more.

Significant improvements are being made in the design of recording systems. Digital sampling rate (typically 2 msec) can be reduced to 0.25 msec, while magnetic tape systems now have increased recording capacity from 1600 to 8000 bpi. The number of channels recorded has increased correspondingly to 256, 512, and even 1024 with certain restrictions. Recent developments in fiber-optics systems may lead to further advances in the recording procedures.

The seismic "master" cable joining the geophone groups ranges from 1 to 2 miles in length. The cable can be reduced to a two-wire system handling several hundred channels by multiplexing signal codes and high-density digital recording.

Large-scale integrated (LSI) electronics can eliminate the miles of heavy master cable entirely by connecting geophone arrays to remote units. Two schemes currently available are (1) wireless, digital telemetry employing long-range FM radio links, and (2) recording on magnetic tape cassettes at each geophone station. The cassettes can be collected at the end of each day to transfer the data to a conventional magnetic tape. Both schemes provide rugged, dependable systems offering greater flexibility in the field with increased productivity. Other advantages are elimination of cross-channel interference inherent in leaky cables and reduction of the 60 Hz power line noise. With the aid of digital computers, efforts are underway to allow increased flexibility in array designs such as space and time variant geophone weighting, summing, and array beam steering. The computers also optimize the effects of changing noise environment, scan incoming signals, digitize signals with increased resolution and dynamic range, store data, and provide field instrument quality control.

Marine data acquisition technology is also improving rapidly. Spatial resolution and signal-to-noise (S/N) ratios are being improved by increasing the number of detector arrays, which may soon approach 1000 or more. Positioning the recording vessel and its receiver arrays continues to be a problem. Satellite surveying is advancing rapidly and may reduce the vessel location errors to less than 30 ft. Inserting azimuth sensors in marine seismic cables has improved the accuracy of receiver station locations.

Many types of energy sources generate seismic waves. Dynamite and other high-energy explosives provide the simplest and most efficient means of releasing energy, but environmental and cost considerations have led to the development of alternate sources such as implosive air guns, electrical sparkers, and vibrating chirp systems. Dynamite and Vibroseis® are the most frequently used sources on land, whereas air

® Trademark of Conoco Inc.

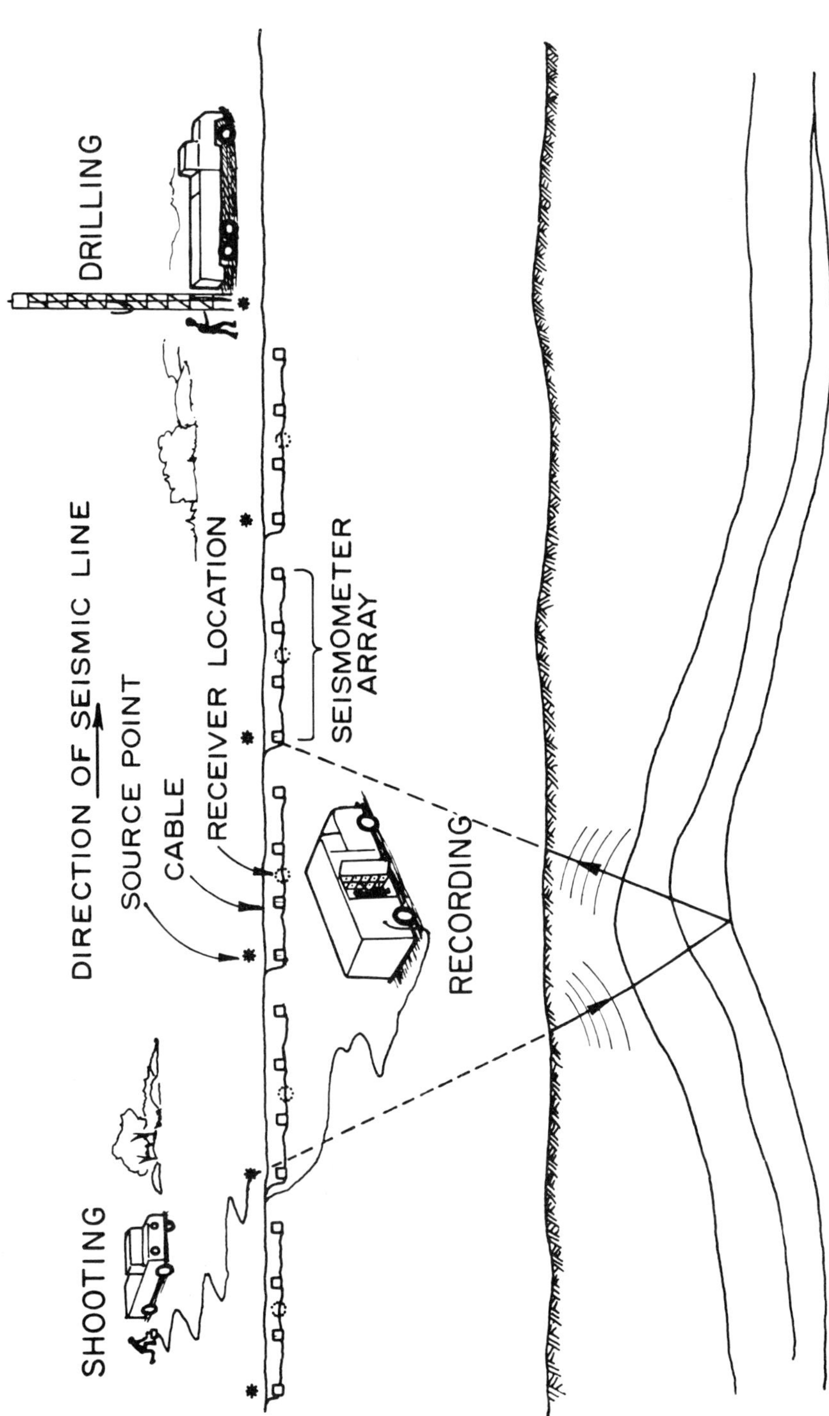

FIG. 4. Configuration of source points and seismometer arrays for CDP survey.

guns and gas exploders are predominantly used in the marine environment.

Digital recording systems have a dynamic range of approximately 80 decibels (dB); a decibel is defined as 20 $\log_{10}(A/A_0)$, where A/A_0 is the amplitude ratio. Signals can rise 100 dB above ambient noise levels, and processing can recover another 20 or 30 dB. Reflection amplitudes decay about 100 dB in the first 4 sec of recording because of spherical spreading and absorption losses. Consequently, amplifier gain levels have to be changed continually during recording in order to preserve the signal amplitudes. Modern recording systems use instantaneous floating point or binary gain controls.

A 48-trace seismic record contains around 2 million bits of information, and a typical crew acquires 100 or more records in a day. These massive amounts of data are processed to enhance the data quality, providing the explorationist with the highest resolution seismic profiles of the subsurface. Data processing technology is another area of complex challenges, some glimpses of which are given next.

Data Processing

Prior to the mid-1950s, seismic data were minimally processed because of the limitations of analog devices. The advent of magnetic tapes gave rise to various filtering and other data enhancement techniques. The "digital revolution" occurred in the 1960s.

Processing begins with the demultiplexing of field records. This results in a work tape with signal traces in sequential order. Each trace is preceded on tape by header information giving elevations, group intervals, sampling rate, trace length, and similar information. The creation of a tape in a format compatible with many different central computing centers is in itself a frustrating task, despite industry attempts to standardize tape formats (Northwood et al, 1967). Adjustments of times and amplitudes to correct various physical phenomena follow demultiplexing and reformatting. We briefly discuss time adjustments before proceeding to the important topic of relative amplitude preservation.

The time adjustments can be divided into static and dynamic corrections (Dix, 1952). A static correction consists of a constant time shift or translation of an entire trace. In other words, a constant time correction is added to or subtracted from all reflection times on a trace. The dynamic corrections, on the other hand, vary with record time and therefore depend upon reflector depth.

Static corrections

The purpose of automatic static computations is to remove time variations caused by elevation changes and near-surface inhomogeneities at the earth's surface. A region of very low velocity extends from the earth's surface to a depth of several tens to hundreds of feet; at this point, due to rock consolidation, the velocities increase from near 2000 ft/sec to 5000 ft/sec or more. Time delays associated with this "weathered layer" disrupt reflection continuity (i.e., trace-to-trace alignment) and pose a major obstacle to correction of the land data. Seismic lines recorded at sea, however, do not usually require static adjustments because of the water layer of constant elevation.

Automatic correction methods (Taner et al, 1974; Wiggins et al, 1976) assume that a simple translation of a trace converts it into a model trace that would have been recorded had sources and receivers been placed on a reference plane with no weathering material present (see Figure 5). This surface consistent time delay is assumed to be the sum of an "initiation" or source related component and a contribution characteristic of a given surface or "receiver" position. The validity of these two assumptions is confirmed by the great success of the automatic static correction programs, although near-surface layers behave as complicated filters whose impulse responses distort amplitude and phase characteristics of seismic wavefronts. Figure 6 shows a typical example of the improved S/N ratios that can be obtained using modern static correction programs.

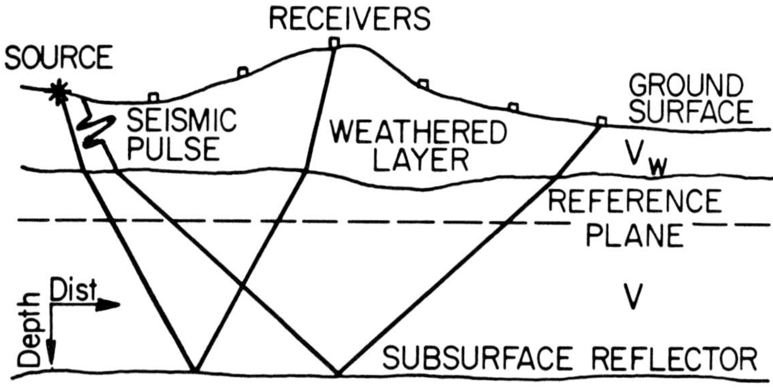

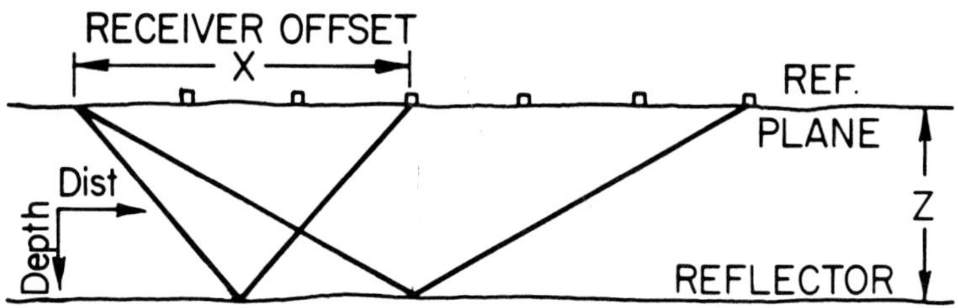

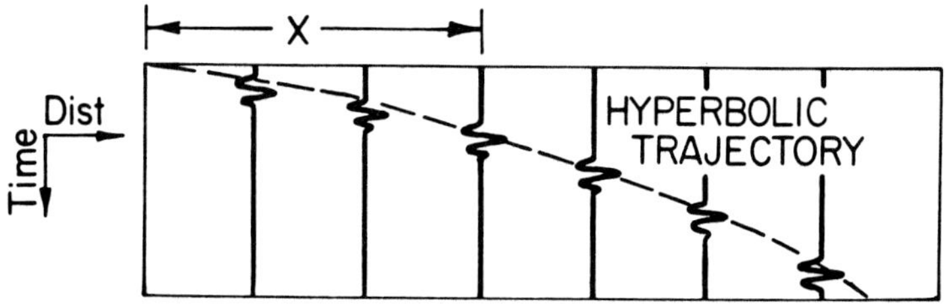

FIG. 5. Static correction model.

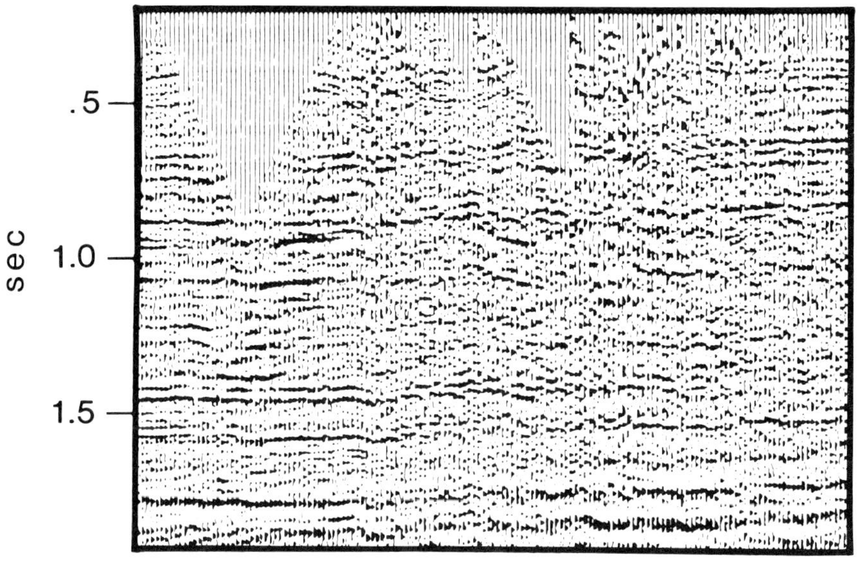

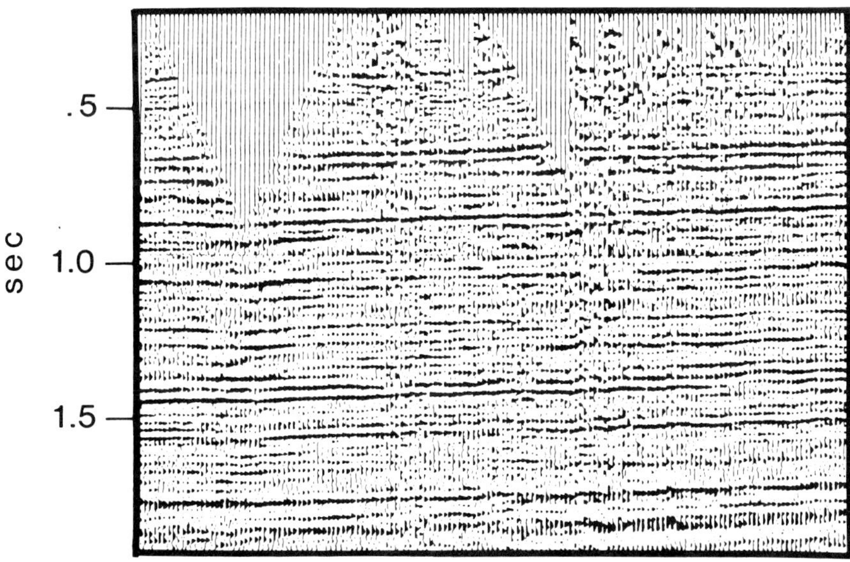

FIG. 6. An example of signal enhancement using modern automatic static correction programs.

Common-depth-point (CDP) stacking

Traces are usually collected into one of four kinds of data sets or "gathers," depending upon the objectives. For this purpose a diagram called a stacking chart is used (see Figure 7). Sources are activated sequentially in the field (Figure 4), with each initiation creating a common-initiation record. Surface positions of sources and receivers are successively rolled along the seismic line; these roll-alongs are shown vertically displaced on the stacking chart for clarity. The diagrams in Figure 7 illustrate four principal trace gathers called "common-initiation," "common-receiver," "common-offset," and "common-depth-point." Common-initiation gathers consist of traces having the same source; common-receiver gathers consist of traces having identical surface locations; common-offset gathers are traces with the same source-receiver distance ("offset"), and common-depth-point gathers have a common surface midpoint between source and receiver (Mayne, 1962).

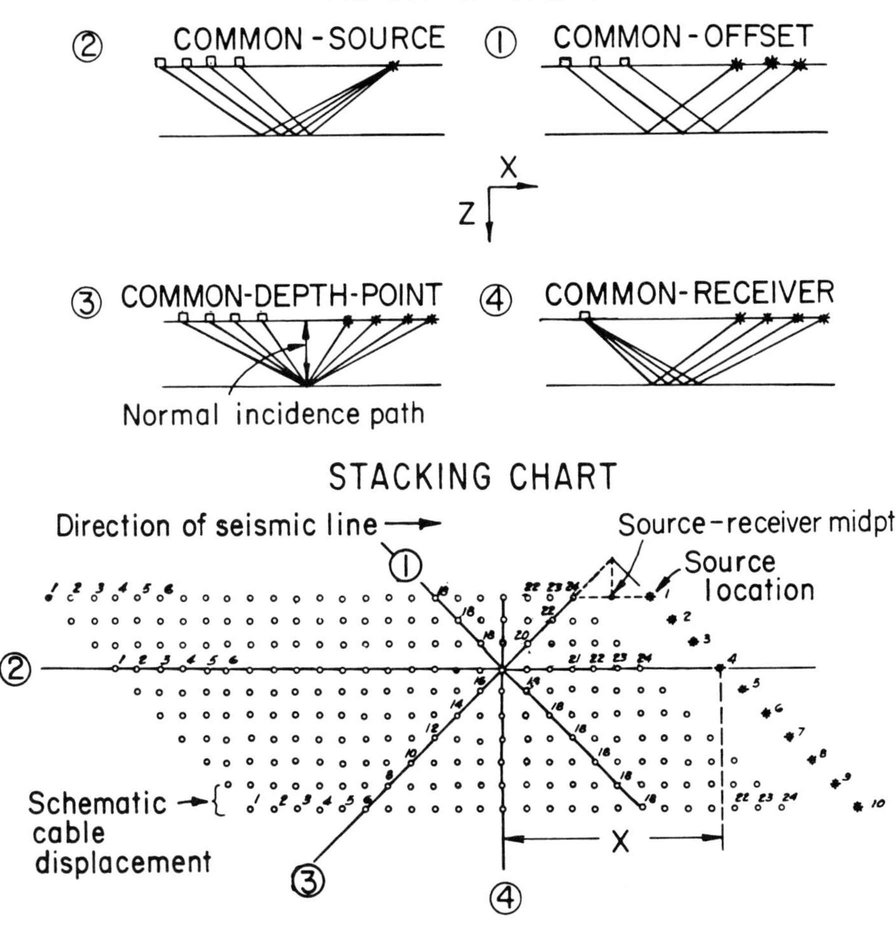

FIG. 7. Source and receiver positions corresponding to four principal planes used for sorting seismic traces. X = source-receiver offset distance; Y = depth flow reference plane.

When the subsurface layers are horizontal (assuming that the velocity is not changing laterally), the CDPs also have common points on reflecting interfaces called "common reflection points" (CRP). The subsurface reflection points are laterally displaced when the reflectors are not horizontal and are spread apart as structures become more complicated. However, because most geometries cause only a small dispersion (Taner et al, 1970; May, 1978) of the CRP locations (Figure 8), CDP stacking is still successful. Work in the last two decades has shown that horizontal (CDP) stacking is the single most important step in seismic signal processing. Stacking consists of the summation of traces from a CDP gather producing a composite trace. The latter's surface position is then equated with the common source-receiver midpoint. Summing CDP traces improves the S/N ratio, assuming that the primary CRP reflections are in phase and add constructively, whereas ambient noise and other seismic signals not in phase cancel. Compositing increases reflection S/N ratios by factors approaching $\sqrt{N}$, where N is the number ("fold") of traces in the CDP gather. CDP stacks of 12, 24, and 48 fold are routinely produced.

Dynamic corrections

Prior to stacking, CDP traces must be corrected for traveltime differences caused by the source-receiver offsets. This correction, called "normal moveout" (NMO), depends upon the depth (record time) to the reflecting horizon and is, therefore, classified as a dynamic correction. NMO is defined by the reflection time of an event, which increases due to an increase in distance from source to receiver for a horizontal reflecting interface in a homogeneous medium of constant velocity. A simple expression for an NMO time increment is (see Figure 5)

$$\Delta t_{\mathrm{NMO}} = t_x - t_0 = \frac{1}{v}\sqrt{4z^2 + x^2} - t_0, \tag{1}$$

where t_0 is the two-way reflection time for a zero-offset trace, t_x is the two-way reflection time for a trace of offset distance x, v is the velocity of the medium, z is the depth to the reflecting horizon, and x is the source-receiver separation (offset).

NMO correction involves subtracting a time increment $\Delta\, t_{\mathrm{NMO}}$ from each record time t_x (interpolating where necessary). This correction converts a trace of offset distance x into a zero-offset trace that would have been initiated and recorded at a common source-receiver midpoint (Figure 7). Equation (1) shows the dynamic nature of the NMO correction, because even in this most elementary case it is a function of depth, velocity, and offset.

Two facts contribute greatly to the success of CDP stacking. First, the reflection time-distance curves (t_x versus x) for complicated structures are approximated by a hyperbolic relationship (Taner et al, 1970) of the form

$$t_x^{\,2} = (t_0 + \Delta t_{\mathrm{NMO}})^2 = t_0^{\,2} + \frac{x^2}{v^2}\,. \tag{2}$$

Second, the purpose of NMO corrections is to align single-bounce ("primary") reflections prior to summing. Multiple-bounce ("multiple") reflections travel at lower average velocities than primary reflections with the same arrival times, since velocity usually increases with depth. Therefore, multiple reflections having greater NMO are misaligned and attenuated in CDP stacking.

NMO corrections and CDP compositing, the whole process known as CDP stacking, create the stacked traces. These correspond to identical source and receiver positions (i.e., zero offset). The zero-offset traces also have identical incident and reflected raypath segments, as shown in Figure 8. The normal incidence raypaths form right angles with reflecting horizons at the points of reflection called normal-incident points. Thus, CDP stacking produces a suite of normal incidence traces with simulated reflection travel paths normal to subsurface horizons (Figure 8).

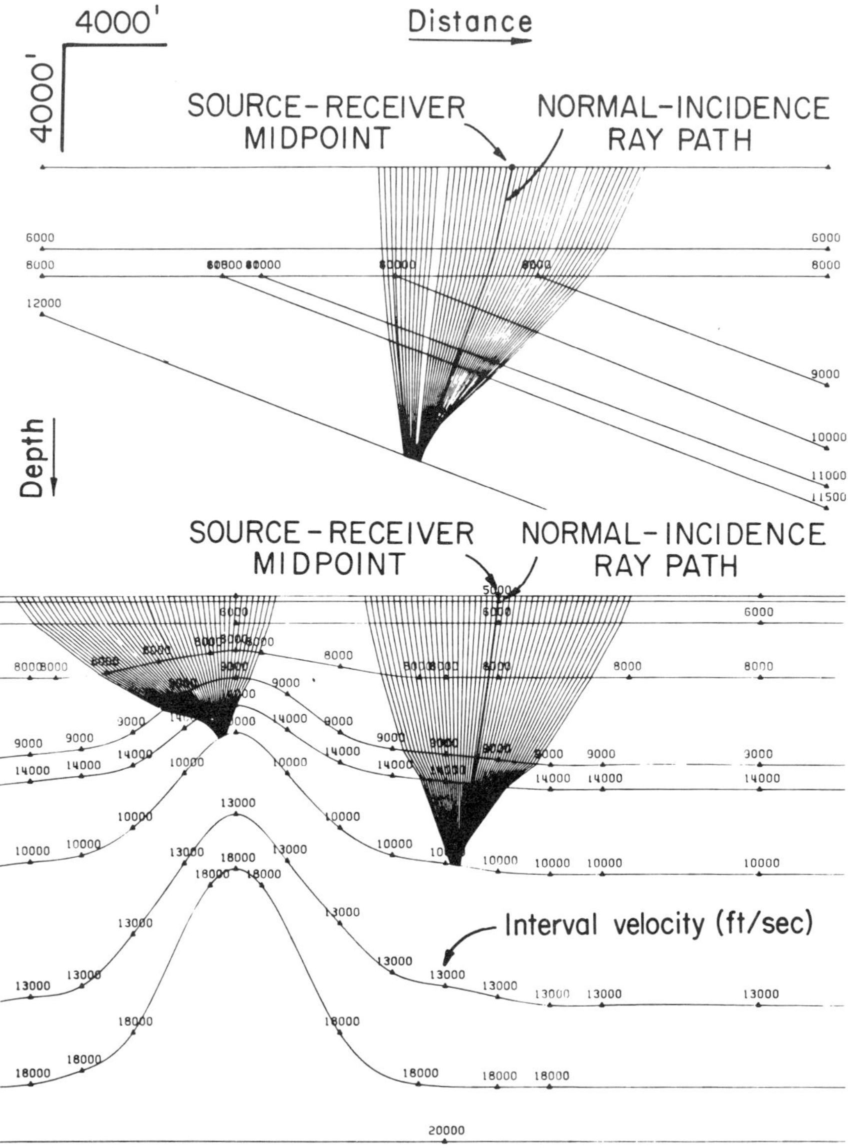

FIG. 8. Raypath diagrams for CDP traces. Normal-incidence raypaths corresponding to ideal CDP compositing.

Velocity analysis

Velocity is an important variable in seismic prospecting, because, among other things, it is needed to calculate the depths of the subsurface reflectors from seismically observed traveltimes. Velocities generally increase with depth, and they vary from 1100 ft/sec in air to values approaching 21,000 ft/sec in deep sedimentary basins. They are measured directly in wells (sonic logs, well velocity surveys) and derived indirectly from the NMO relationships of the seismic reflections. In well velocity surveys, the seismometers and/or sources are placed at varying depths in the well (Dix, 1952).

The objective of CDP stacking is to increase S/N ratios through in-phase summing of the primary events, for which velocity must be known for proper NMO corrections prior to summing. The hyperbolic form of the $t - x$ curves [equation (2)] provides a means for estimating velocity when the CDP gathers are scanned along hyperbolic trajectories for signal coherence. These scans establish a velocity function for NMO corrections. The hyperbolic relationship is not quite valid for more than one subsurface layer; nevertheless, the approximation works well even in areas of relatively complex geologic situations, provided parameters are correctly determined (Taner et al, 1970). They are accurate within 2 to 5 percent in geologic areas of simple structural deformations, where inclination angles of interfaces do not exceed 15 degrees. The interval velocities are often estimated from the NMO velocities within 5–10 percent for use in stratigraphic studies and for detection of hydrocarbon accumulations.

A velocity-versus-time display called a ''velocity spectrum'' (Taner and Koehler, 1969) can be used to estimate the velocity function. It is calculated from CDP gathers by scanning various hyperbolic trajectories for maximum reflection coherency. As shown in Figure 9, the spectrum is generated as a function of time t_0 [equation (1)], by incrementally changing the velocity between minimum and maximum values. This spectrum may be displayed as contour lines which represent the intersection of level planes of constant coherency with the coherency surface. Interpretation of velocity spectra requires skill and experience, because multiple reflections and other seismic events in addition to primary reflections tend to align themselves along hyperbolic trajectories. The idea is to locate the peaks on the coherency surface that correspond to primary reflections. These peaks are then joined to obtain the stacking velocity.

Coherence is a measure of similarity (Neidell and Taner, 1971). The following normalized coherency function employing zero-lag values of autocorrelation and crosscorrelation functions is not sensitive to rms signal amplitude variations between traces:

$$S = \frac{2}{M(M-1)} \sum_{i=1}^{M} \sum_{i>j} \frac{R_{ij}(0)}{\sqrt{R_{ii}(0)R_{jj}(0)}}, \tag{3}$$

where M is the number of CDP traces, R_{ii} (0) is the zero-lag value of the autocorrelation function of the ith trace, and R_{ij} (0) is the zero-lag value of the crosscorrelation function between the ith and jth traces. This normalized crosscorrelation function varies between -1 and 1, where 1 corresponds to perfect signal coherency.

Another useful quantity for measuring multichannel coherence is semblance coefficient (S_c), which is defined as the normalized output/input energy ratio in terms of zero-lag values of correlation functions. Output energy is measured on a composited time gate obtained by summing input time gates:

$$S_c = \frac{\sum_{i=1}^{M} \sum_{j=1}^{M} R_{ij}(0)}{M \sum_{i=1}^{M} R_{ii}(0)}. \tag{4}$$

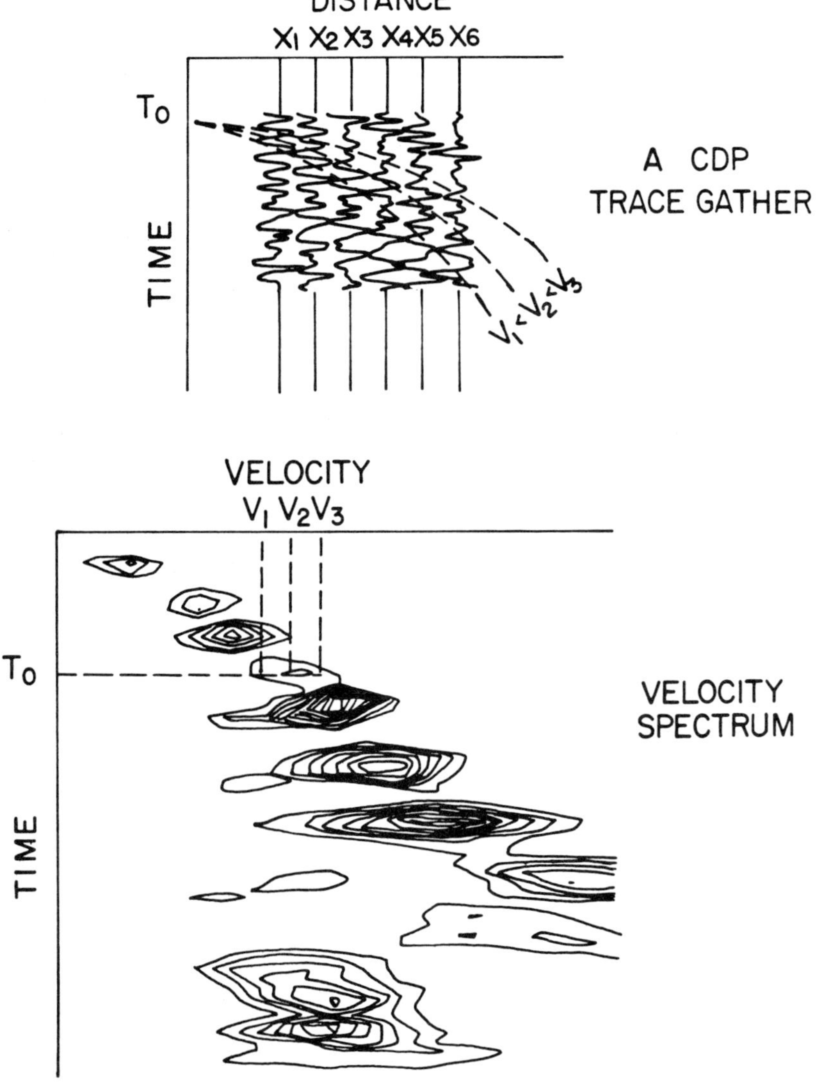

FIG. 9. A velocity spectrum displaying coherency as a function of reflection time and rms velocity.

The S and S_c coherence measures are closely related. The semblance coefficient S_c is sensitive to channel amplitude differences and varies between 0 and 1, with 1 denoting identical signals.

Relative amplitude preservation

Until the late 1960s, seismic amplitudes were used almost exclusively for identifying seismic events. The "bright spots" technology, which relates large-amplitude changes with the presence of hydrocarbons, has added a new and significant dimension to exploration. Seismic amplitudes can pinpoint changes in rock composition, layer thickness, and stratigraphic conditions. The approach has had 70 percent success (Flowers, 1976) in locating gas accumulations in young, unconsolidated Plio-Pleistocene deposits (not over 12 million years in age) in offshore Nigeria, Indonesia, and the United States Gulf Coast.

The improper use of automatic gain control (AGC) and trace average-amplitude procedures can destroy amplitude information. Proper time-dependent adjustments in amplitudes are required after the recording gain functions have been recovered, because the human eye cannot assimilate dynamic ranges of 80 dB. Both statistical and deterministic approaches for inverse gain functions are used to correct trace amplitudes for average attenuation rates while preserving instantaneous variations caused by changes in subsurface acoustical impedances. Deterministic approaches define general models to describe the factors affecting amplitudes such as spherical spreading of the wavefront, frequency-dependent absorption losses, reflection and transmission losses, and source and receiver array effects. Statistical approaches, on the other hand, produce average gain functions based on collections of traces sorted by common offset, source, receiver, etc. Figure 10 shows a record of common-depth traces both before and after gain correction. The statistical approach gives poor results in areas of low S/N ratio, where estimates of the gain functions tend to be affected by noise. Deterministic models may yield better results in such noisy areas.

In general, both compressional (P) and shear (S) wave modes propagate in an elastic medium. Snell's law and Fermat's principle of minimum-time paths govern reflection and refraction at an interface, where incident energy splits into reflected and transmitted P and S modes. Mode conversion does not occur for normal incidence waves. The normal incidence reflection coefficient, relating incident and reflected amplitudes, is given by

$$r = \frac{\rho_2 V_2 - \rho_1 V_1}{\rho_2 V_2 + \rho_1 V_1}, \tag{5}$$

where subscripts 1 and 2 denote the media for incident and transmitted waves, ρ is the density, and V is the interval velocity (Grant and West, 1965). Appropriate equations for oblique incidence are much more complicated (Ewing et al, 1957) because of mode conversion. The above equations for reflection and transmission coefficients are very useful in seismic work and quite accurate for stacked traces in the areas of simple geologic structures.

Porous rocks at depth are usually filled with water, oil, or gas. The gas-filled rocks tend to have lower velocities than either oil- or water-saturated rocks. In many situations, this difference can cause significant lateral change in seismic amplitudes, which is the basis of so-called "bright spot" interpretation.

In unconsolidated sandstones, gas has a much larger effect on formation velocities than oil, so that amplitude anomalies associated with gas/brine contacts are greater than those related to oil/brine and gas/oil interfaces. Normal incidence reflection coefficients for gas-filled sandstones encased in shales may approach 40 percent, as compared to 10 percent or less for brine-charged sandstones. Coefficients for oil-bearing sands exhibit intermediate values. Thus hydrocarbons may produce amplitude anomalies around 12 dB. Polarity reversals may also characterize hydrocarbon accumulations if acoustic impedances across the upper interface of gas and, to a lesser extent, oil-bearing sands encased in shale decrease (negative $\rho_2 V_2 - \rho_1 V_1$). In contrast, velocities and densities associated across the upper interface of a brine-

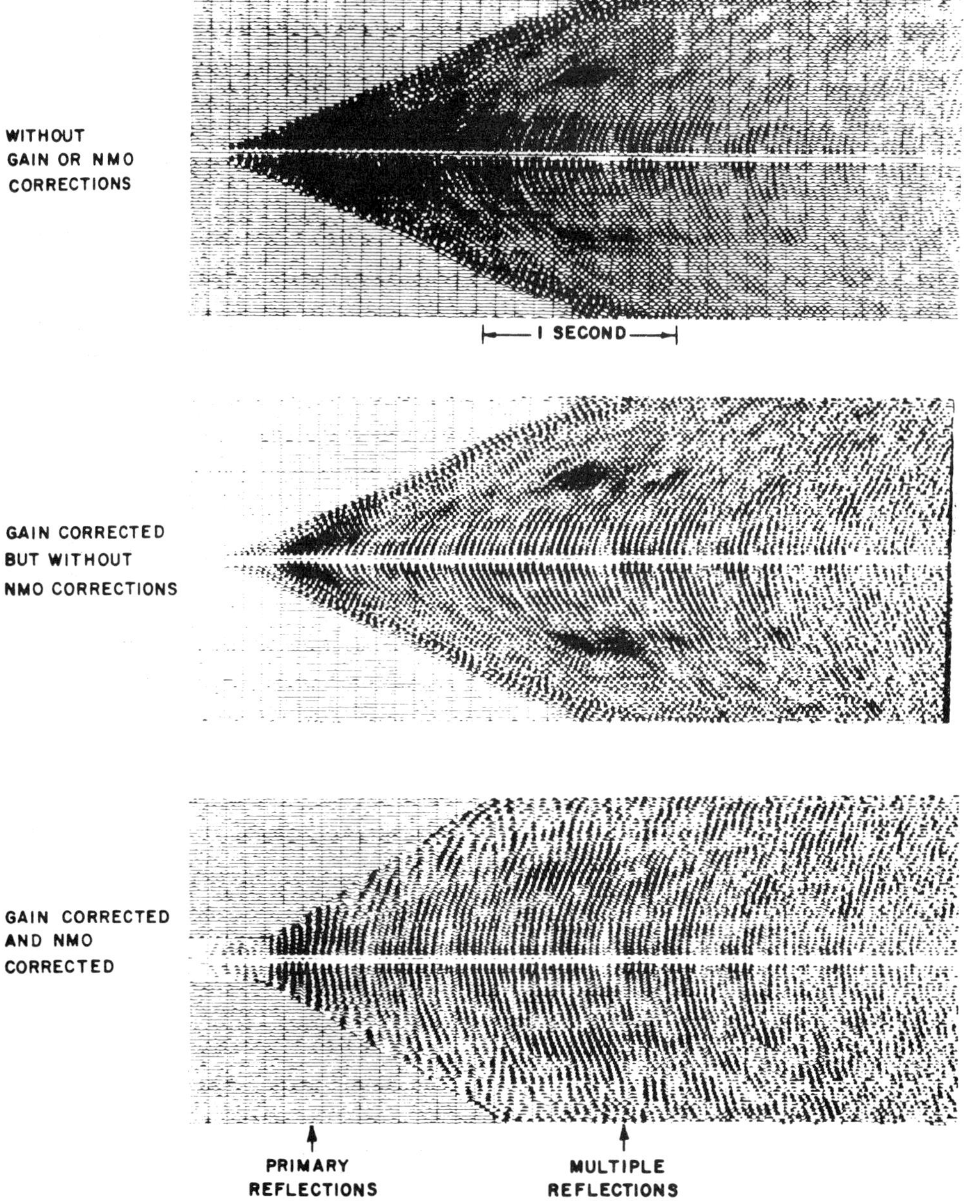

FIG. 10. Two CDP trace gathers showing the effects of gain and NMO corrections.

filled reservoir sand encased in shale usually increase with depth (positive $\rho_2 V_2 - \rho_1 V_1$). Thus rapid lateral increases in amplitude and sudden changes in polarity as shown in Figure 11 may indicate a hydrocarbon accumulation at depth.

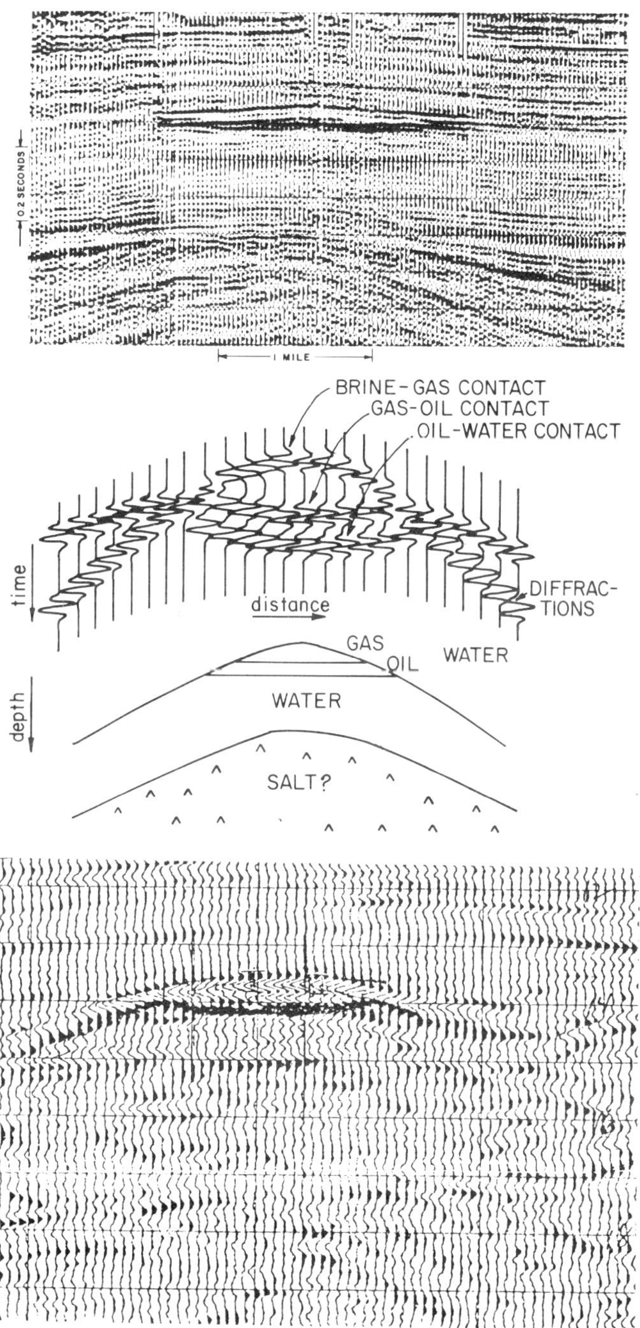

FIG. 11. An amplitude anomaly exhibiting many seismic features of an idealized bright spot associated with a hydrocarbon reservoir.

The bright spot concept, when applicable, allows the explorationist to locate hydrocarbons, determine reservoir dimensions (Lindsey and Craft, 1973), and identify fluid content. The method is based on the assumption that seismic amplitudes are proportional to normal incidence reflection coefficients, and therefore lateral changes in reflection amplitude correspond to lateral changes in rock properties such as velocity and/or density. After the traces are appropriately gain compensated, the common practice is to estimate lateral and vertical amplitude variation of the background noise and to compensate the amplitude variation of an anomaly for changes in the background noise. This correction is usually based on noise estimates obtained from two windows encompassing, but not including, the anomaly.

Compensated trace amplitudes are then converted to reflection coefficient values either by direct calibration with a known control point, or by normalization so as to provide relative changes in reflection coefficient. The seismic amplitudes are related to the reservoir rock properties by calculating densities and velocities as functions of fluid saturation, and then substituting these velocities and densities into the reflection coefficient relationship [equation (5)]. Bulk density and compressional velocity of shale surrounding a sand reservoir can be obtained by averaging data from well logs in the area. The bulk density of a sand reservoir, however, can be computed for various fluid contents using a weighted-by-volume average of the brine, gas, or oil, and sand members (Domenico, 1974):

$$\rho = \phi S_w \rho_w + \phi\,(1-S_w)\,\rho_h + (1-\phi)\,\rho_s, \tag{6}$$

where

ρ = bulk density of the reservoir,

ρ_w = brine density,

ρ_h = hydrocarbon density (gas or oil),

ρ_s = sand grain density,

ϕ = fractional porosity,

S_w = fractional brine-water saturation.

The compressional velocity of the sand reservoir can be obtained by an equation derived by Geertsma (1961) from a theory developed by Biot (1956) for propagating elastic waves in a porous elastic solid containing a compressible viscous fluid (Domenico, 1976):

$$V_p^2 = \left[\left(\frac{1}{c_b} + \frac{4\mu}{3}\right) + \frac{(1-\beta)(1-\beta-2\phi/K) + \phi\rho_b/K\rho_f}{\phi c_f + c_s(1-\phi-\beta)}\right]\left[\frac{1}{\rho_b(1-\rho_f\phi/\rho_b K)}\right], \tag{7}$$

where

μ = rigidity of dry rock,

K = fluid-matrix coupling factor; $K = 1$ means no coupling and $K = \infty$ is perfect coupling,

ϕ = porosity,

c_f = pore fluid compressibility,

c_s = matrix compressibility,

c_b = compressibility of dry rock,

$\beta = c_s/c_b$,
ρ_s = matrix density,
ρ_f = fluid density,
ρ_b = density of fluid-filled rock.

The normal incidence reflection coefficient equation [equation (5)] is valid only for plane wavefronts. Consequently, the approach outlined here for relating seismic amplitudes to rock properties may not be applicable in areas of geologic complexity where incidence angles are large and wavefronts are curved.

Bright spots also possess diagnostic features. Large-amplitude events of limited lateral extent, which are flat (no dip) on a stacked section, sometimes correspond to reflections from gas/brine, gas/oil, or oil/brine interfaces (Figure 11). These contact events, often called "flat spots," may be important hydrocarbon indicators. They are horizontal because the fluids align themselves along gravitational equipotential surfaces regardless of the complexity of geologic structures. Such events may also help to define the reservoir dimensions. In thick reservoirs, such events can have a convex downward appearance ("velocity pulldown") on a stacked section. Thus, some bright spots may have a "fisheye" appearance, as shown in Figure 11. This effect occurs because reflections from the top of the reservoir are convex upward in appearance with the geologic structure, whereas slower reservoir velocities cause contact events to be concave downward. Diffracted wavefronts from edges of reservoirs where hydrocarbons terminate add to this fisheye effect and provide an additional hydrocarbon indicator. Another criterion is the marked attenuation of amplitudes of reflections originating from horizons beneath reservoirs. Large transmission losses and strong reverberations associated with shallow accumulations attenuate or "mask" reflections from underlying strata and deeper reservoirs.

Amplitude anomalies do not always indicate hydrocarbon accumulations. Reflected signal strengths depend upon subsurface impedance contrasts, and many factors other than hydrocarbon accumulation can cause large impedance contrasts. Thin lenses of lava and tightly cemented layers of silt, lime, and lignite give rise to bright spots similar to those associated with hydrocarbon reservoirs (McNabb, 1974). The top cross-section in Figure 12 shows both shallow and deep bright spots (Flowers, 1976). The shallow anomaly is a brine-filled sand, whereas the deeper anomaly is productive. It is very easy to misinterpret the upper anomaly because both sands produce similar seismic responses. Low-saturation gas sand and rocks deposited in shallow water environments also produce large-amplitude reflections. Domenico (1974, 1976) showed that a small percentage of free gas in a reservoir (less than 10 percent) can give nearly as large a velocity change as full gas saturation. This effect is shown in Figures 13 and 14, and it is believed to account for the nonproductive anomalies shown in Figure 12. By no means are all hydrocarbon deposits commercial; hence, careful interpretations must be made to establish thicknesses, fluid content, saturation levels, and areal extent.

The bright spot method works best in outlining gas reservoirs in unconsolidated sand reservoirs at depths not exceeding 6000 ft. Amplitude anomalies associated with older rocks at greater depths are exceedingly difficult to interpret because rocks are more indurated, have less pore space and, therefore, smaller impedance changes across elastic interfaces. Multiple reverberations as well as geologic complexities also tend to become more bothersome with depth. Onshore surface conditions further complicate interpretations because of changes in the shallow layers, topography, and

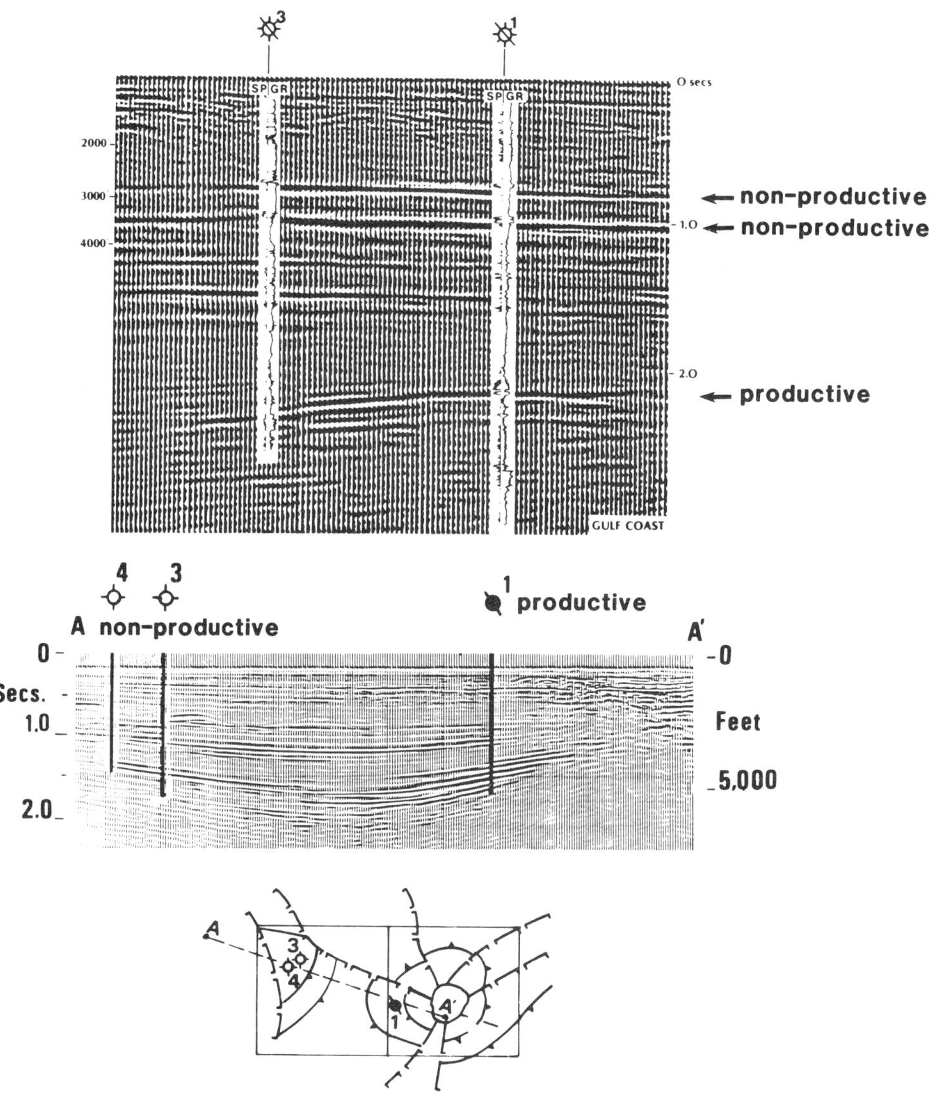

FIG. 12. Examples of pitfalls in interpreting bright spots on seismic lines.

variations in source and receiver coupling. Despite many difficulties, amplitude anomalies have led to many new oil and gas fields throughout the world, and many unexpected benefits have resulted from interpretation of seismic amplitudes.

Digital filtering

A major benefit of the digital revolution of the 1960s has been the application of digital filters to seismic data. The filtering applications in exploration seismology are virtually unlimited; some examples are improvement of S/N ratios, system design with linear and nonlinear elements, pattern recognition, image enhancement, and inverse operators. Band-pass and 60-Hz notch filters have been used since the early 1920s. Until the late 1950s, these filters were by necessity analog devices with

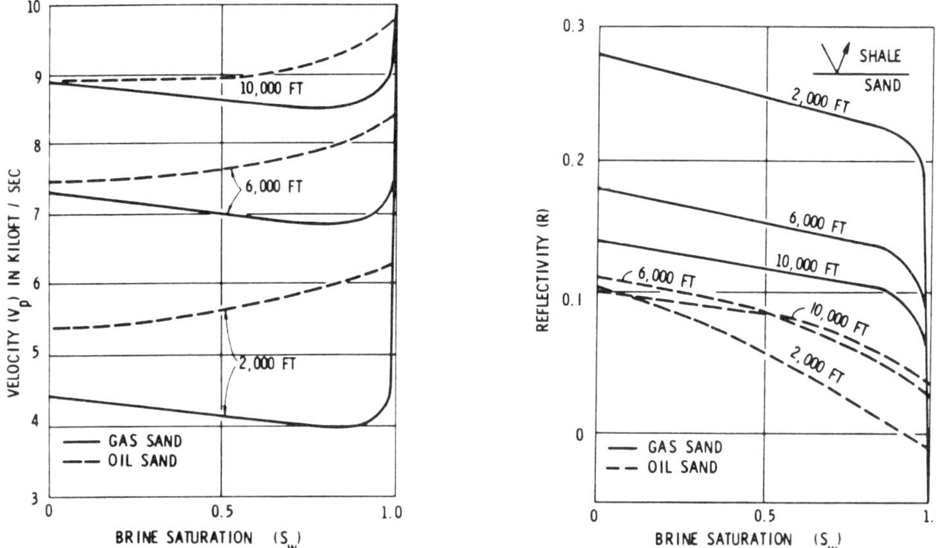

FIG. 13. Theoretical calculations of compressional velocity and reflectivity as functions of brine saturation for gas and oil sands at various depths in a Gulf Coast environment.

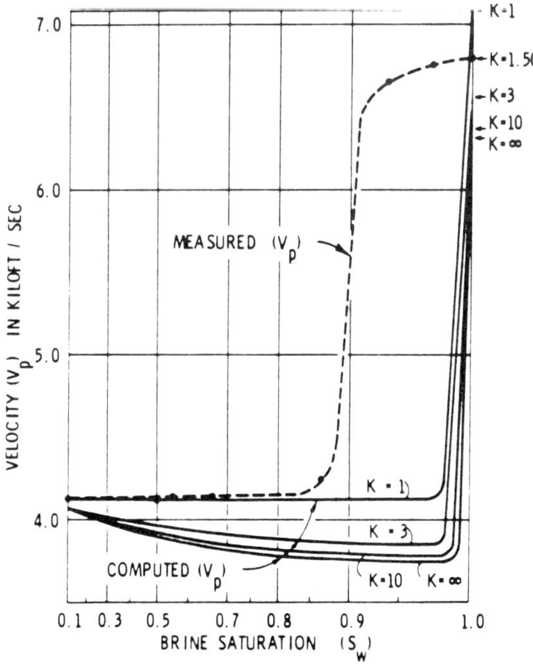

FIG. 14. Measured and computed theoretical compressional wave velocities as a function of brine saturation.

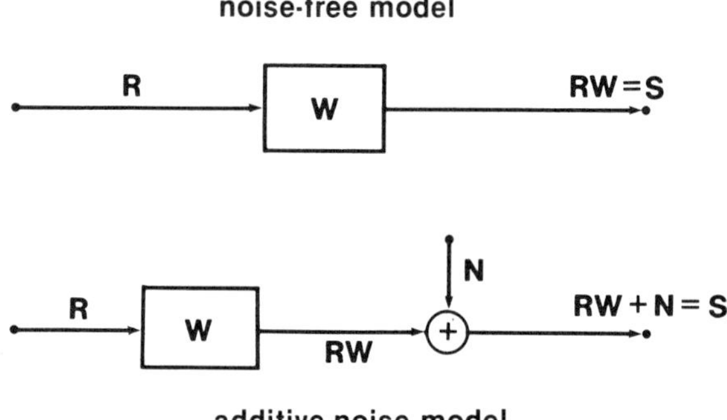

FIG. 15. The convolutional system model of seismic data in the frequency domain, with and without additive noise. Functions R, W, N, and S are functions of frequency. R = reflectivity sequence, W = basic wavelet, N = additive noise, and S = seismic model.

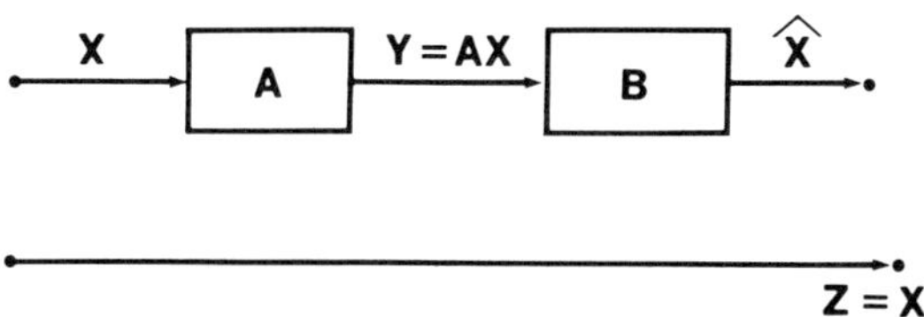

FIG. 16. A frequency domain representation of fundamental convolution and deconvolution operations. Note that while system A is defined to be a linear system, system B is not constrained to be linear and may be either a linear or nonlinear system. A = linear system, X = input to A, Y = output of A and input to B, B = deconvolution process, Z = desired output from B, and X = actual output from B.

small roll-off rejection rates. Research conducted by the Geophysical Analysis Group (GAG) at the Massachusetts Institute of Technology during the mid-1950s (Flinn et al, 1967) led to new developments that have carried through to the present.

A common use of filters, both analog and digital, is to increase S/N ratios by passing certain frequency bands. The noise is often constrained to relatively narrow bandwidths, and therefore it can be attenuated significantly by judicious choice of bandwidth. Examples are 60- and 120-Hz highline pickup, surface waves (ground roll), and wind noise. Band-limited filters can be designed and applied in either time or frequency domains. Ormsby (1961) filters serve adequately as convolutional operators, while time-domain recursive implementations are also possible (Shanks et al, 1972).

The convolutional model of the reflection seismogram (Robinson, 1957) is the basis of a large number of other important digital filters. In its simplest form the convolutional model assumes that a seismic trace is the result of convolving a "basic wavelet," attributed to the impulse response of the seismic energy source, with a sequence of spikes whose amplitudes correspond to subsurface reflection coefficients as described by equation (5), and whose time separation equals the appropriate two-way traveltime between source and receiver. Figure 15 illustrates convolutional models with and without additive noise. The basic wavelet generated from the source pulse includes other physical effects such as frequency-variant attenuation of energy

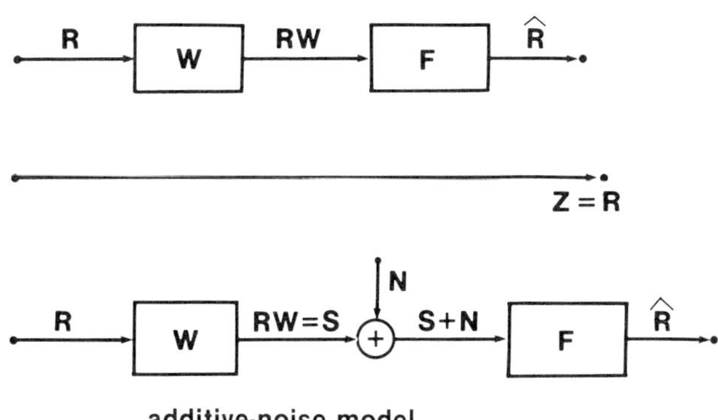

FIG. 17. Basic convolutional models for designing simple inverse filters, where *R*, *W*, *N*, *Z*, and *S* are functions of frequency. *R* = reflectivity sequence, *W* = basic wavelet, *N* = additive noise, and *S* = *RW* = seismic model.

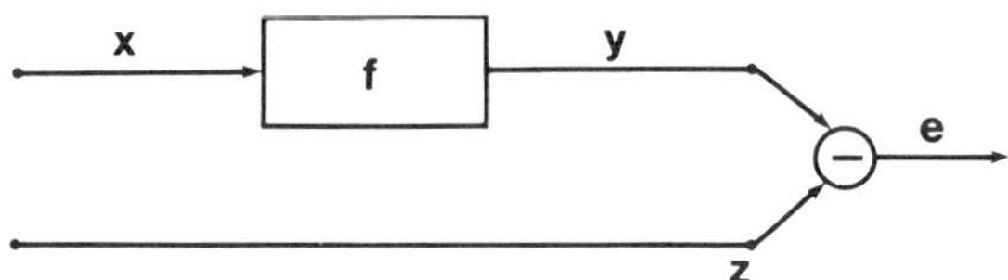

FIG. 18. Convolutional model for a generalized discrete Wiener shaping filter where **x, f, y, z,** and **e** are discrete time-domain vectors. **x** = input vector, **f** = discrete Wiener filter, **y** = actual output, **z** = desired output, and **e** = error between actual and desired outputs.

due to inelastic absorption, interbed (peg-leg) multiple reflections, residual NMO and static errors, geophone array effects, recording instrument filters, etc. These effects are often lumped together in one linear operator.

A number of the physical effects on propagating seismic waves mentioned above can be approximated as a linear filter or convolution operation. The inversion procedure (see Figure 16) to undo the convolution operation, known as "deconvolution," has led to a wide variety of techniques, which are discussed next.

A simple model for the noise-free case is convolution of a reflectivity sequence (R) with a basic wavelet (W). Then the seismic signal (S) is the frequency domain product of R and W. The goal is to deconvolve the wavelet W from S to obtain the reflectivity sequence, which represents the subsurface geology. Thus, the deconvolution operation in this case is to design linear systems F, as shown in Figure 17, such that their outputs are estimates of the reflectivity sequence R:

$$F = \frac{1}{W} . \tag{8}$$

In the presence of noise where $S = RW + N$, the filter F is given by

$$F = \frac{R}{RW + N} . \tag{9}$$

A discrete formulation of a Wiener shaping filter is shown in Figure 18. Wiener (1948) suggested designing analog filters based on the criterion of minimizing the mean-square-error between actual and desired outputs. In matrix form, the error can be written as

$$\mathbf{e} = \mathbf{X}\mathbf{f} - \mathbf{z}, \tag{10}$$

where $\mathbf{e}$ and $\mathbf{z}$ are $m \times 1$ column vectors; $\mathbf{f}$ is an $n \times 1$ column vector, and $\mathbf{X}$ represents an $m \times n$ convolution matrix (Figure 19). The squared error E, given by

$$\mathbf{E} = \mathbf{e}^T\mathbf{e} = \mathbf{f}^T\mathbf{X}^T\mathbf{X}\mathbf{f} - 2\mathbf{f}^T\mathbf{X}^T\mathbf{z} + \mathbf{z}^T\mathbf{z}, \tag{11}$$

is Wiener's objective function to be minimized. Wiener's normal equations are obtained by differentiating with respect to the filter coefficients $\mathbf{f}$, yielding

$$\frac{dE}{d\mathbf{f}} = 0 = 2\mathbf{X}^T\mathbf{X}\mathbf{f} - 2\mathbf{X}^T\mathbf{z}. \tag{12}$$

Thus we obtain a matrix formulation for the time-domain solution of a Wiener shaping filter:

$$\mathbf{f} = [\mathbf{X}^T\mathbf{X}]^{-1}\,\mathbf{X}^T\mathbf{z}. \tag{13}$$

Matrix $\mathbf{X}^T\mathbf{X}$ has Toeplitz symmetry with its elements consisting of autocorrelation coefficients of the input time series x. Similarly, vector $\mathbf{X}^T\mathbf{z}$ has elements corresponding to crosscorrelation coefficients obtained by crosscorrelating input time series x with desired output vector $\mathbf{z}$.

The first example deals with a Wiener inverse filter where the desired output $\mathbf{z}$ is a Dirac delta function (''spike''), as shown in Figure 19, consisting of elements containing all zeros except a single 1 denoting the spike. The spike location in vector $\mathbf{z}$ affects the magnitude of error vector and has been the subject of intensive investigation (Treitel and Robinson, 1966). The spiking filter is used, among other things, for deconvolving the basic seismic wavelet W (see Figure 17), for source signature deconvolution, and for removing the effects of recording instruments filter. An efficient algorithm for solving Wiener equations [equation (13)] was described by Levinson (1947), which corresponds to minimum-delay wavelets.

Another application of the Wiener filter involves the problem of bubble oscillations generated by some energy sources in marine recording. An impulse source in a water environment causes acoustic oscillations created by expansions and contractions of gas bubbles; the attenuation of these reverberations by means of a digital filter is called debubbling (Wood et al, 1978). Figure 20 shows the result of a filtering procedure involving matched as well as least-squares inverse components; here the source signature is known through field measurements. The objective is to shape the narrow-band, reverberating source signature into a broadband, highly compressed waveform. This example illustrates the ability of Wiener filters to shape input signals into desired output waveforms.

The third example utilizes a prediction error operator to remove reverberations caused by the marine water layer (Peacock and Treitel, 1969). Multiple reflections arise in the water layer because both the air-water and water-rock interfaces are strong reflectors with large acoustic impedance contrasts of -1 and about one-half, respectively. The water layer acts as a strong waveguide giving rise to slowly decaying signals whenever seismic pulses pass through it, as illustrated in Figure 21. Thus, the source pulse generates a reverberating wavetrain, which causes the reflections from the subsurface to be reverberatory. One approach to solving this prob-

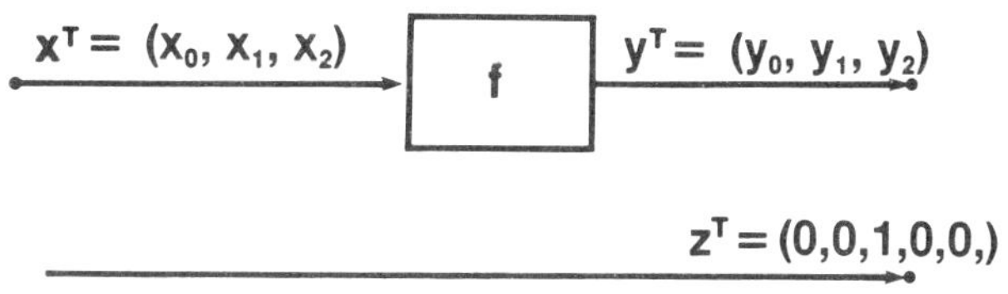

$$x^T = (x_0, x_1, x_2) \rightarrow f \rightarrow y^T = (y_0, y_1, y_2)$$

$$z^T = (0,0,1,0,0,)$$

$$\begin{bmatrix} x_0 & 0 & 0 \\ x_1 & x_0 & 0 \\ x_2 & x_1 & x_0 \\ 0 & x_2 & x_1 \\ 0 & 0 & x_2 \end{bmatrix} \begin{bmatrix} f_0 \\ f_1 \\ f_2 \end{bmatrix} = \begin{bmatrix} 0 \\ 0 \\ 1 \\ 0 \\ 0 \end{bmatrix} + \begin{bmatrix} e_1 \\ e_2 \\ e_3 \\ e_4 \\ e_5 \end{bmatrix}$$

$$X\,f = z + e$$

$$f = [X^T X]^{-1} X^T z \qquad \text{L.M.S.}$$

$$\begin{bmatrix} f_0 \\ f_1 \\ f_2 \end{bmatrix} = \begin{bmatrix} (x_0^2 + x_1^2 + x_2^2) & (x_1x_0 + x_2x_1) & (x_0x_2) \\ (x_1x_0 + x_1x_2) & (x_0^2 + x_1^2 + x_2^2) & (x_1x_0 + x_2x_1) \\ x_0x_2 & (x_0x_1 + x_1x_2) & (x_0^2 + x_1^2 + x_2^2) \end{bmatrix}^{-1} \begin{bmatrix} x_2 \\ x_1 \\ x_0 \end{bmatrix}$$

FIG. 19. An example of a simple Wiener inverse filter calculation for a three-sample input wavelet.

lem assumes that primary reflections from the subsurface arrive randomly in time (Robinson, 1957) and are therefore unpredictable. Reverberations associated with the water layer, on the other hand, are predictable. Therefore, a digital filter may be designed to remove the predictable component in the recorded signal while preserving the unpredictable component, which constitutes a prediction-error filter.

The first step in designing a prediction-error filter is to estimate the predictable component of a seismic trace. A model for obtaining Wiener normal equations for a prediction filter is shown in Figure 22. The desired output in this case simply becomes the input signal advanced in time, leading to the following set of normal equations corresponding to equation (13):

$$\begin{bmatrix} p(0) \\ p(1) \\ \cdot \\ \cdot \\ \cdot \\ p(L-1) \end{bmatrix} = \begin{bmatrix} R_{xx}(0) & R_{xx}(1) & \cdot & R_{xx}(L-1) \\ R_{xx}(1) & \cdot & & \cdot \\ \cdot & \cdot & & \cdot \\ \cdot & \cdot & & \cdot \\ \cdot & \cdot & & R_{xx}(1) \\ R_{xx}(L-1) & \cdot & R_{xx}(1) & R_{xx}(0) \end{bmatrix} \begin{bmatrix} R_{xx}(\alpha) \\ R_{xx}(\alpha+1) \\ \cdot \\ \cdot \\ \\ R_{xx}(\alpha+L-1) \end{bmatrix}. \tag{14}$$

Parameters α and L denote the prediction interval and prediction-operator length in number of samples, respectively. In actual practice, choice of parameter α depends upon water depth and the source pulse. The prediction operator coefficients $p(0)$, $p(1), \ldots, p(L-1)$ are used to construct the prediction-error operator p_e for prediction interval α,

$$p_e^T = [1,0,0,\ldots 0, -p_0, -p_1,\ldots, -p_{L-1}]. \tag{15}$$

This prediction-error operator is then convolved with the marine trace to yield the "dereverberated" (deconvolved) output.

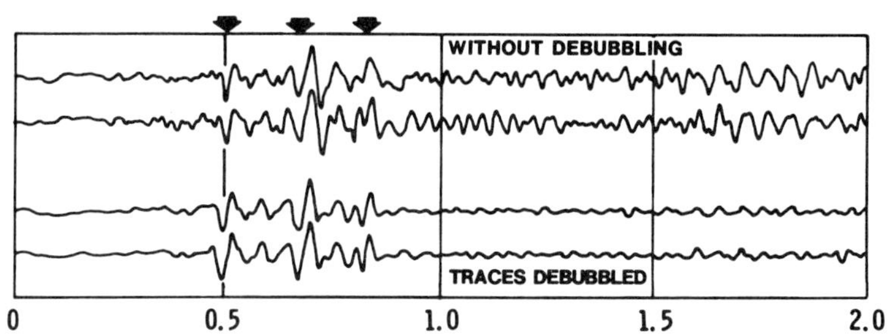

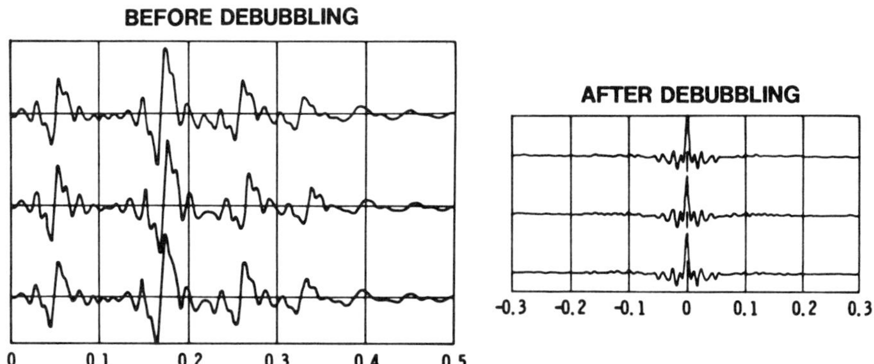

FIG. 20. An example of debubbling marine source signatures using a matched filter cascaded with a Wiener shaping filter.

An alternate approach to the problem consists of deriving a system response of the water-generated reverberation spike train based on physical principles. The z-transform expression of this system response can be shown to be (Wood and Treitel, 1975)

$$W(z) = \frac{1}{1 + cz^n}.$$

Here $c < 1$ is the normal incidence reflection coefficient at the water bottom, and integer n represents the two-way traveltime in the water layer as shown in Figure 21. A deterministic inverse filter to remove water reverberations is therefore

$$W^{-1}(z) = \frac{1}{W(z)} = 1 + cz^n. \tag{16}$$

In actual practice the reverberations encountered in marine environments are attenuated best using a prediction-error filter as given by equation (15); however, the simple two-sample operator of equation (16) provides insight into the physics of the problem. A comparison of Figures 23 and 24 shows that a properly designed prediction-error operator removes a significant amount of reverberating energy. A prediction-error operator with a unit sample prediction interval ($\alpha = 1$) is equivalent

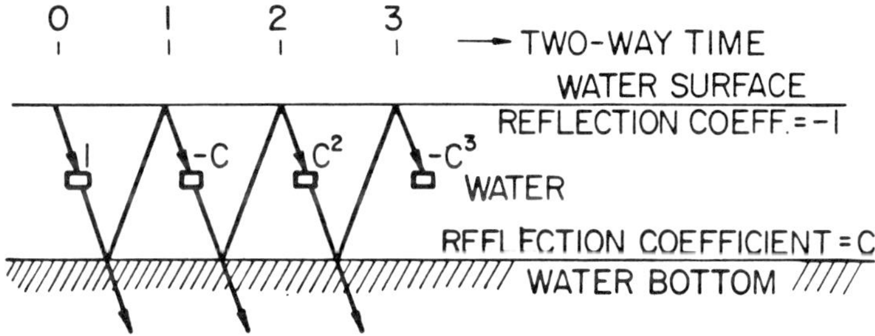

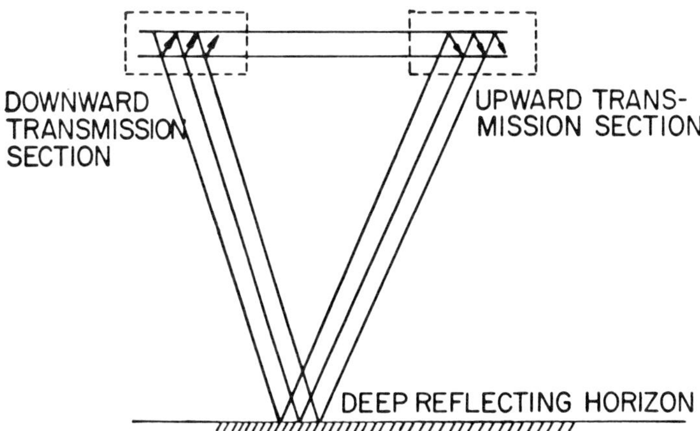

FIG. 21. The water layer in a marine environment acts as a waveguide giving rise to reverberations that decay slowly with time due to large reflection coefficients at the air-water and water-rock interfaces.

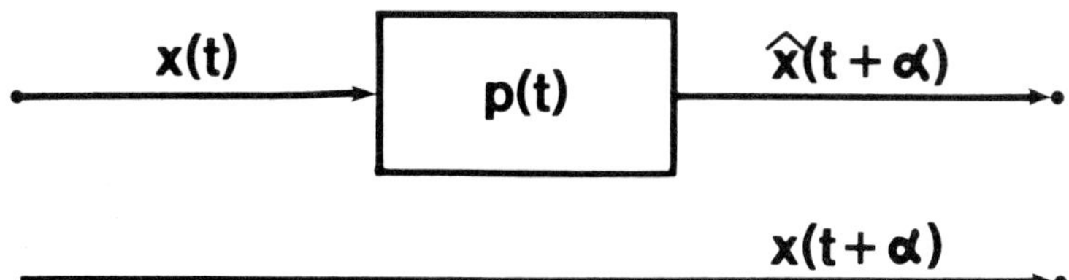

FIG. 22. A system model for designing a prediction filter using Wiener's least-mean-square criterion.

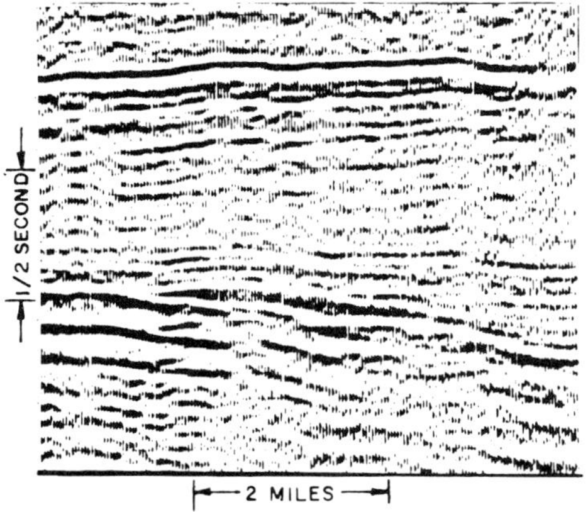

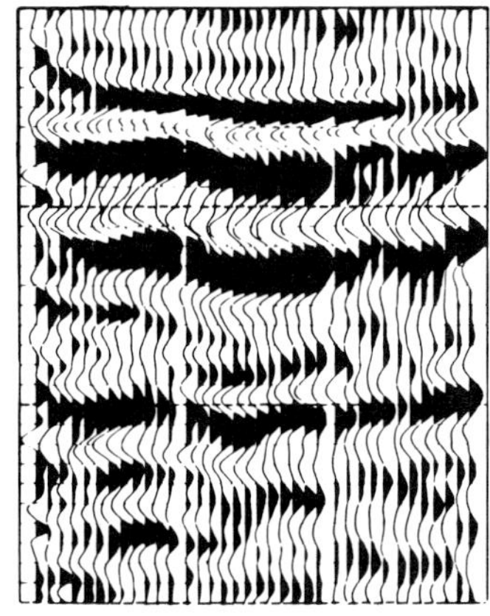

FIG. 23. CDP marine seismic data.

(within a constant scale factor) to a Wiener spiking filter. Significantly, these normal equations occur frequently in many linear prediction problems such as speech compression (Makhoul, 1975).

The above discussion of digital filtering has omitted many other techniques used in seismic data processing. The material given here should provide a good starting point for the reader interested in pursuing the knowledge further. Kalman filtering algorithms, as well as the homomorphic deconvolution approach (Ulrych, 1971), have been applied with some success. Alternate objective functions and minimization criteria using nonlinear techniques based on statistical theory have been developed and are under further investigation. Two-dimensional filters are also used extensively in seismic data processing (Robinson, 1967).

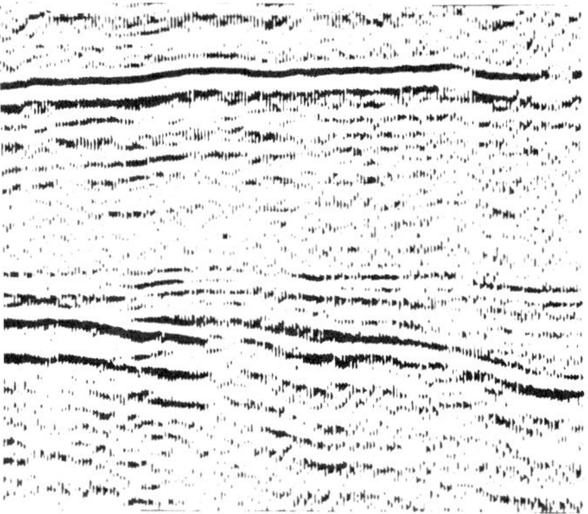

PREDICTIVE DECONVOLUTION (SECOND ZERO CROSSING)

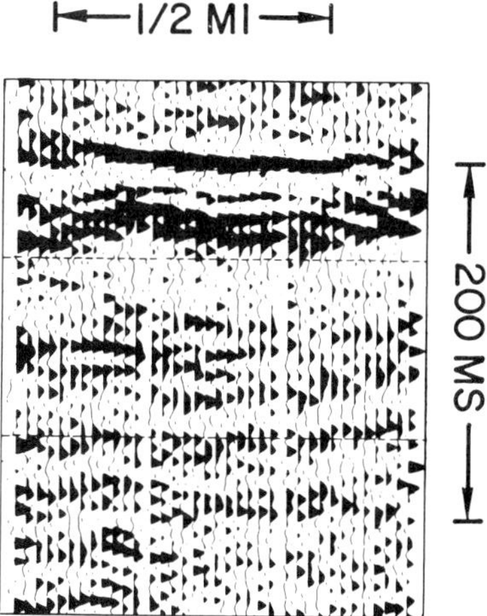

WIENER-LEVINSON (UNIT PREDICTIVE) INVERSE

FIG. 24. Deconvolution of the seismic data shown in Figure 23 using least-mean-square, prediction-error filters.

Modeling and Interpretation

Successful interpretation depends upon the ability to combine the knowledge gained from seismic data with geologic information derived from data such as well logs and cores. Two general methods are forward and inverse approaches.

Forward approach

The forward approach employs a deterministic seismic model of the earth constructed from the geologic model of the subsurface. A comparison between the actual seismic data and seismic model results in modification of the geologic model to reduce any discrepancies. An iterative comparison process continues until a satisfactory fit between synthetic and actual data is obtained. Models may be altered many times before an interpreter is convinced that he has developed a geologically valid solution. Small-scale physical models of the geologic subsurface can be constructed from materials that simulate the earth's acoustic properties. Electronic sensors measure the model's sonic response to a scaled seismic source. Theoretical models involve construction of mathematical models of the earth's acoustic properties. The seismic response of these mathematical models is simulated by solution of the elastic differential equations of motion subject to boundary conditions imposed by geologic criteria. Mathematical models constructed in one and two spatial dimensions are in common use; interest in and need for 3-D modeling studies is growing rapidly. A separate chapter treats the subject of seismic models in some detail.

Synthetic seismogram.—The one-dimensional model is a single synthetic seismic trace simulating a normal incidence seismic trace at the well location. Sonic and density logs are used to construct a seismic model of the subsurface. The model consists of a series of horizontal acoustic impedance boundaries as shown in Figure 25, from which a reflectivity time series is computed using equation (5). This time series consists of the arrival times and amplitudes of normal incidence, compressional reflected plane waves produced by an ideal, impulsive source (spike) from the acoustic boundaries in an elastic medium. The energy from a field seismic source is band-limited compared to the Dirac delta (spike) function, so direct comparison between a reflectivity series and seismic data is difficult. The shape of the impulse contained in processed seismic data represents a combined effect of a field seismic source, the lumped impulse response of the earth, and any acquisition-processing distortions. Figure 26a shows how an ideal impulse is modified by the earth filter as well as by the acquisition and processing effects to produce a basic seismic wavelet. Figure 26b shows the result of combining the basic seismic wavelet with the reflectivity time series to produce a synthetic seismogram that can be compared with the actual data.

Figure 27 shows a comparison of the synthetic seismic trace and sonic log with field data in the vicinity of the well. Such an exercise is useful in correlating the reflection events on the field data with the subsurface reflecting boundaries.

The acquisition and processing effects on the basic seismic wavelet cause a wide variety of phase distortions in the seismic data. In actual practice, the wavelet rarely has a zero-phase characteristic, thereby creating a severe problem in the proper correlation of field data with zero-phase synthetic seismograms. If the wavelet in the field data is zero-phase, reflection arrival times correspond accurately to wavelet peak-and-trough times. The interpreter picking peaks and troughs as reflection arrival times is accurate only to the extent that it can be assumed the data contain zero-

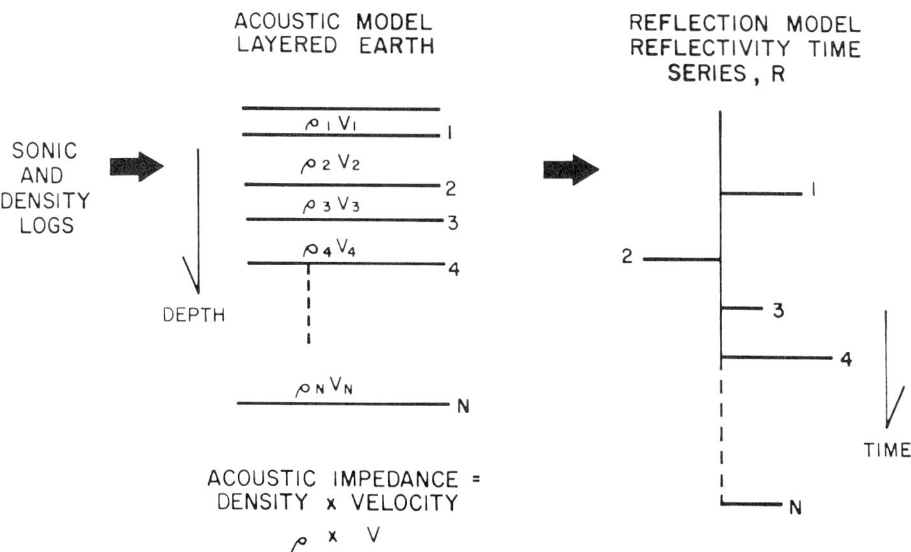

FIG. 25. A 1-D acoustic model of a stratified earth is derived from log measurements. The model then is used to construct a reflectivity times series, which is a broadband synthetic seismogram.

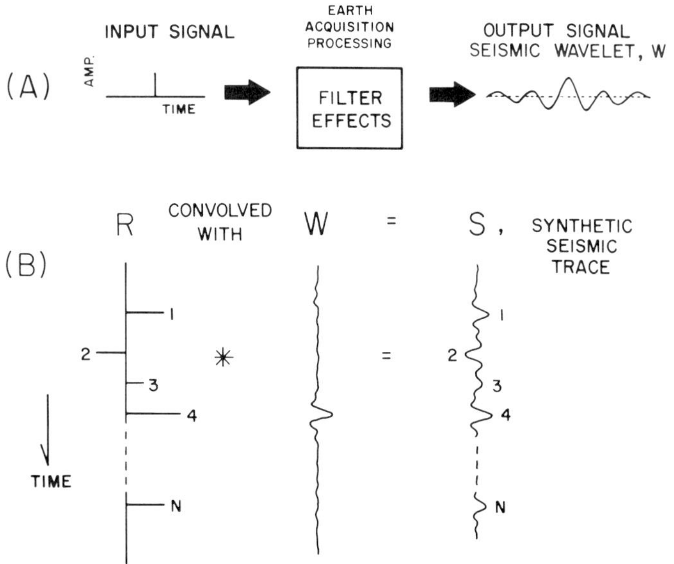

FIG. 26. (a) The convolutional system model shows the basic seismic wavelet resulting from earth, acquisition, and processing filter effects. (b) Construction of normal-incidence synthetic seismograms using the convolutional system model.

phase wavelets. Thus, the symmetry inherent in zero-phase wavelets when combined with short pulse length can help simplify the interpretation and improve its accuracy. Figure 28 illustrates the effect that phase distortions can have on the basic appearance of seismic data and on the peak-and-trough times that are interpreted as reflection arrival times. The procedure of applying a digital filter to remove the phase distortions is called "wavelet processing" or "dephasing" (Lindsey et al, 1978). It is usually accomplished with a simple inverse filter when an estimate of the wavelet is available.

Numerous statistical and deterministic algorithms exist for estimating wavelets, but no particular one is applicable in all situations. Wavelet estimation remains an

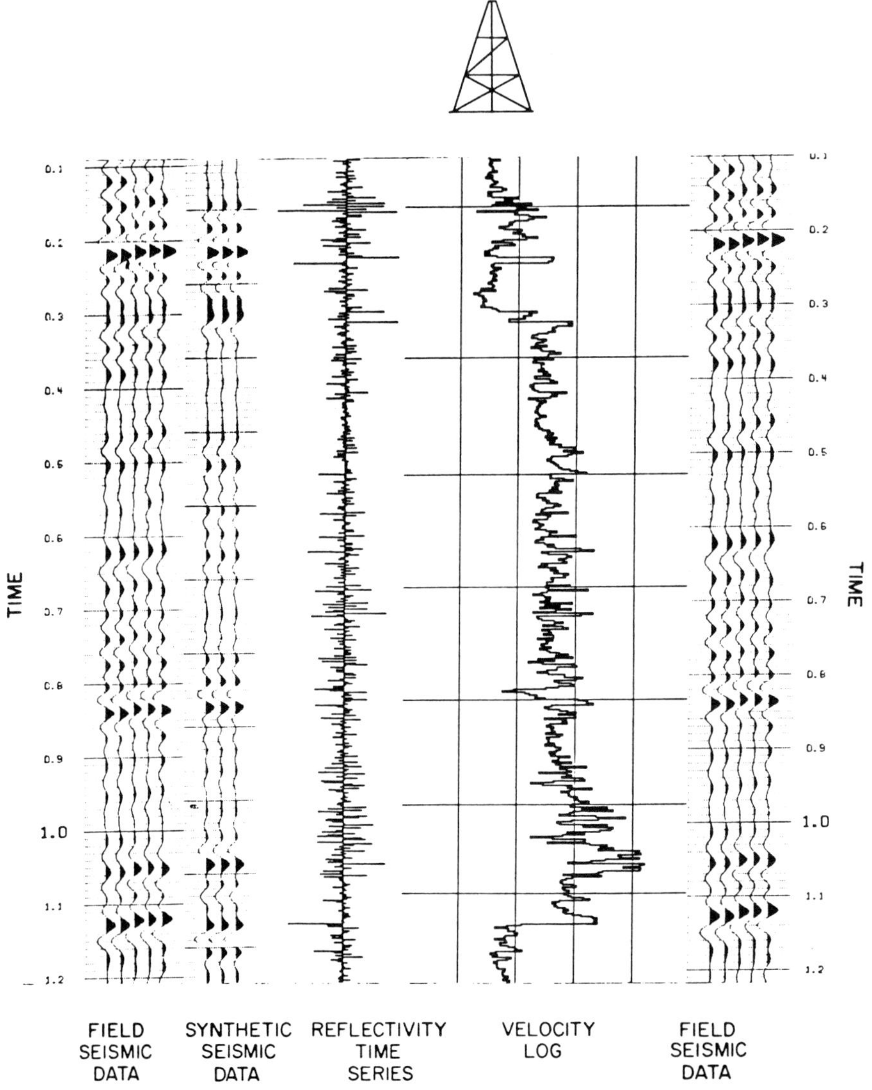

FIG. 27. A synthetic seismogram (band-limited reflectivity series) derived from well logs, compared with real seismic data. (Data courtesy of Venex Corporation)

area of active research and development. A deterministic approach combines the use of sonic and density logs and field seismic data in close proximity to the well location. The reflectivity time series is correlated with the field data traces nearest the well to extract the wavelet. Once the wavelet is estimated, an inverse filter can be designed and applied to the field data to convert the wavelet's phase to zero, as shown in Figures 29 and 30.

Two- and three-dimensional modeling studies provide synthetic seismic profiles for spatially varying geologic models. The interpreter is thus able to see the acoustical response to both structural and stratigraphic changes. Figure 31 shows a 2-D geologic model and the resulting synthetic seismic profile. Model parameters such as layer thicknesses and velocities can be varied iteratively until event arrival times and amplitudes match the actual field data within specified tolerances. Results of

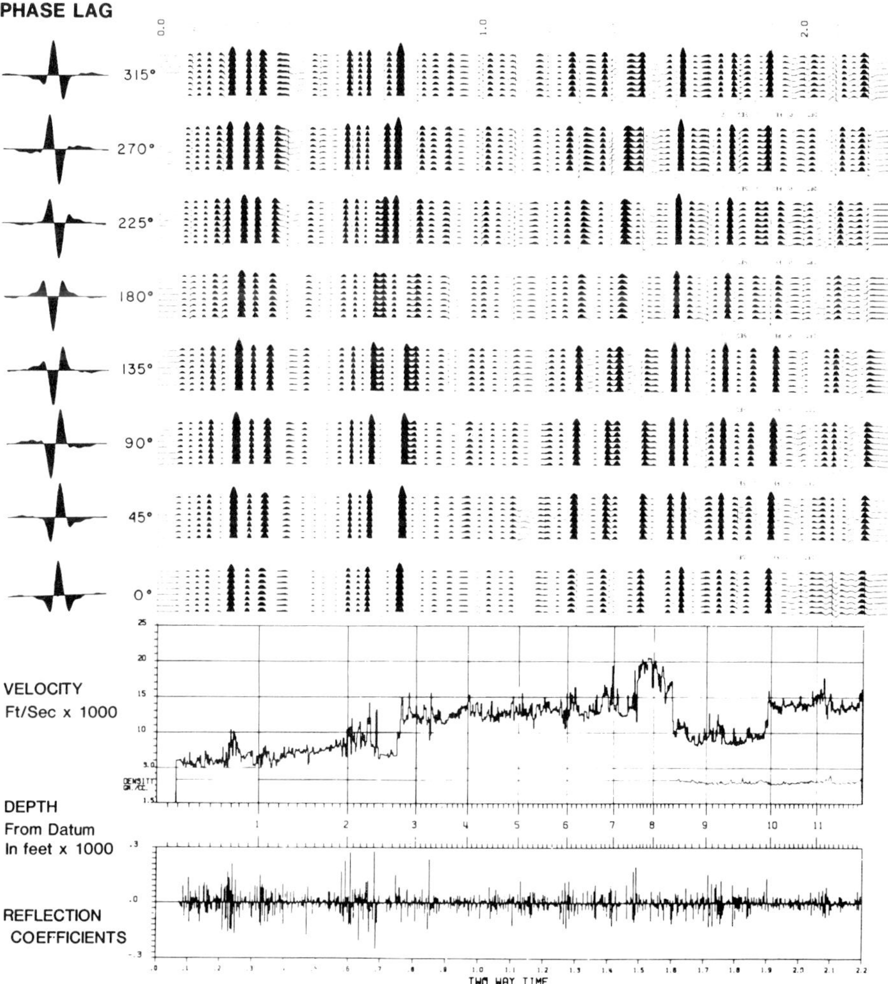

FIG. 28. Synthetic seismograms obtained by convolving a reflectivity sequence with a series of constant-phase basic wavelets. (Data courtesy of Venex Corporation)

such model studies often emphasize the importance of acquiring high-frequency data in the field and subsequently using deconvolution and wavelet processing techniques to achieve the resolution necessary to observe thin layers (Kallweit and Wood, 1982). For hydrocarbon reserve estimates in a field, an explorationist needs to know the vertical thickness and lateral extent of the reservoir. Many reservoirs, especially those that are stratigraphic in nature, have relatively small dimensions for seismic resolution. Consequently, determining the limits of seismic resolution is a problem of major importance. For vertical resolution, the simplest situation is to identify reflections from the top and bottom of a thin bed (Widess, 1973) and to measure the time interval. Zero-phase wavelets are required for accurate interpretations. The convolutional model, in this case, is a zero-phase wavelet convolved with two spikes (of equal or unequal amplitudes as the case may be) separated by the time interval

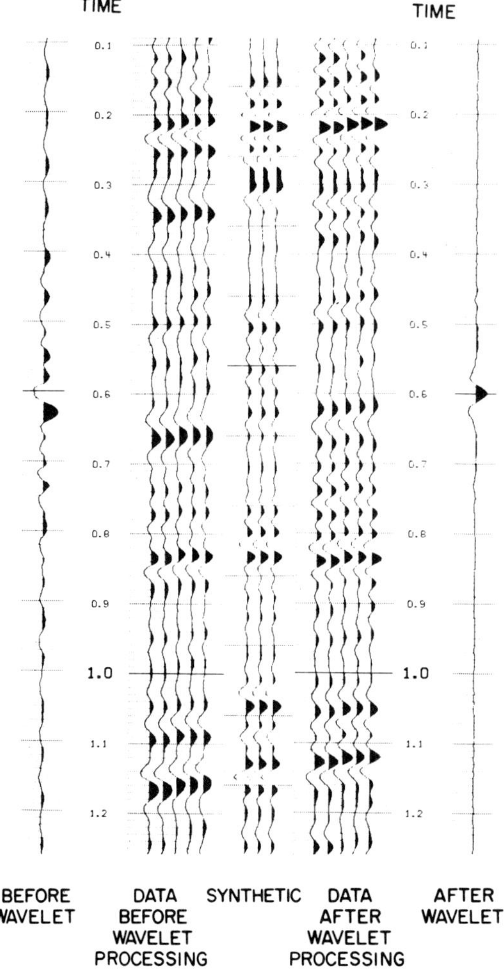

FIG. 29. An example of wavelet processing with a basic wavelet derived from well logs in the close vicinity of the seismic field line. (Data courtesy of Venex Corporation)

corresponding to the two-way vertical traveltime through the bed as shown in Figure 32. Constructive and destructive tuning of the band-limited wavelets cause distortions resulting in a discrepancy between the measured and actual times. This phenomenon is illustrated in Figure 33 for equal spikes of the same and reverse polarities. In the noise-free case, the mathematical relationship is simple and the limits of temporal resolution can be analytically established. The practical limit of resolution occurs at a separation where the central lobe of a symmetrical wavelet tunes with its adjacent side lobe. The tuning thickness resolution limit corresponds to a one-quarter wavelength condition and approximates $1/1.4 f_u$ for wavelets with white spectra, where f_u represents the upper terminal frequency of a broadband wavelet in

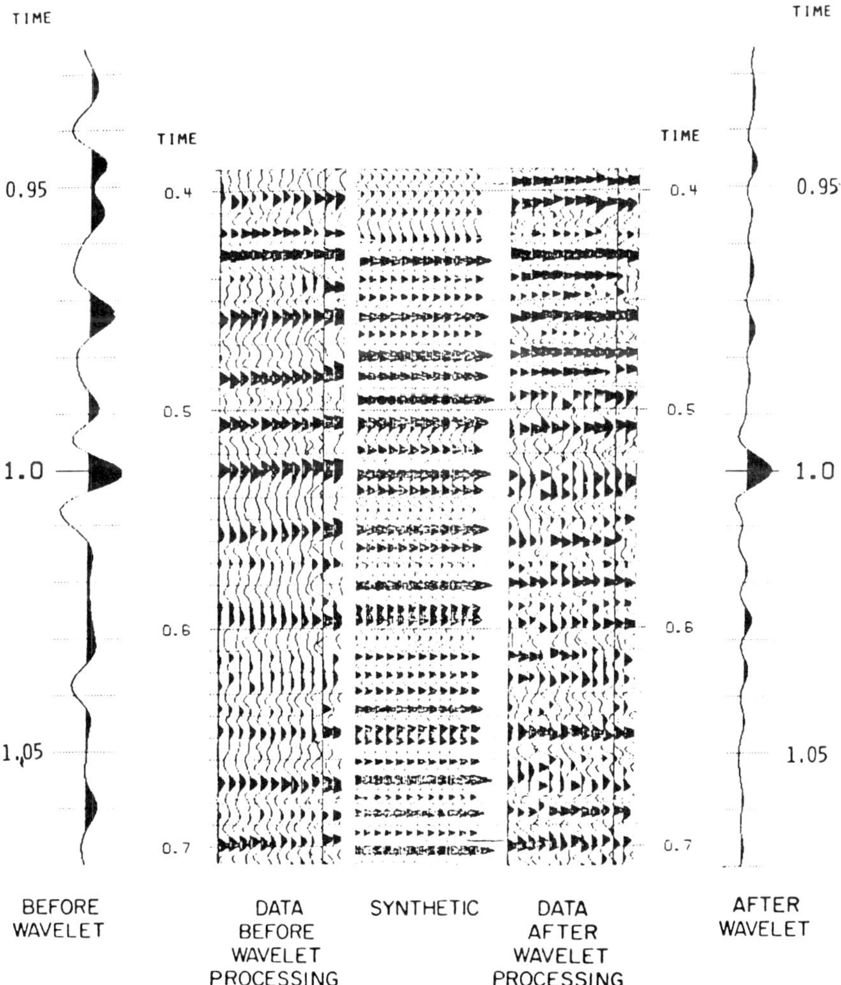

FIG. 30. An example of wavelet processing with a basic wavelet derived from well logs in the close vicinity of the seismic field line.

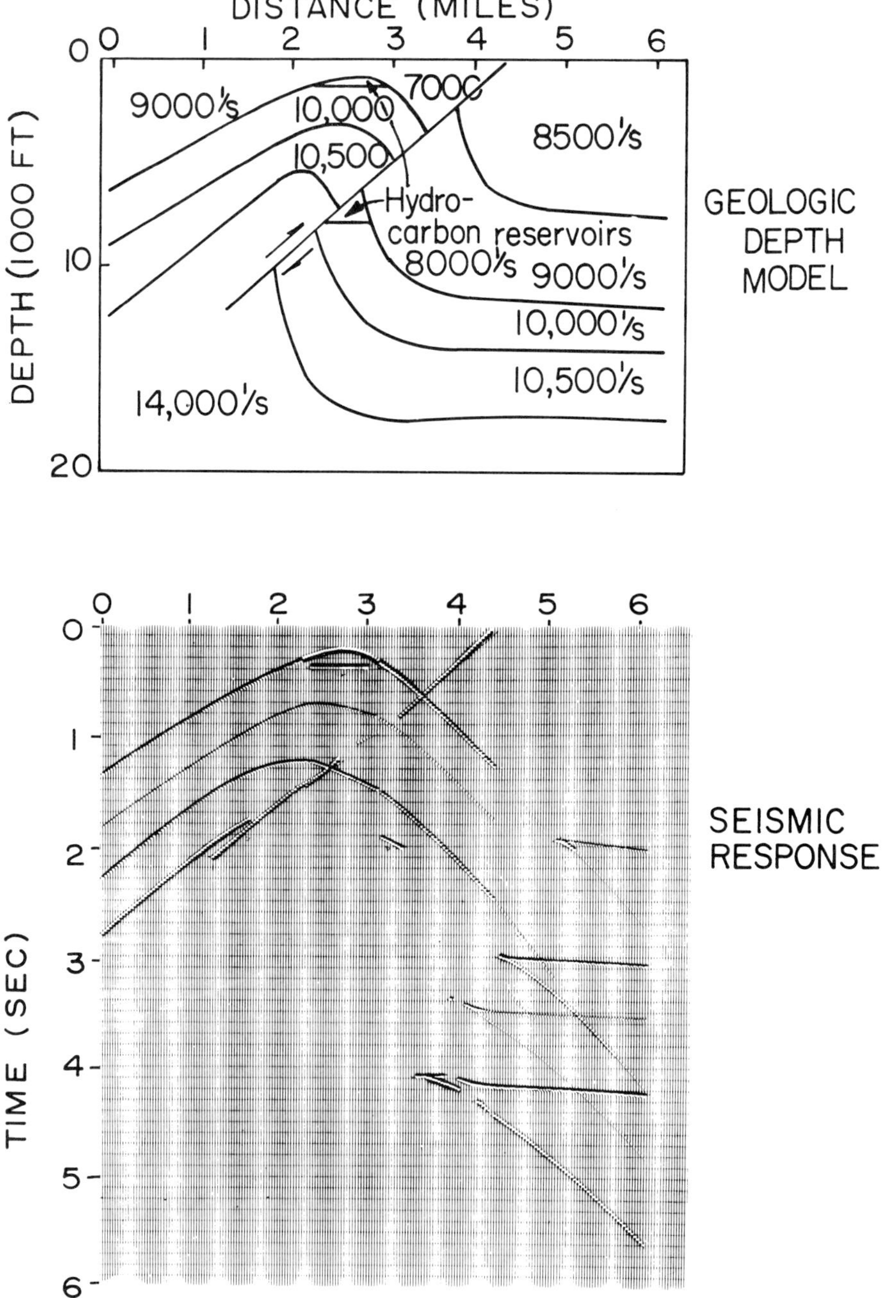

FIG. 31. A 2-D model of a faulted anticline structure with hydrocarbon reservoirs.

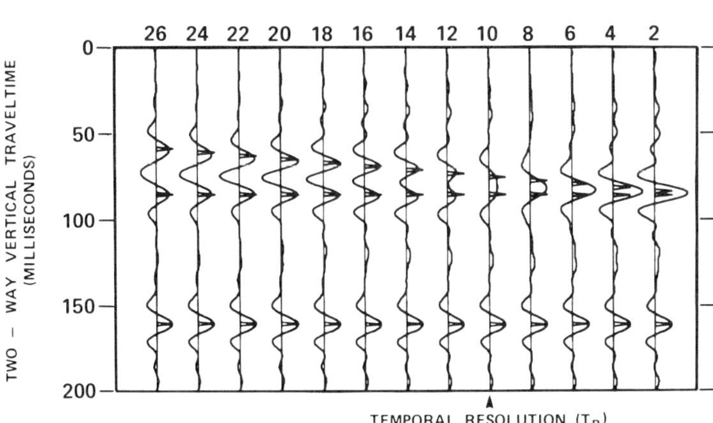

FIG. 32. The seismic response for a wavelet convolved with two spikes of equal amplitude and equal polarity is a composite wavelet that varies as a function of spike separation, i.e., layer thickness.

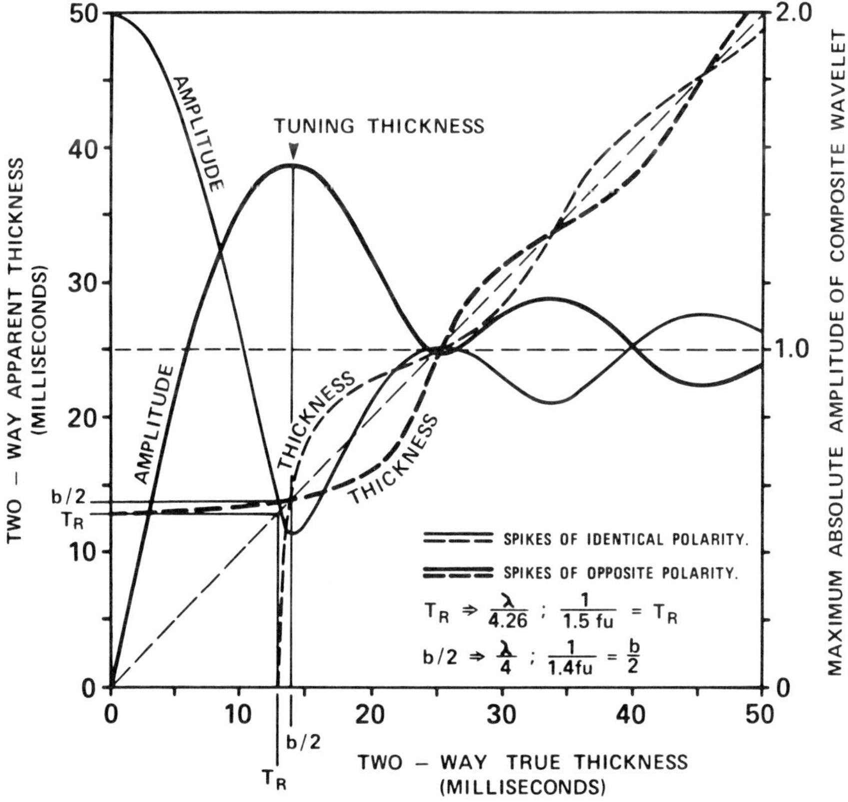

FIG. 33. Resolution and detection graphs for two-termed reflectivity spike models consisting of a wedge of material bounded above and below by dissimilar material. These curves show tuning thickness and temporal resolution limits for a 12.5–50 Hz sinc wavelet.

excess of two octaves. In the tuning range, the time interval of the spikes is encoded in the seismic amplitude, assuming that the entire amplitude variation is caused by the tuning effects (Neidell and Poggiagliolmi, 1979). Amplitude calibration then permits net-pay thickness calculations for arbitrarily thin beds.

Inverse approach

In the forward approach, the geologic model was assumed to be an acoustic impedance sequence, which was then operated on to produce a synthetic seismic model for comparison with actual seismic data. The inverse approach starts with the actual seismic profiles and proceeds to calculate the model or physical situation, namely, the acoustic impedance sequence. In the highest resolution limits, this approach allows an interpreter to obtain information on physical rock properties from seismic data which is used conventionally to provide structural and stratigraphic information. The amplitudes of the reflected events on the seismic data are assumed to be proportional to the actual reflection coefficients. The following inversion equation is obtained:

$$a_{i+1} = \frac{a_i(1 + r_i)}{1 - r_i}, \tag{17}$$

where a_i represents the acoustic impedance $\rho_i v_i$. A solution for equation (17) can be obtained through recursion.

$$\frac{a_{i+1}}{a_0} = \prod_{j=0}^{i} \frac{1 + r_j}{1 - r_j}. \tag{18}$$

Approximate solutions for equation (18) can be developed by using the fact that acoustic impedances usually vary slowly as a function of depth. For example, equation (17) can be written as

$$r_i = \frac{a_i + \Delta a - a_i}{a_i + \Delta a + a_i}, \text{ where } a_{i+1} = a_i + \Delta a,$$

$$r_i = \frac{\Delta a}{2a_i + \Delta a} \cong \frac{\Delta a}{2a_i},$$

or (19)

$$r_i \cong \lim_{\Delta t \to 0} \frac{1}{2} d\,(\ln a).$$

Thus,

$$r \cong \frac{1}{2} d\,(\ln a).$$

Integrating both sides of equation (19) with respect to time yields

$$\frac{1}{2} \ln \frac{a(t)}{a(0)} \cong \int_0^t r(\tau) d\tau. \tag{20}$$

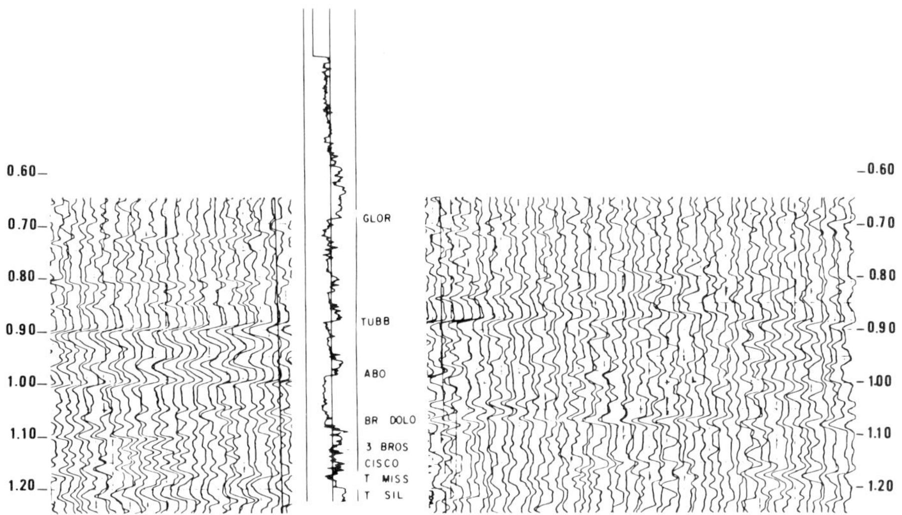

FIG. 34. A comparison of a band-limited acoustic impedance log with seismic derived acoustic-impedance profiles.

This mathematical relationship is the reason exploration seismologists frequently use the term "integration" interchangeably with "inversion."

Recent improvements in the acquisition and processing technology can provide seismic data of sufficient quality in many areas to produce acoustic impedance profiles of the subsurface (Becquey et al, 1979; Lindseth, 1979). Bandwidths exceeding two octaves and S/N ratios greater than 12 dB are generally required. A typical example is shown in Figure 34.

Migration.—A problem of fundamental importance in seismic interpretation is the proper positioning of reflected events in space and time. Interpreters must map reflections observed in the time domain into the depth domain. This transformation is not simple even in the 2-D case, as may be noted in Figure 31.

In space, a stacked trace is associated with the source-receiver midpoint. However, a subsurface reflection point will not be vertically below this midpoint if the reflecting point is not a part of a horizontal plane (see Figure 8). Migration of seismic data is necessary whenever subsurface geologic structures deviate from a transversely homogeneous, horizontally stratified medium. Migration also brings diffraction events, resulting from subsurface discontinuities such as faults and truncations, into focus. In such situations, the seismic profiles must be "migrated" to position reflection events at proper subsurface location, both in space and time. The migration problem for linear segments is illustrated in Figure 35. A segment P_1P_2 of true dip angle β appears on an unmigrated CDP profile as linear segment $P_1' \, P_2'$ with apparent dip angle α. Simple geometrical considerations show that apparent and true dip angles bear the following relationship:

$$\sin \beta = \tan \alpha. \tag{21}$$

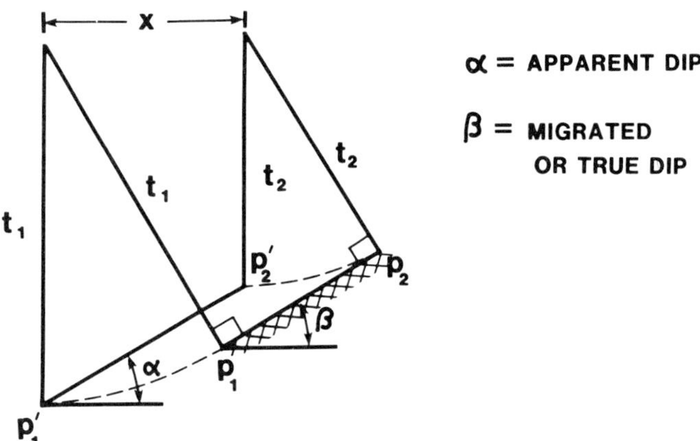

FIG. 35. An example of how linear segments migrate on CDP seismic sections.

Numerous computer algorithms have been developed for migration; they have improved significantly in the last decade due to application of scalar wave-equation theory. These techniques fall under the broad category of wave-equation migration and can be further subdivided into finite-difference, integral, and Fourier transform solutions. Each method has its advantages and disadvantages, and all of them are used with good success. Finite-difference techniques, pioneered by Claerbout (1971), led the way in the early 1970s. The Kirchhoff integral-equation techniques in two and three dimensions were discussed by Schneider (1978), while Stolt (1978) described a Fourier transform approach. An example of wave-equation migration is shown in Figure 36 (pp. 88–89) to illustrate that migration is a necessary tool for seismic interpreters. The general concepts associated with the migration of seismic data are discussed by Johnson and French in Chapter 5.

Summary

This chapter outlines how seismic data are acquired, processed, and interpreted. New developments are continually improving the seismic technology. Many developments are undergoing intensive operational evaluation. For example, the interactive terminals are beginning to provide viable interpretive systems due to the increased power of mini- and microcomputers. Three-dimensional seismic surveys offer promise in the near future for exploration in difficult areas as higher capacity recording systems with 500–1000 channels become an economic reality. Vertical profiles conducted within well bores, as contrasted with conventional surface surveys, are providing accurate velocity control, higher frequency data near the well bore and information to help distinguish the signals from certain types of noise. The use of shear waves in conjunction with compressional waves has been an active area of research; this would lead to seismic techniques for directly measuring the Poisson's ratio of the subsurface strata. Direct measurement of rock properties will significantly improve stratigraphic mapping capabilities.

A continuing trend within the industry has been to increase the resolving power of the reflection data through improved acquisition and processing procedures, since resolution is crucial for mapping subtle stratigraphic traps. Specialized field procedures, under favorable field conditions, allow recovery of frequencies up to 200 Hz, as opposed to the more usual 50 Hz, to depths of 8000 ft. Innovations such as those currently in progress represent a never-ending quest of geophysicists involved in hydrocarbon exploration.

REFERENCES

Anstey, N. A., 1970, Seismic prospecting instruments: Signal characteristics and instrument specifications, *in* Geoexploration Monog., v. 1, Berlin, Gerbruder Borntraeger.

Becquey, M., Lavergne, M., and Willm, C., 1979, Acoustic impedance logs computed from seismic traces: Geophysics, v. 44, p. 1485–1501.

Biot, M. A., 1956, Theory of propagation of elastic waves in a fluid-saturated porous solid, I & II: J. Acoust. Soc. Am., v. 28, p. 168–191.

Claerbout, J. F., 1971, Toward a unified theory of reflector mapping: Geophysics, v. 36, p. 467–481.

Crook, N. T., 1979, Sharpening our tools: Geophysics, v. 44, p. 1635–1636.

Dix, C. H., 1952, Seismic prospecting for oil: Harper and Brothers.

Domenico, S. N., 1974, Effect of water saturation on seismic reflectivity of sand reservoirs encased in shale: Geophysics, v. 39, p. 759–769.

——— 1976, Effect of brine-gas mixture on velocity in an unconsolidated sand reservoir: Geophysics, v. 41, p. 882–894.

Evenden, B. S., and Stone, D. R., 1971, Seismic prospecting instruments: Instrument performance and testing, *in* Geoexploration Monog., v. 2, Berlin, Gerbruder Borntraeger.

Ewing, W. M., Jardetsky, W. S., and Press, F., 1957, Elastic waves in layered media: New York, McGraw-Hill.

Flowers, B. S., 1976, Overview of exploration geophysics—Recent breakthrough and challenging new problems: AAPG Bull., v. 60, p. 3–11.

Geertsma, J., 1961, Velocity-log interpretation: The effect of rock bulk compressibility: SPE J., v. 1, p. 235–248.

Grant, F. S., and West, G. F., 1965, Interpretation theory in applied geophysics: New York, McGraw-Hill.

Heiland, C. A., 1940, Geophysical exploration: Englewood Cliffs, New Jersey, Prentice-Hall.

Jakosky, J. J., 1950, Exploration geophysics: Trija.

Kallweit, R. S., and Wood, L. C., 1982, The seismic resolution of zero-phase wavelets: Geophysics, v. 47, no. 7, p. 1035–1046.

Kramer, F. S., Peterson, R. A., and Walter, W. C., Eds., 1968, Seismic energy sources 1968 handbook: United Geophysical Corp.

Levinson, N., 1947, The Wiener rms error criterion in filter design and prediction: J. Math. Phys., v. 25, p. 261–278.

Lindseth, R. O., 1979, Synthetic sonic logs—A process for stratigraphic interpretation: Geophysics, v. 44, p. 3–26.

Lindsey, J. P., and Craft, C. I., How hydrocarbon reserves are estimated from seismic data: World Oil, v. 177, p. 23–25.

Lindsey, J. P., Dedman, E. V., and Ausburn, B. E., 1978, New seismic techniques define stratigraphic, lithologic details: World Oil, v. 5, p. 56–60.

Makhoul, J., 1975, Linear prediction: A tutorial review: Proc. IEEE, v. 63, p. 561–580.

May, B. T., and Hron, F., 1978, Synthetic seismic sections of typical petroleum traps: Geophysics, v. 43, p. 1119–1147.

Mayne, W. H., 1962, Common reflection point horizontal data stacking techniques: Geophysics, v. 27, p. 927–938.

McNabb, D., 1974, Bright-spot warning: It's not infallible: Oil and Gas J., v. 72, p. 50–51.

Neidell, N. S., and Taner, M. T., 1971, Semblance and other coherency measures for multichannel data: Geophysics, v. 36, p. 482–497.

Neidell, N. S., and Poggiagliolmi, E., 1977, Stratigraphic modeling and interpretation—Geophysical principles and techniques: AAPG Special Memoir 26, p. 389–416.

Nettleton, L. L., 1940, Geophysical prospecting for oil: McGraw-Hill.

Northwood, E. J., Weisinger, R. C., and Bradley, J. J., 1967, Recommended standards for digital tape formats: Geophysics, v. 32, p. 1073–1084.

Ormsby, J. F. A., 1961, Design of numerical filters with applications to missile data processing: J. Assoc. Comp. Mach., v. 8, p. 440–466.

Peacock, K. L., and Treitel, S., 1969, Predictive deconvolution: Theory and practice: Geophysics, v. 34, p. 155–169.

Robinson, E. A., 1957, Predictive decomposition of seismic traces: Geophysics, v. 22, p. 767–778.

——— 1967, Multichannel times series analysis with digital computer programs: San Francisco, Holden-Day.

Schneider, W. A., 1978, Integral formulation for migration in two and three dimensions: Geophysics, v. 43, p. 49–76.

Shanks, J. L., Treitel, S., and Justice, J. H., 1972, Stability and synthesis of two-dimensional recursive filters: IEEE Trans. Audio and Electroacoust., v. AU-20, p. 115–128.

Stolt, R. H., 1978, Migration by Fourier transform: Geophysics, v. 43, p. 23–48.

Taner, M. T., and Koehler, F., 1969, Velocity spectra-digital computer derivation and application of velocity functions: Geophysics, v. 34, p. 859–881.

Taner, M. T., Koehler, F., and Alhilali, K. A., 1974, Estimation and correction of near-surface time anomalies: Geophysics, v. 39, p. 441–463.

Taner, M. T., Cook, E. E., and Neidell, N. S., 1970, Limitations of the reflection seismic method: Lessons from computer simulations: Geophysics, v. 35, p. 551–573.

Treitel, S., and Robinson, E. A., 1966, The design of high-resolution digital filters: IEEE Trans. Geosci. Electr., v. GE-4, p. 25–38.

Ulrych, T. J., 1971, Application of homomorphic deconvolution to seismology: Geophysics, v. 36, p. 650–660.

Widess, M. B., 1973, How thin is a thin bed?: Geophysics, v. 38, p. 1176–1180.

Wiener, N., 1948, The extrapolation, interpolation, and smoothing of stationary time series: New York, John Wiley and Sons.

Wiggins, R., Larner, K., and Wisecup, R. D., 1976, Residual statics analysis as a general linear inverse problem: Geophysics, v. 41, p. 922–938.

Wood, L. C., and Treitel, S., 1975, Seismic signal processing: Proc. IEEE, v. 63, p. 649–661.

Wood, L. C., Heiser, R. C., Treitel, S., and Riley, P. L., 1978, The debubbling of marine source signatures: Geophysics, v. 43, p. 715–729.

REFERENCES FOR GENERAL READING

Data acquisition

Crawford, J. M., Doty, W. E. N., and Lee, M. R., 1960, Continuous signal seismograph: Geophysics, v. 25, p. 95–105.

Cunningham, A. B., 1979, Some alternate vibrator signals: Geophysics, v. 44, p. 1901–1921.

Geyer, R. L., 1970, The Vibroseis system of seismic mapping: J. Can. SEG, v. 6, p. 39–57.

Goupillaud, P. L., 1976, Signal design in the Vibroseis technique: Geophysics, v. 41, p. 1291–1304.

Gray, R. L., Leitinger, J. H., and Hollister, J. C., 1968, Determination of seismic system distortion and its compensation using digital filters: Geophysics, v. 33, p. 285–301.

Hermont, A. J., 1956, Design principles for seismic reflection amplifiers: Geophys. Prospect., v. 4, p. 279–293.

Holzman, N., 1963, Chebyshev optimized geophone arrays: Geophysics, v. 28, p. 145–155.

Lombardi, L. V., 1955, Notes on the use of multiple geophones: Geophysics, v. 20, p. 215–226.

Parr, J. O., and Mayne, W. H., 1955, A new method of pattern shooting: Geophysics, v. 20, p. 539–564.

Smith, M. K., 1956, Noise analysis and multiple seismometer theory: Geophysics, v. 21, p. 337–360.

Data processing

Claerbout, J. F., 1976, Fundamentals of geophysical data processing: New York, McGraw-Hill.

Craft, C. I., 1973, Detecting hydrocarbons—For years the goal of exploration geophysicists: Oil and Gas J., v. 71, p. 122–125.

Hammond, A. L., 1974, Bright spot: Better seismological indicators of gas and oil: Science, v. 185, p. 515–517.

Digital filtering

Derusso, P. M., Roy, R. J., and Close, C. M., 1965, State variables for engineers: New York, John Wiley and Sons.

Oppenheim, A. V., Schaefer, R. W., and Stockham, T. G., 1968, Non-linear filtering of multiplied and convolved signals: Proc. IEEE, v. 56, p. 1264–1291.

Papoulis, A., 1962, The Fourier integral and its application: New York, McGraw-Hill.

FIG. 36. An example of an unmigrated (a) and migrated (b) seismic profile.

MIGRATED

Robinson, E. A., 1962, Random wavelets and cybernetic systems: Charles Griffin, England.

——— 1966, Multichannel Z-transforms and minimum-delay: Geophysics, v. 31, p. 482–500.

Stoffa, P. L., Buhl, P., and Bryan, G. M., 1974, The application of homomorphic deconvolution to shallow-water marine seismology—Part I: Models: Geophysics, v. 39, p. 401–416.

——— 1974, The application of homomorphic deconvolution to shallow-water marine seismology—Part II: Real data: Geophysics, v. 39, p. 417–426.

Treitel, S., and Robinson, E. A., 1964, The stability of digital filters: IEEE Trans. Geosci. Electr., v. G.E. 2, p. 6–18.

Wiggins, R., 1978, Minimum entropy deconvolution: Geoexpl., v. 16, p. 21–35.

Forward approach

Brown, A. R., 1979, 3-D seismic survey gives better data: Oil and Gas J., November, p. 57–71.

French, W. S., 1974, Two-dimensional and three-dimensional migration of model-experiment reflection profiles: Geophysics, v. 39, p. 265–287.

——— 1975, Computer migration of oblique seismic reflection profiles: Geophysics, v. 40, p. 961–980.

Hautefeville, A., and Cotton, W. R., 1979, Three-dimensional seismic surveying aids exploration in the North Sea: Oil and Gas J., November, p. 72–79.

Hiltermann, F. J., 1970, Three-dimensional seismic modeling: Geophysics, v. 35, p. 1020–1037.

Kelly, K. R., Ward, R. W., Treitel, S., Alford, R. M., 1976, Synthetic seismograms: A finite-difference approach: Geophysics, v. 41, p. 2–27.

Peterson, R. A., Fillippone, W. R., and Coker, F. B., 1955, The synthesis of seismograms from well log data: Geophysics, v. 20, p. 516–538.

Sengbush, R. L., Lawrence, P. L., and McDonald, F. J., 1961, Interpretation of synthetic seismograms: Geophysics, v. 26, p. 138–157.

Inverse approach

Claerbout, J. F., and Doherty, S. M., 1972, Downward continuation of moveout corrected seismograms: Geophysics, v. 37, p. 741–768.

Gevers, E. C. A., and Watson, S. W., 1978, Quantitative interpretation of seismic data using well logs: S.P.E. of AIME, 53rd Annual Conf., paper no. SPE 7439, October.

Nath, A. K., Meckel, L. D., and Wood, L. C., 1977, Synergistic interpretation of the convolution model: SEG Cont. Educ. Symp.

Chapter 4

MODELING—THE FORWARD METHOD

Introduction

Oil and gas deposits are found in porous formations in sedimentary basins. Under normal conditions, the reservoirs occur at locations where the appropriate porous formation is at a higher elevation than the surrounding region. The task of the exploration geophysicist is to locate such occurrences. The most common means of doing this is the proper interpretation of seismic data recorded for the region of interest. As with any physical procedure of this nature, it becomes highly desirable to simulate the data collection process in the laboratory to gain insight by the examination of known situations. This process forms the concept behind seismic modeling for exploration purposes.

The results from modeling of seismic wave propagation are used in many ways by the geophysicist to predict and understand seismic wave phenomena. Uses fall in the broad categories of (1) demonstrating the consistency of interpretation of real data; (2) providing synthetic data sets for testing processing techniques, acquisition parameters, and new ideas; and (3) educating the geophysicist in wave propagation.

Perhaps the most obvious use of modeling is in the generation of synthetic data sections for a proposed model for comparison with the actual data. Although it would be unusual for the synthetic data to match the real data precisely, the interpreter can, by the degree of likeness, increase his confidence in the proposed geologic cross-section. This comparison of model data with field data can be performed at various stages (e.g., common source records, stacked sections, migrated sections) by applying similar processing to both the synthetic and real data. In the process of modeling, the geophysicist can often distinguish between primary events, multiple reflections, diffractions, etc., thereby adding to his knowledge of the data. Modeling also provides insight into the portions of the model from which there are no primary returns, resulting from steeply dipping structure or shadow zones. With the insight gained from the modeling process, the geophysicist can often refine the interpretation, and the process may be repeated until a suitable match is obtained between the synthetic data from the model and the actual field data.

Synthetic data sets from modeling provide excellent test data. The research geophysicist can use the synthetic data set in developing and testing new data processing or acquisition techniques. Similarly, the operations geophysicist can use the synthetic data to develop and test data acquisition parameters and the data processing sequence to be used in a new exploration region.

Modeling is also used to generate displays which can provide some understanding of the basic phenomena associated with seismic wave propagation. The mathematics associated with wave propagation is sometimes formidable, and extrapolation into physical understanding is usually difficult. Through modeling, wave propagation phenomena can be readily demonstrated and displayed in the form of examples to support the mathematical theory. Unfortunately, the use of modeling as an educational aid was not extensive until the recent advent of powerful computer modeling systems. The educational aspects of modeling are certainly not limited to academia. Many exploration areas of current interest are associated with extremely complex geologic environments. A detailed understanding or intuitive insight into the wave phenomena associated with such situations is understandably beyond the capabilities of even the most experienced geophysicists without the aid of model results. Even relatively simple situations can be deceptive. Excellent examples of this were given by Tucker and Yorston (1973) who used model results to demonstrate the correct interpretation of numerous phenomena occurring in seismic sections which are not as straightforward as they might initially appear.

General Discussion

There are many methods of modeling seismic wave propagation. The approaches include analog models in the laboratory, numerical evaluation of various analytical solutions, and the wave equation or approximations to the wave equation.

Analog simulations of the wave equation were one of the earliest means of generating synthetic data sets for complex geologic models. One of the more commonly used methods for acoustic models is to record data in a water tank. Models of varying complexity may be built of various rubber-like materials which support few shear waves (or none at all); these models thus approximate acoustic wave propagation. Typical experimental methods and results are in Hilterman (1970) and French (1974). This technique has recently seen renewed interest and is one of the easiest methods of obtaining three-dimensional (3-D) data for complicated geometries. Other experimental techniques have been used where the propagation medium is elastic and effects of shear waves are included. Some examples of such modeling are photoelastic modeling described in Riley and Dally (1966) and modeling with bonded plates discussed in Angona (1960).

Exact analytical solutions to the wave equation are known for a few simple geometries. Many of the geologic models of exploration interest may be considered, to a first order of approximation, to consist of a series of flat parallel layers, that is, models which vary along only one spatial coordinate. The analytical solution for such flat-layer models may be obtained without approximation. This is not to say, however, that the evaluation of the solutions to provide numbers is trivial. If the source is assumed to be a plane wave propagating normal to the boundaries, the solution can be obtained fairly readily. A general discussion of this approach was given in Robinson and Treitel (1980). Many times, even the one-dimensional (1-D) model is simplified to include only primaries from the hypothesized reflecting layers. A reflectivity series is then created and convolved with the wavelet to produce a "primaries only" record which does not include multiples.

The solution for flat layers can be extended analytically to point and line sources by transform techniques, although the evaluation of the solution becomes challenging. One method of evaluating the integrals for a line source parallel to flat layers stems from the work of Cagniard [Cagniard (1962)]. The extension to a point source requires more work and was dissussed by Perkeris et al (1965). Although the solution to flat-layer geometry provides valuable information concerning wave propagation, it is obviously inadequate to explain many of the effects occurring in geometries of exploration interest.

Ray tracing, based on geometrical optics, is probably the most widely used modeling procedure. A discussion of ray-tracing methods and numerous references were given in Hubral and Krey (1980). With this approach seismograms can be generated which predict the arrival times and general amplitudes of waves from complex models. One thereby obtains plots of the events occurring in time as a function of source-receiver separation. A knowledge of the arrival time of events, coupled with the general amplitude of that event, is often adequate for a preliminary interpretation of the subsurface. One commonly used specialization of ray theory is the generation of so-called "zero offset" sections where each trace corresponds to events recorded as a function of time by a receiver at the same location as the source. The composite of such traces for incremental source-receiver locations along the surface gives the zero-offset section which provides an approximation to common midpoint sections (CMP).[1]

Ray tracing by geometrical optics is based on the assumption that all geometric dimensions in the model are large with respect to the wavelength of the incident wave. The method, therefore, cannot account for phenomena resulting from rapid changes of curvature, diffractions produced by faults, and other wave phenomena associated with the curvature of the wavefront and/or boundary. Two of the more common approaches for extending geometrical optics were presented by Červený et al (1977) and Keller (1962). A number of other approaches using Kirchhoff or Huygen's integral representations based on Green's theorem have been investigated which give more general results than those obtainable by simple geometrical optics. Typical examples are the diffraction study conducted by Trorey (1970) and the work of Hilterman (1975).

As more complex models become of interest, the demand for more sophisticated modeling approaches increases. This has led to the development of numerical methods employing finite-difference and finite-element techniques as a means of obtaining complete solutions to the wave equation. In the remainder of this chapter, application of the finite-difference method to produce synthetic solutions to the acoustic wave equation is discussed. The finite-element method is an alternate approach and yields results quite similar to the finite-difference method. A general treatment of the finite-element method was given by Strang and Fix (1973). Recent investigations by Bamberger et al (1980) show the finite-element method has definite advantages for elastic modeling where high Poisson ratios are encountered. These advantages are lost for lower Poisson ratios. A comparable analysis by Marfurt (1981, personal communication) indicates the finite-difference method is preferable for acoustic modeling.

In the following treatment, modeling of the two-dimensional (2-D) acoustic wave equation using finite-difference techniques is discussed. In many applications, acoustic models provide realistic synthetic seismograms while at the same time avoiding

[1]"Common depth point" is often used in the industry; however, "common midpoint" is more informative and refers to the compositing of all source receiver pairs with a common midpoint. For flat-layered geometry, this also results in events which are reflected from a common depth point (CDP).

the complexity of the full elastodynamic wave-equation solutions. After formulating and discretizing the heterogeneous acoustic wave equation, the modeling system will be demonstrated through generation and processing of a synthetic CMP data set.

Formulation and Discretization of the Acoustic Wave Equation

Preliminary considerations

An acoustic wave equation valid in a heterogeneous medium, i.e., a medium in which the velocity and density are allowed to vary with position, is used. This is in contrast to the more classical approach of specifying a model as a combination of homogeneous units with appropriate boundary conditions between the units. Whereas the classical homogeneous formulation requires boundary conditions to be satisfied explicitly at all boundaries, the heterogeneous formulation implicitly satisfies the boundary conditions at every point in the model. Although the governing partial differential equation for the heterogeneous formation is much more complicated than the corresponding equation for a homogeneous medium, the heterogeneous approach provides a technique which can be used routinely to model complex geometries automatically.

The desired heterogeneous acoustic wave equation may be easily derived following Brekhovskikh (1960, p. 171). The equation of continuity and Euler's relation are

$$\frac{\partial p}{\partial t} + \rho c^2 \nabla \cdot \mathbf{V} = 0, \tag{1}$$

and

$$\frac{\partial \mathbf{V}}{\partial t} + \frac{1}{\rho} \nabla p = 0, \tag{2}$$

where p is the acoustic pressure, $\mathbf{V}$ is the particle velocity, ρ is the density, and c is the local velocity of propagation in the medium. Differentiating equation (1) with respect to time, applying the divergence operator to equation (2), recalling that ρ and c may be functions of position, and solving the resulting system for a partial differential equation in terms of the pressure yields the heterogeneous acoustic wave equation

$$\frac{\partial^2 p}{\partial t^2} - \rho c^2 \nabla \cdot \left[\frac{1}{\rho} \nabla p \right] = \delta(t)\, \delta(\mathbf{r}), \tag{3}$$

where $\delta(\cdot)$ is the Dirac delta distribution and $\mathbf{r}$ is the cylindrical position vector.

Introducing the source term on the right-hand side of equation (3) creates an inhomogeneous partial differential equation. It arises formally by introducing its time integral into the continuity equation. Its inclusion permits any time variation or space variation to be synthesized from the solution to equation (3). As we shall see later, a band-limited time function is generally employed in order to obtain more interpretable displays of the waves as they propagate through the model.

A complete statement of the problem requires boundary conditions at infinity, commonly referred to as a radiation condition. This condition requires that the media eventually become uniform, thereby producing no incoming waves. In practice, artificial boundary conditions are introduced at the side and bottom boundaries to reduce unwanted reflections from those boundaries. This will be discussed in a later section.

For the subsequent numerical solution, only 2-D Cartesian geometries (exhibiting no variation along the transverse direction) are considered. This will by necessity force the source to be a line source, oriented along the transverse direction. The governing equation [equation (3)] may be approximated by a number of discretization methods, and the resulting algebraic system may be solved by various numerical algorithms. The algorithm presented here follows Kelly et al (1976). The accuracy of the method in a homogeneous medium was investigated by Alford et al (1974). This straightforward approach results in an explicit finite-difference system of second-order accuracy in both the spatial and temporal sampling intervals.

Difference approximations

A standard second-order finite-difference approximation for the time derivative is

$$\frac{\partial^2}{\partial t^2} P = \frac{1}{\Delta t^2} [P(m,n,k + 1) - 2P(m,n,k) + P(m,n,k - 1)] + 0(\Delta t^2), \tag{4}$$

where $P(m,n,k) = p(m\Delta x,n\Delta z,k\Delta t)$. The spatial derivatives are in a form considered in Mitchell (1969 p. 24), and they may be approximated by

$$\frac{\partial}{\partial x}\left[a(m,n)\frac{\partial}{\partial x}P(m,n,k)\right] = \frac{a(m + 1/2,n)}{\Delta x^2}[P(m + 1,n,k) - P(m,n,k)] - \frac{a(m - 1/2,n)}{\Delta x^2}[P(m,n,k) - P(m - 1,n,k)], \tag{5}$$

where a sufficient approximation for the parameter $a(m \pm 1/2,n)$ is

$$a(m \pm 1/2,n) = \frac{a(m \pm 1,n) + a(m,n)}{2}. \tag{6}$$

The approximation for the terms containing the partials with respect to z follows directly from equation (5) by symmetry.

The source distribution term appearing on the right-hand side of equation (3) appears worrisome at first. Upon closer examination, however, straightforward replacement of the Dirac delta function by a Kronecker delta function results in the desired properties. The dependent variable appears only linearly in the finite-difference equations; hence, the introduction of a Kronecker delta function permits any other source distribution to be synthesized by a discrete convolution of the "discrete Green's function" with the desired source distribution. This replacement is motivated by considering which discrete function $D(n)$ satisfies the requirement

$$F(0) = \sum_n F(n)\, D\,(0), \tag{7}$$

where $F(n)$ is an arbitrary discrete function. In practice, a band-limited time function is usually used. However, for the present a Kronecker delta will be retained.

Combining the approximations for the individual terms occurring in equation (3) results in an arithmetic equation in terms of the pressure P at three sequential time levels ($k - 1$, k, and $k + 1$). Because the spatial derivatives have been performed at the kth time step, the governing finite-difference equation can be solved explicitly for the field at the future time step ($k + 1$) in terms of the previous two time levels. Specifically, the algorithm to solve the finite-difference governing equation is

$$P(m,n,k+1) = 2P(m,n,k) - P(m,n,k-1) + \frac{r(m,n)^2}{2\rho(m,n)} \tag{8}$$

$$\cdot \Big\{ [\rho^{-1}(m+1,n) + \rho^{-1}(m,n)][P(m+1,n,k) - P(m,n,k)]$$

$$- [\rho^{-1}(m,n) + \rho^{-1}(m-1,n)][P(m,n,k) - P(m-1,n,k)]$$

$$+ \frac{(\Delta x)^2}{(\Delta z)^2} [\rho^{-1}(m,n+1) + \rho^{-1}(m,n)][P(m,n+1,k)$$

$$\cdot - P(m,n,k)] - [\rho^{-1}(m,n) + \rho^{-1}(m,n-1)]$$

$$\cdot [P(m,n,k) - P(m,n-1,k)] \Big\} + \delta^{m,n,k}_{m_o,n_o,1},$$

where the Kronecker delta function is defined as

$$\delta^{m,n,k}_{m_o,n_o,k_o} = \begin{cases} 1 \text{ where } m = m_0,\ n = n_0,\ k = k_0 \\ 0 \text{ otherwise} \end{cases},$$

and $r(m,n) = c(m,n)\Delta t/\Delta x$, where $c(m,n)$ is the discrete value of the velocity at the grid point (m,n).

Stability

If equation (8) is to be a suitable numerical algorithm for solving the finite-difference governing equation, it must produce bounded solutions (a requirement of stability). The necessary conditions for stability in heterogeneous media are not obvious. The stability requirement for a homogeneous region is well known, and we select the stability ratio r, such that the algorithm is stable in each and every homogeneous unit in the model. Our experience indicates that this serves as an adequate condition for physically realistic problems. The stability requirement is

$$r = \frac{C\Delta t}{\Delta x} \le \frac{1}{\sqrt{2}}, \tag{9}$$

where C is the maximum velocity in the model.

The stability requirement does not impose intractable conditions on the allowable relative sizes of Δx and Δt. This is in contrast to parabolic governing equations where implicit algorithms are typically used to remove severe stability restrictions. Fortunately, we can use an explicit scheme for solution of the finite-difference equations since the explicit schemes are amenable to very rapid and repetitive application of a simple numerical algorithm. This considerably simplifies the optimization of the computer programs and software.

Boundary conditions

Implementation of the finite-difference algorithm requires that a geologic model be restricted to a finite region of space. At the surface of the model, a free surface is generally imposed by explicitly setting the pressure to zero. The model is bounded laterally and in depth by simply terminating the model. This introduces reflections from the sides and bottom which can corrupt the results. Various solutions have been suggested to suppress the undesired edge reflections. The most straightforward solution is simply to extend the model laterally by approximately a factor of two, thereby removing the interference of the edge reflections from the events of interest. This approach is very expensive in terms of computer resources.

Another commonly used method is based on the work of Lysmer and Kuhlemeyer (1969) who impose a viscous boundary condition to "dissipate" incident energy mathematically. Another approach suggested in Smith (1974) attempts to cancel the first edge reflections by solving the problem twice, using complementary boundary conditions, thereby producing edge reflections of opposite polarity and subsequently adding the two solutions together. Again this increases the computing expense of the model. In still another approach discussed in Kelly et al (1975), the undesired edge reflections are removed in certain situations through filtering in the frequency-horizontal wavenumber domain. This approach is not viable for complicated models. The most satisfactory solution to the edge reflection problem found to date involves imposing artificial boundary conditions at the model boundaries designed to minimize reflections. Such artificial boundary conditions were developed independently in Clayton and Engquist (1977) and Reynolds (1978). We use the boundary conditions described by Reynolds, and we have found them to be satisfactory.

A method has thus been formulated for obtaining a numerical solution to the acoustic wave equation which is tractable for an essentially arbitrary 2-D model. The solution to the difference equation $P(m,n,k)$ will differ to some degree from the solution to the continuous equation $p(x,z,t)$. The difference between the discrete solution and the continuous solution arises from truncation of the expansion for the continuous partial derivative operator which produced the finite-difference operator. The error is a function of the size of the discretization intervals with respect to a typical wavelength in the medium. This truncation or discretization error has been analyzed in terms of grid dispersion (Alford et al, 1974). For acoustic models, grid dispersion manifests itself by the shorter wavelengths propagating with slower velocity. This tends to produce a wave with a ringing tail which becomes more pronounced as the wave propagates to greater distances. Experimentally, we have found that spatial sampling of approximately 10 grid points/wavelength will produce acceptable results, where the wavelength is selected as the wavelength at the upper half-power wavelength of the band-limited source (Alford et al, 1974). As we begin to run larger and larger models, involving larger propagation distances, it becomes advisable to use finer grids of perhaps 12 or more grid points/wavelength.

Initial state and solutions

The model is specified by the velocity and density distributions defined on a discrete grid. The system is assumed to be undisturbed prior to the source initiation, and the solution is generated recursively, using equation (8), for all times at every spatial location on the grid. By recording the results at the surface for all time, one obtains a synthetic seismogram for the model. If the field is recorded at one instant of time, at all points on the grid one generates a "snapshot" of the wave disturbance at that time. If enough snapshots are saved, they can be made into a movie showing the progression of the waves in the model (Alford et al, 1977).

The snapshots provide a very necessary interpretational aid. The synthetic seismograms resulting from a complex model can become complicated enough to tax the interpreter's ability to identify the events. Fortunately, the snapshots provide a method of following the events in the model, thus permitting their identification.

The desire for snapshots makes it advisable to use a band-limited source time distribution, thereby reducing the artifacts of dispersion to a manageable level. Otherwise, it would be expedient to use a Kronecker time distribution from which any desired source distribution could be obtained by the appropriate convolution.

Example

The model

A diagram of the geologic model is shown in Figure 1. It consists of two flat layers with a dipping layer terminating on the second horizontal layer. Beneath this unconformity, there is an anticlinal layer containing a simulated hydrocarbon reservoir at the top. The indicated velocities were chosen to be typical for this type of geologic model, and corresponding densities were obtained from the Gardner relationship (Gardner et al, 1974). The model width is 20,000 ft and depth extends to 12,000 ft. The finite-difference model consisted of 840 grid points horizontally and 480 grid points vertically. Two seconds of output data at a sample rate of 4 msec were desired. The stability requirement given in equation (9) forced the actual computation time increment (Δt) to be approximately 0.88 msec. This resulted in a total of approximately 2300 time steps to be calculated for each model.

A typical acquisition spread to be simulated consists of a shot located in the middle of a split spread with 50 seismometer groups on either side of the shotpoint. The groups extend for a maximum offset of 5000 ft, with a group interval of 100 ft. Since the grid size was selected to be 25 ft, it was possible to simulate an array of four seismometers per group by simply summing the results at four adjacent grid points. The seismometer groups were located at the surface of the model, and the shots were located at a depth of 100 ft beneath the surface.

In order eventually to simulate ten-fold CMP acquisition, 42 sets of data were computed with the shot moving from left to right in increments of 500 ft per shot. A standard roll-in roll-out of the shots across the seismometer spread was simulated at the ends of the model.

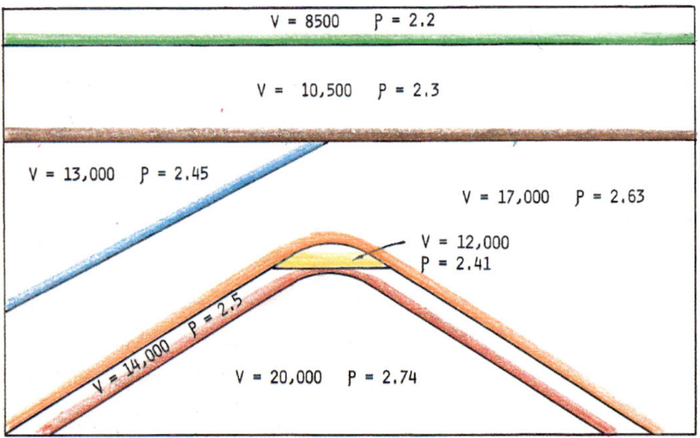

FIG. 1. The model.

To facilitate interpretation and discussion of the results, the events are color coded with respect to the horizon from which they were generated. As indicated in Figure 1, events from the upper layer are shown in green, those from the second layer in brown, etc.

Simulated data

The finite-difference output for the shot at the center of the model is discussed next. The data are presented in the form of snapshots (Figure 2), a common source (CS) gather (Figure 3), and a vertical seismic profile (Figure 4). [The vertical banded appearance of the traces in the background is a result of photographic processes and is not of physical significance.] Snapshots are obtained by storing the wave field at every spatial grid point for a fixed instant in time. The vertical seismic profile (VSP) was obtained by recording the data along a vertical line passing through the source, thereby simulating the acquisition of data down a borehole with a shot at the surface directly above the borehole. For convenience, the VSP data have been plotted with one-half of the CS seismogram joined along the common receiver location. On the left side of the display (Figure 4), the abscissa represents surface location, and on the right side of the display, the abscissa represents depth down the borehole. The time scale is common to both and extends downward.

Analysis

We will begin our discussion by using the VSP-CS display to correlate the location of the events with the structural boundaries from which they were generated. At a time of approximately 100 msec, the direct event can be seen progressing downward on the VSP with the slope of the event indicating the velocity of propagation in the media at that time. On the CS portion of the display, the direct event propagates horizontally across the model essentially at the velocity at the upper media. At a time of approximately 600 msec the direct event is no longer clearly visible on the CS display, due in part to interference with the reflection from the first interface and due to discrimination of the four-seismometer array group against horizontally propagating waves and the free surface effect.

At approximately 200 msec, an event is observed propagating upward on the VSP which arrives at the surface at approximately 360 msec and correlates with the event on the CS portion of the display. This event is the primary reflection from the first interface. The wave then reflects from the surface and propagates downward as evident on the VSP. It once again encounters the first boundary at approximately 450 msec, at which time the first multiple from the layer is generated. This multiple can be observed arriving at the surface at approximately 575 msec, and it correlates with the reflection on the CS display. The multiply reflected wave then once again reflects from the surface and propagates downward through the model; however, we shall no longer consider that wave in detail.

If we now return to the direct event propagating downward through the model as visible in the VSP, we observe another reflection being generated at approximately 450 msec which propagates upward through the model, arriving at the surface at approximately 850 msec. This event is the reflection of the downward moving direct wave from the second interface. It also reflects from the surface at the model, propagates downward, encounters the first boundary, and produces an upward moving wave which arrives at the surface at approximately 1150 msec. This event is an example of what is commonly termed a "peg-leg multiple."

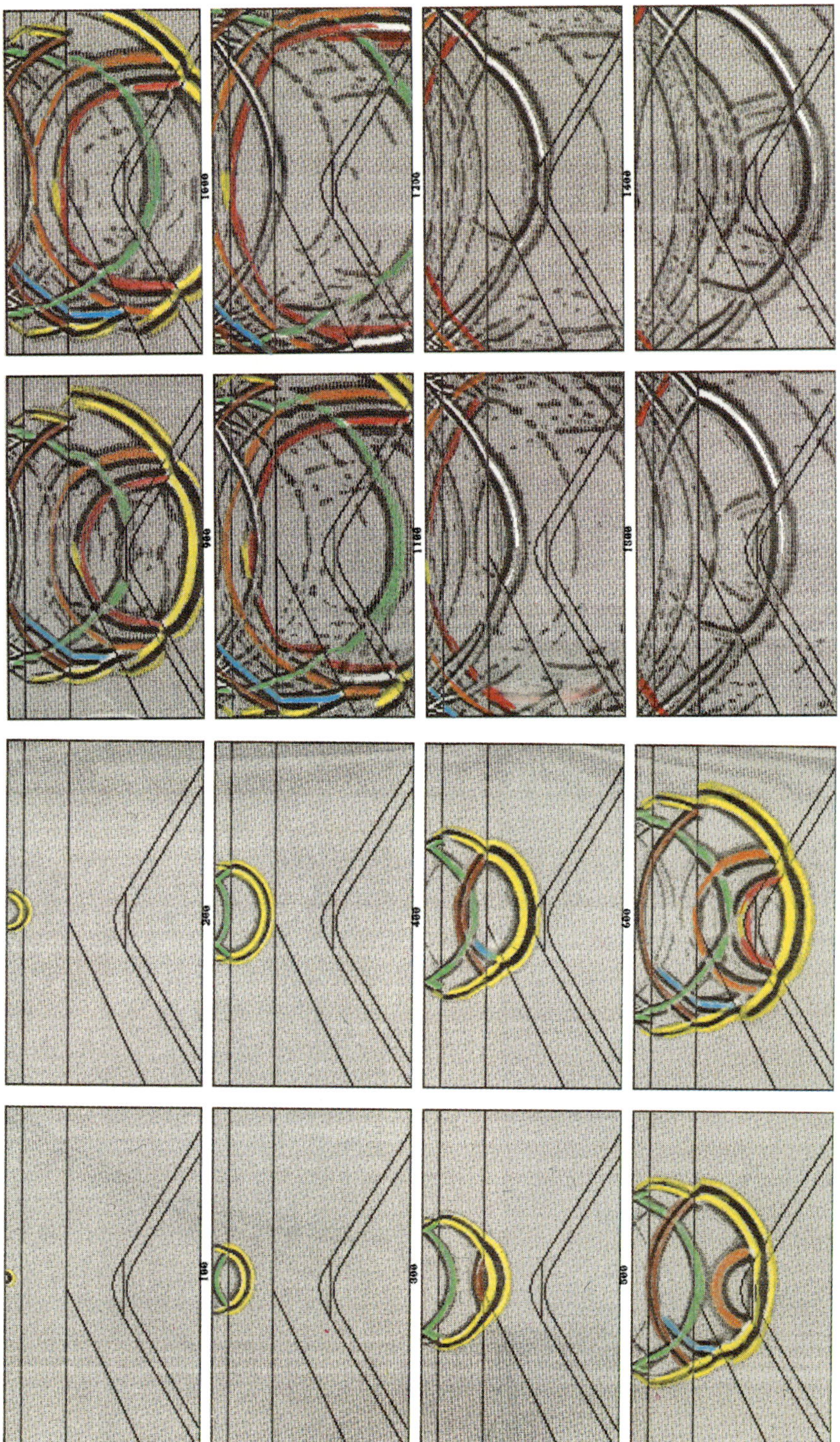

FIG. 2. Snapshots for center shotpoint.

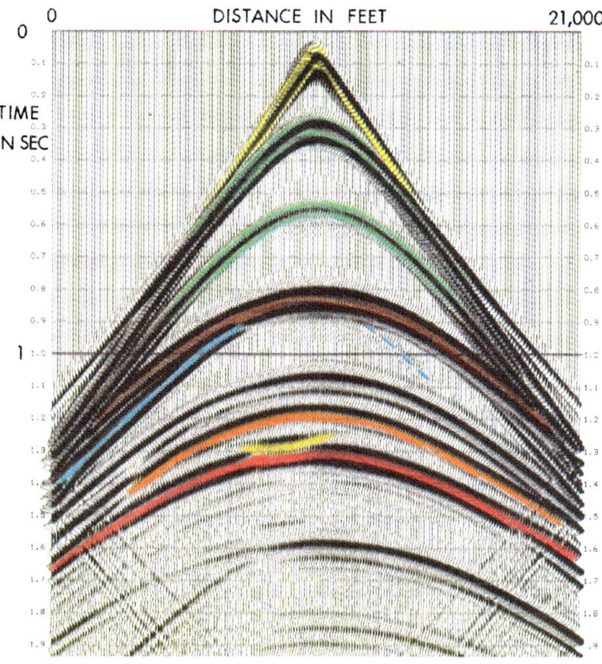

FIG. 3. Seismogram for full-width spread about the center shotpoint.

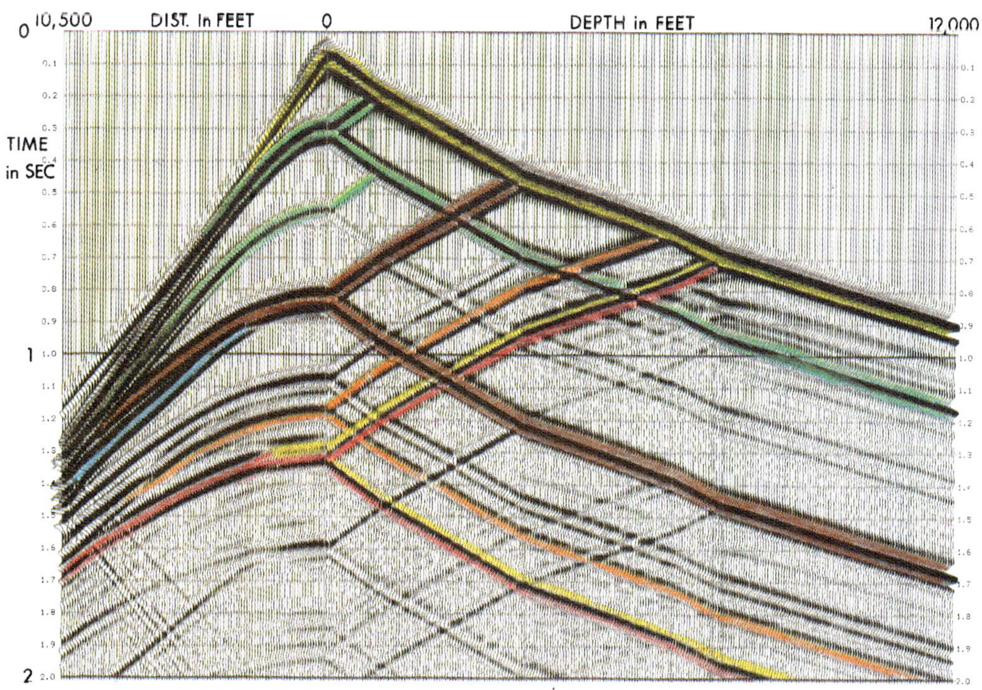

FIG. 4. VSP-CS display.

If one follows a similar procedure, the reflections from the anticlinal top, the reservoir contact event, and anticlinal bottom can be followed to the surface on the VSP and their respective arrival times can be determined.

Alternatively, one could have used the snapshots to deduce the preceding information, although the VSP is perhaps more straightforward to use for the original space-time correlation of the events. The snapshots, on the other hand, provide a method of tying various events spatially in the model. An example of such an event is the reflection from the dipping interface identified in blue on Figure 4 (but not visible on the VSP). If we consider the snapshots displayed in Figure 2, the contribution from a dipping interface can be identified and followed to the surface. The events previously discussed are indicated in Figure 2 by the color correlation. They can be followed as they propagate upward and downward throughout the model on the snapshots.

In the snapshot taken at 500 msec, the outgoing wave has encountered the corner of the pinchout, and in the following frame at 600 msec an event (indicated in blue) can be observed being generated at the dipping interface and connecting that boundary with the reflected wave from the flat layer. This wavefront can be followed through the snapshots and observed to arrive at the surface at between 900 and 1000 msec. In the seismogram of Figure 3, the blue event breaking away from the second primary (in brown) can be identified as the reflection from the dipping layer.

The CS seismograms for the odd numbered shotpoints are displayed in Figure 5. (The seismogram considered in Figure 3 corresponds to shotpoint no. 23 in this data set, although the maximum far-trace offset has now been limited to 5000 ft on either side of the source corresponding to the CMP acquisition simulated.) Now that the events have been identified for one seismogram with the aid of the VSP and the snapshots, interpretation of the remaining CS seismograms comprising the CMP data set is straightforward. The events from the two upper layers, since they are horizontal, are invariant from shotpoint to shotpoint. The multiples of these events occurring in the upper layer are similarly readily identified and are also invariant from shotpoint to shotpoint. The event in purple, corresponding to the reflection from the dipping interface, can be identified and traced on adjoining panels as indicated in the display. The diffractions from the pinchout can similarly be identified on the panels corresponding to shotpoints 19–27.

CMP Stacking, Processing, and Interpretation

CMP stacking

In the preceding section, the events on the individual common source seismograms were identified. We can now attempt to optimize the processing in order to accentuate the primary reflections. Probably the most powerful tool available to geophysicists for the attenuation of multiple reflections and other undesirable "noise" is the CMP stacking procedure. Because we have synthetic noise-free data, noise will not be a consideration, and we will concentrate our attention on multiple attenuation.

In Figure 6, the common source seismograms are displayed with NMO corrections applied. All of the data sets were corrected using the RMS velocity for an assumed well passing through the center of the model; horizontal variations in velocity were not taken into account. This approach was chosen solely for its simplicity, since better results would have been obtained if a careful velocity analysis of the CMP gathers had been employed. Since the events occurring at early times experience

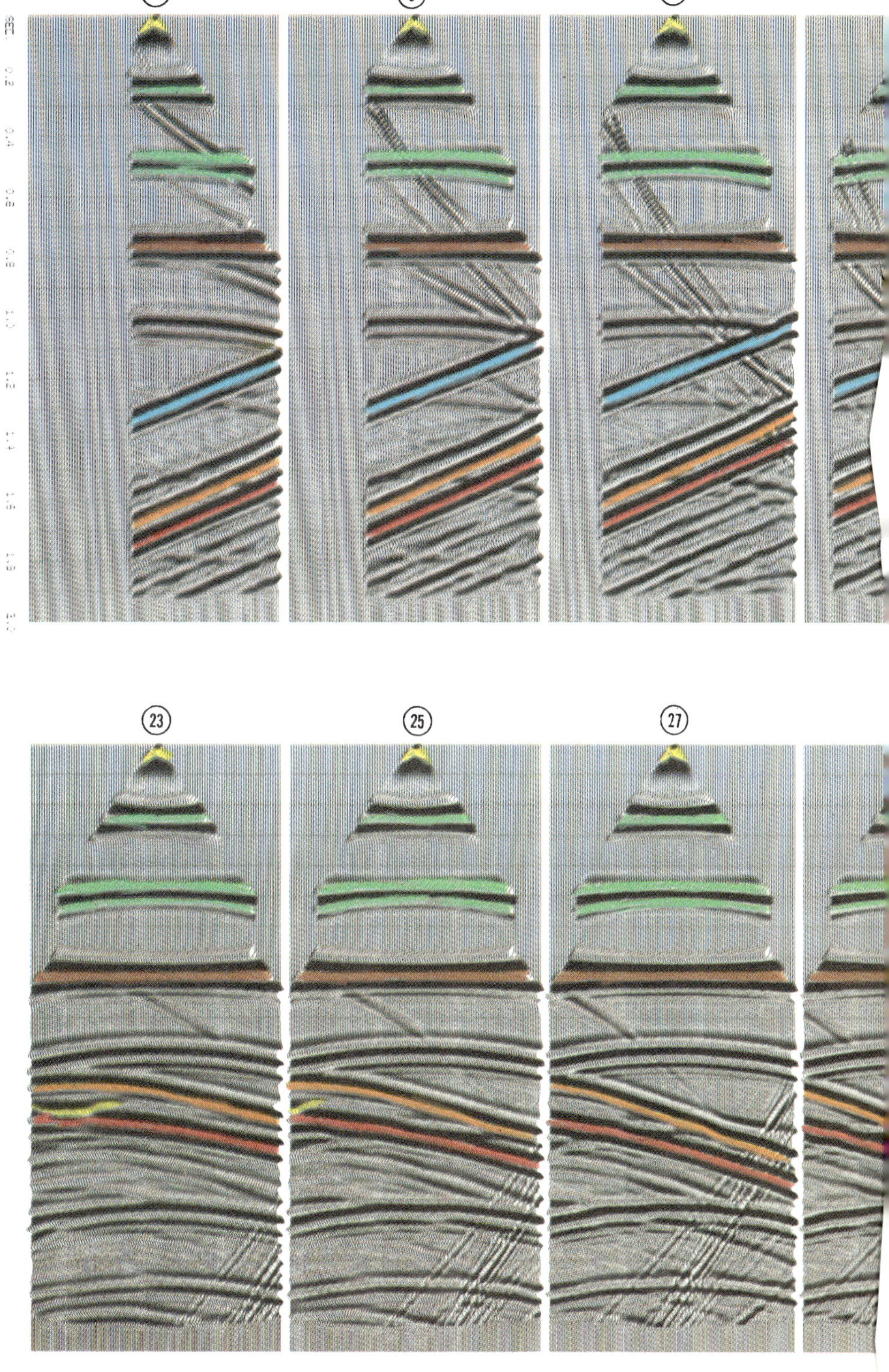

FIG. 6. CS seismograms after NMO correction.

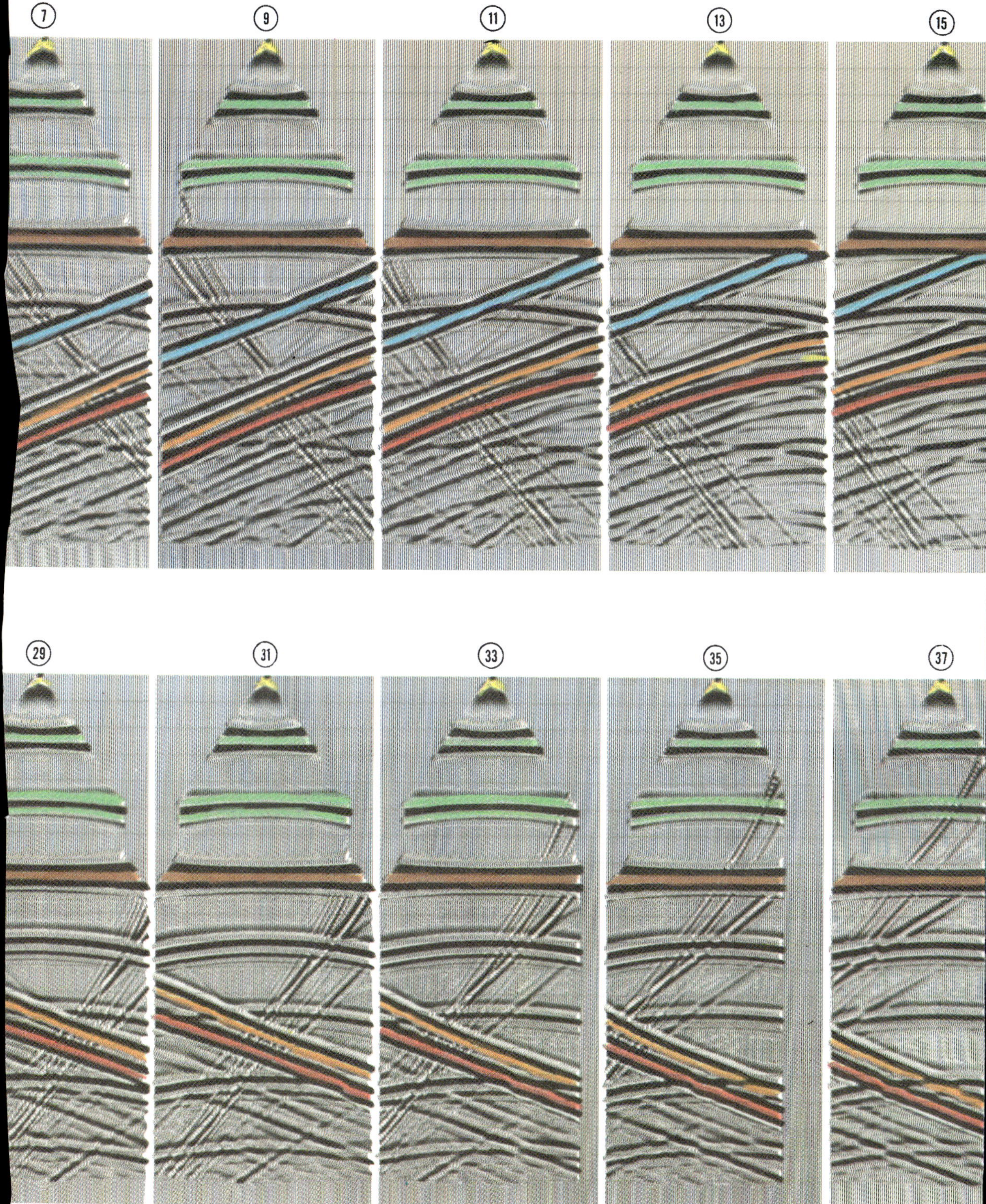

7
9
11
13
15
29
31
33
35
37

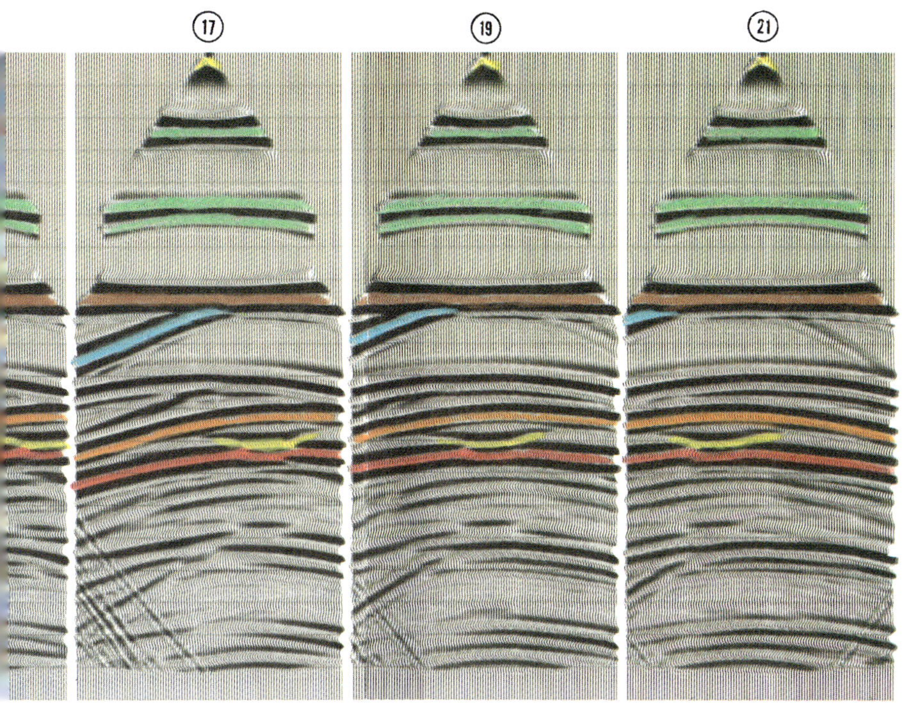
17
19
21

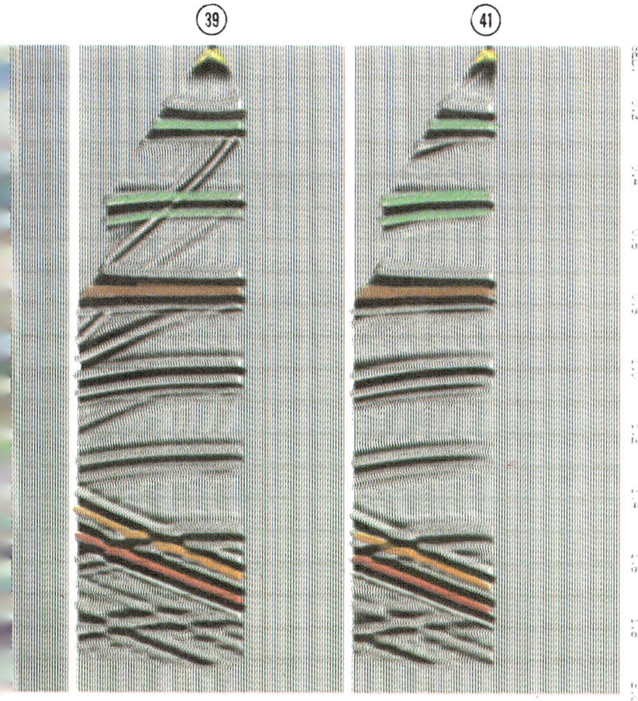
39
41
SEC.
0.2
0.4
0.6
0.8
1.0
1.2
1.4
1.6
1.8
2.0

more severe NMO stretch at the greater ranges, a mute has been applied to suppress the data from large ranges. The mute also suppresses the direct arrival.

One observes on the NMO corrected CS seismograms that the primaries from the first interface (occurring at approximately 200 msec) and the primary from a second interface (occurring at approximately 800 msec) have been flattened. The multiples from these events, occurring in the upper layer, have not been flattened as much since they experience greater moveout than a primary occurring at the same time. Observing Figure 6, we suspect that the maximum offset, which is included in a CMP gather, will be the dominant factor in multiple attenuation. Conversely, NMO stretch becomes more severe as greater ranges are included, and we suspect that artifacts due to this effect may be present in the CMP sections.

A single-fold CMP section (shown in Figure 7) was obtained by restricting the maximum offset included in the stacking process to one source interval, resulting in only one trace at each common depth point. Since noise has not been included in the model, a single-fold CMP stack is sufficient to give a very close resemblance of the stack section to the original model.

The reflection from the first interface occurs at approximately 200 msec; the multiple occurs at approximately 500 msec. The reflection from the second interface and the dipping layer occur at approximately 750 msec, and their multiple occurs at approximately 1000 msec. As one would expect, the multiples have not been attenuated on the single-fold section. The events originating at the sides of the seismogram are edge reflections from the model boundary. The multiples are colored in this display; in the following displays, the multiples will not be colored to permit a more careful examination of their variation with different approaches to stacking.

Figure 8 is a display of a five-fold CMP section. It was generated by including ranges in the CMP gather up to a maximum offset of 2500 ft. This simulates using 42 shots with a split spread with a maximum offset of 2500 ft on either side of the shot, which results in five traces in the sum for each output CMP trace. One can observe very little attenuation of the multiples over the single-fold section. On the contrary, the primary reflection from the first interface appears to be degraded on the five-fold stack. This is a result of including longer ranges in the stacking process, thereby introducing more NMO stretch distortion in the stacked section. This produces the observed rippling in the early events. Had random noise been included in the synthetic data, one would expect to improve the signal-to-noise (S/N) ratio by $\sqrt{5}$ or approximately 7 dB. To understand why the multiples have been largely unaffected by the stacking process, we can refer back to the NMO corrected seismograms in Figure 6. If we limit our attention to the center portions of the NMO corrected seismograms, we see there is little differential moveout between the zero-offset trace and the far trace at 2500 ft. There will, therefore, be very little cancellation of the multiples between the near and far traces. The relative amplitudes of the edge reflections have also been decreased.

In Figure 9 an alternate method has been employed to obtain a five-fold CMP section. In this case, the data from all 42 shots were used assuming a single ended geophone spread extending from the shotpoint to the right for a maximum offset of 5000 ft. This acquisition scheme results in the same number of groups per spread (50) as in the previous case; however, now the data correspond to groups at ranges up to twice the previous far-trace distance. Multiples are now greatly attenuated with respect to primaries since there is more differential moveout from near trace to far trace; therefore there is more destructive interference when the CMP traces are

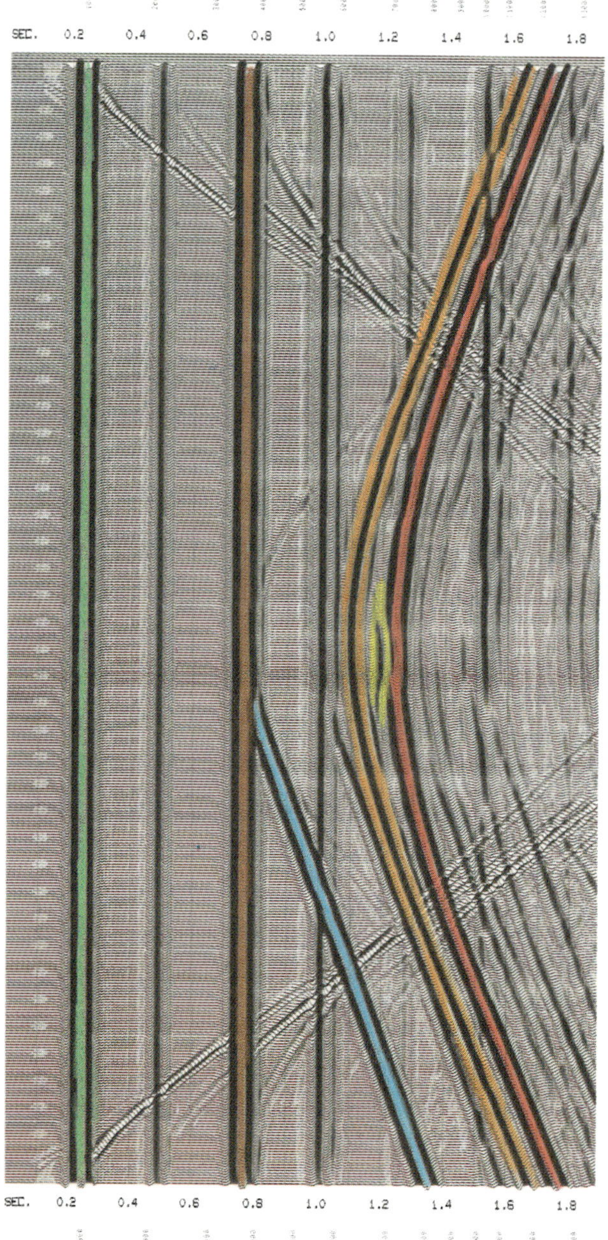

FIG. 7. Single-fold CMP section. 42 shots, 500 ft maximum offset, split spread.

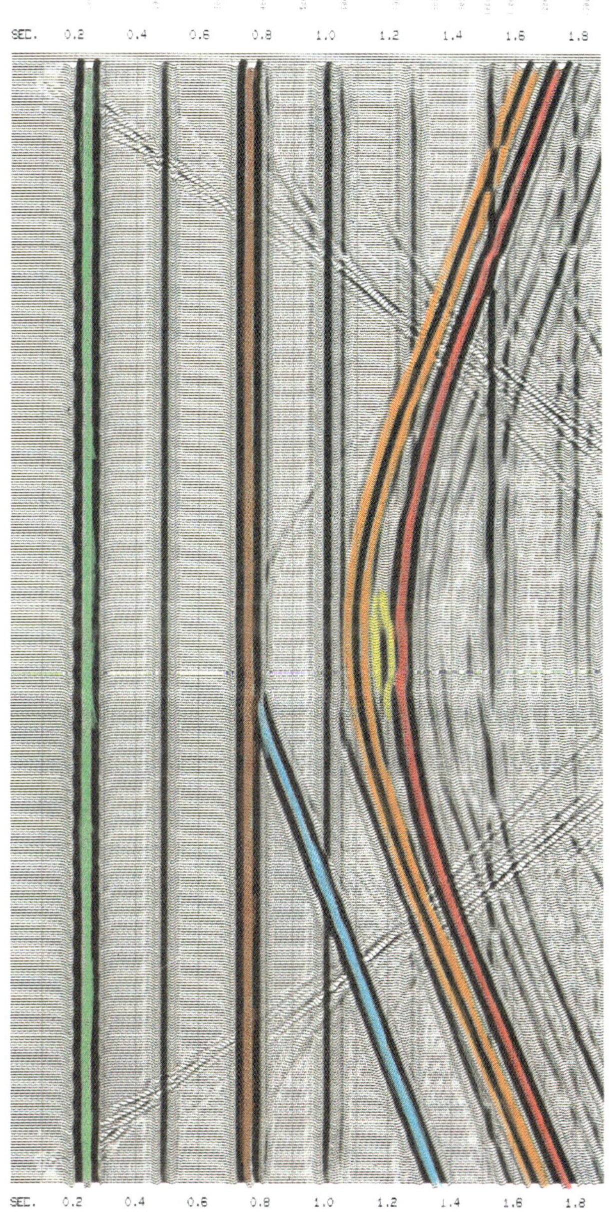

FIG. 8. Five-fold CMP section. 42 shots, 2500 ft maximum offset, split spread.

summed in the stacking process. The flat primary events exhibit more distortion due to residual NMO when the 5000-ft maximum offsets are included than they did when only 2500-ft maximum offsets were included (Figure 8).

Figure 10 shows yet another method of obtaining a five-fold CMP section. In this case, a split spread has been used containing the maximum offset of 5000 ft on either side of the shot, but only 21 shotpoints have been included—every other shot has been suppressed. There is substantially no difference between this section and the previous section. This is expected since the fold and the maximum offset are the same. The decision as to whether to employ 21 shots with a 10,000-ft split spread, or 42 shots with a 5000-ft single ended spread, would be based upon considerations of field implementation and economics.[2] One would have to decide between a longer spread containing twice the number of geophone groups but half as many shots, or a shorter spread containing half the number of geophone groups and twice the number of shots. Other factors also have to be considered; for example, one may be able to obtain denser geophone spacing with a single ended spread for the same number of groups thereby offering the possibility of higher frequency (resolution) data.

In Figure 11 data from all the shotpoints in all the ranges have been included, resulting in a ten-fold CMP section. The increase in the fold has had little or no effect on the multiple attenuation since we are still including the same maximum far-trace offset. The ripple seen on the primary event as a result of the NMO distortion has been reduced. Had random noise been included in the CS seismograms, one would expect an increase in the S/N ratio by approximately 3 dB over the five-fold data set, and of approximatedly 10 dB over the single-fold CMP section. This ten-fold CMP section is representative of the optimum CMP stack one can obtain for this model without resorting to unified velocity and migration techniques appropriate for complex structural situations.

Other processing

In Figure 12, predictive deconvolution techniques (Peacock and Treitel, 1969) have been applied to the ten-fold CMP stack in an attempt to reduce the multiple content further. One observes considerable improvement on the multiple from the first primary event, and not quite as much of an improvement on the peg-leg multiple from a second interface. This is a result of how predictive deconvolution was used. The design window contained only the first event and its multiple, resulting in essentially a deterministic deconvolution of that multiple event. The multiple event from the second interface contains two kinematic analogs, i.e., events arriving by two different wave paths but at the same time. This produces a multiple from the second interface with twice the relative amplitude as the multiple from the first interface, and the deconvolution process does not have sufficient information to remove the event. In normal field procedures, the operator would be designed for a larger window and would contain more multiple statistics.

Now that a simulation of a noise-free "final stacked section" has been generated, it is timely to discuss the events from the anticlinal layer and the fluid contact event at the bottom of the reservoir.

---

[2]If one had obtained a shot at every group location, reciprocity could be invoked to demonstrate that split-spread data result in gathering redundant information for the acoustic case. Given appropriate source and receiver considerations, it also results in redundant information for the elastic case. A discussion of reciprocity for the general case of a heterogeneous, anisotropic, elastic medium was given in Knopoff and Gangi (1959).

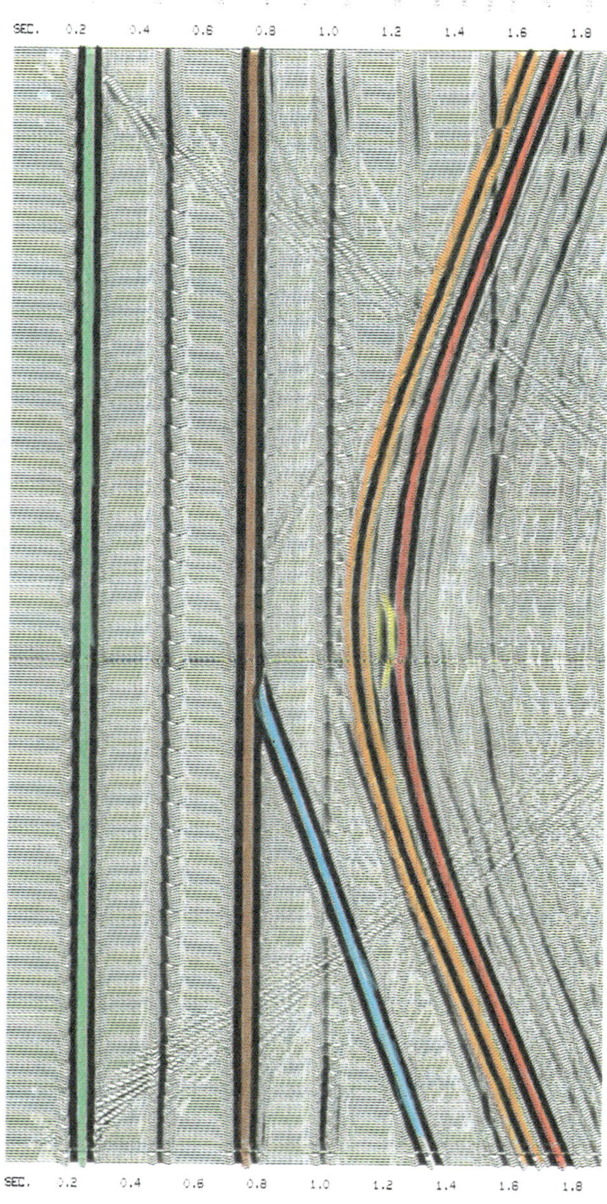

FIG. 9. Five-fold CMP section. 42 shots, 5000 ft maximum offset, single-ended spread.

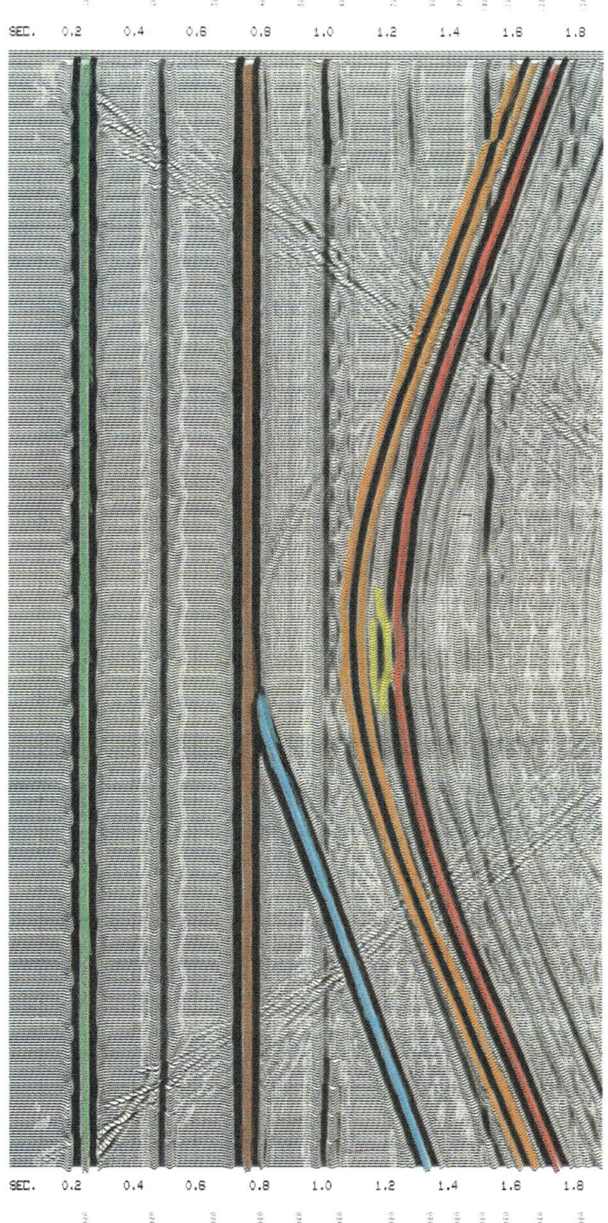

FIG. 10. Five-fold CMP section. 21 shots, 5000 ft maximum offset, split spread.

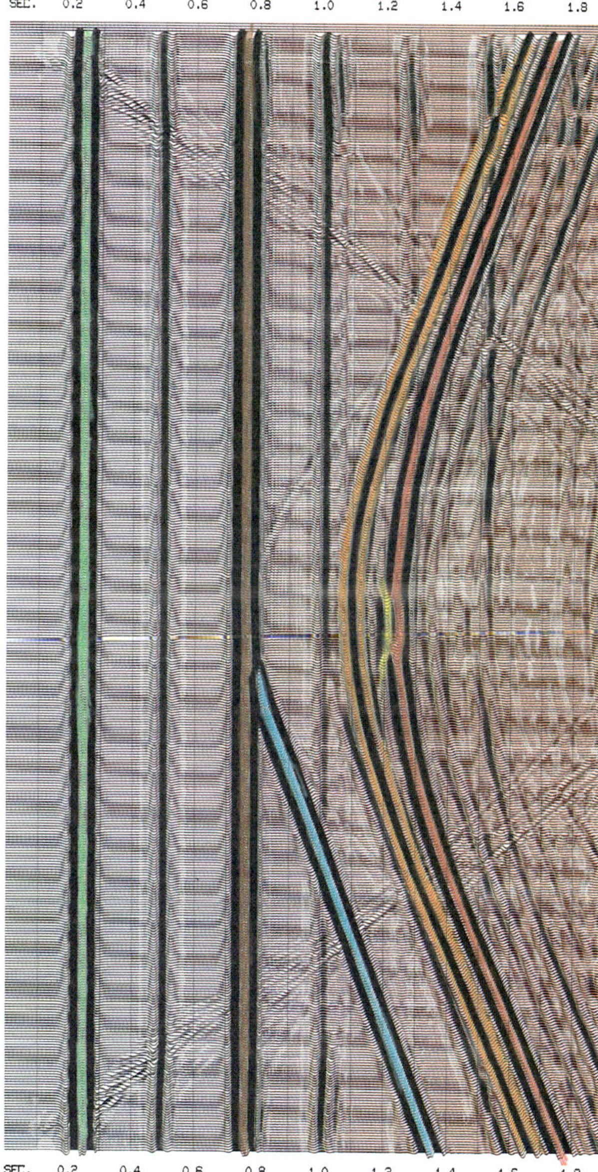

FIG. 11. Ten-fold CMP section. 42 shots, 5000 ft maximum offset, split spread.

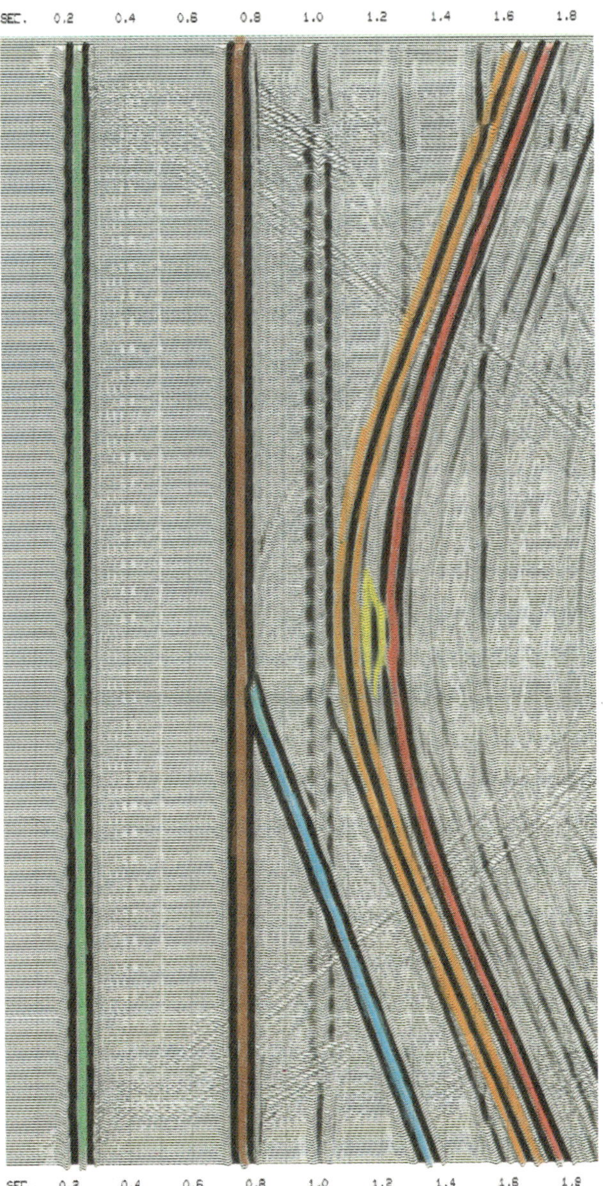

FIG. 12. Ten-fold CMP section after predictive deconvolution.

Figure 13 is a display of the results of migrating the deconvolved, ten-fold CMP section in Figure 12 using a constant-velocity (12,000 ft/sec) frequency-domain migration process (Stolt, 1978). The resulting migrated section bears a much closer resemblance to the original model than the CMP section. The flat layers have not been moved; however, the reflection from the dipping reflector has been moved updip giving an indication of the actual portion of the dipping layer from which information was gathered at the surface. The curvature of the anticline has also been increased as expected.

Interpretive aspects

There are two curious features associated with the contact event. If we follow the reflection from the top of the anticlinal layer, it is fairly continuous, consisting of a small negative trough, followed by a prounounced positive peak, followed again by a small negative trough. There is some decrease of the amplitude occurring at the apex at the anticline due to spreading of the events, although it is hard to discern on the displays. The reflection from the lower boundary of the anticline consists of a small peak followed by a pronounced trough and then another small peak, suggesting a reversed polarity from the upper boundary. It appears to be quite continuous on either side but almost appears to split at the top of the apex. This is the result of destructive interference of the event from the lower anticlinal layer with the reflection from the bottom of the reservoir. The first portion of the anticline reflection tends to cancel with the later part of the reflection from the contact event. This, coupled with the velocity pull down, results in the apparent discontinuity of the event.

Another interesting observation is that the reflection from the reservoir lower boundary (contact event) extends for only a small portion of the actual extent of the reservoir. This is a result of a lens effect as the wave passes through the low-velocity reservoir. Consider for a moment the zero-offset approximation to a CMP stack. Reflected rays from the horizontal contact event leave the reflector perpendicular to the boundary and pass upward from the low-velocity region into a high-velocity region through a concave lens. As the rays pass from the low velocity to the high velocity, they are bent away from the normal, causing the rays to focus beneath the surface of the model. The rays then diverge from the focus slightly before actually arriving at the surface. This results in a contact event that is considerably narrower than the actual reflector at the base of the reservoir.

There is also a fortuitous artifact introduced on the contact event observed on the migrated section of Figure 13. It is extremely tempting to identify the event tying the upper and the lower anticlinal reflectors as being the reflection from the contact at the lower boundary of the reservoir. The resulting "fish eye" shape is what would be expected from a low-velocity lens. The horizontal extent of the "reflection" even approximates the actual reservoir width, which in this case is known. Unfortunately, as previously discussed, the reflection from the contact boundary was focused on its path upward through the model. In a constant-velocity migration process, however, there is no conceivable mechanism by which to explain the redistribution of the reflected energy along the horizontal reservoir boundary. Such a migration process cannot account for the ray bending experienced by the reflection from the reservoir contact and cannot spread the flat event across a flat surface. We must conclude, therefore, that the contact event must be an artifact of the migration process.

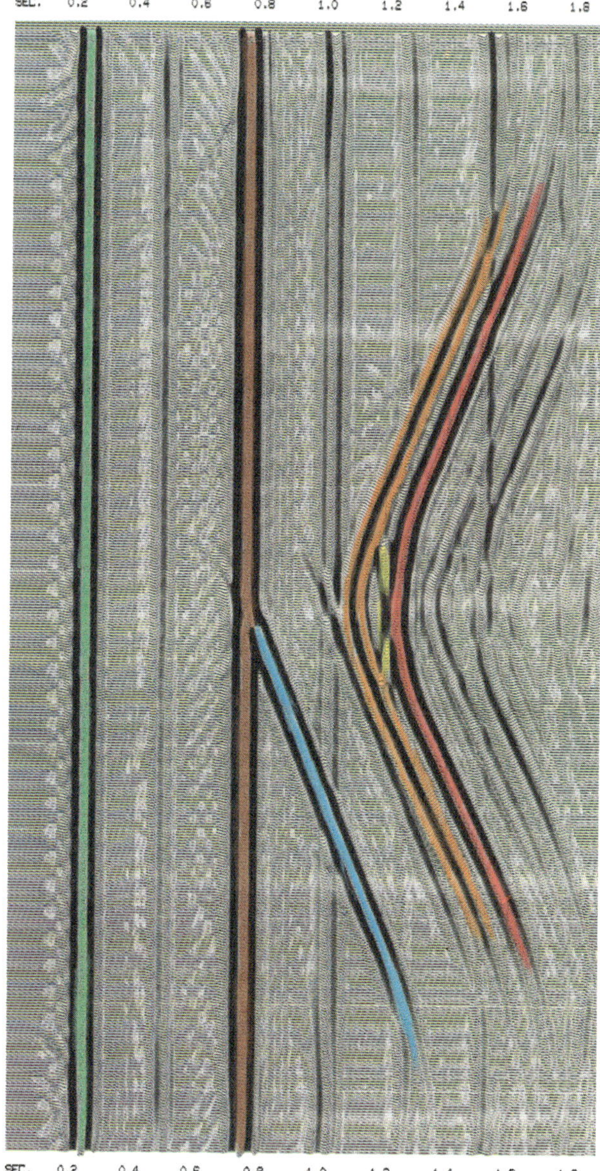

FIG. 13. Migrated ten-fold CMP section. Constant velocity, 12,000 ft/sec.

One possible explanation for the preceding phenomenon is that it is a result of inappropriately migrating the diffractions from the corners of the reservoir. Without further investigative modeling, one can only speculate as to the true mechanism behind the artifact. The irony is that an erroneous interpretation of the artifact results in a more accurate picture of the reservoir.

Discussion and Conclusions

In the process of generating a numerical solution for the wave equation by a finite-difference process, the solution is calculated at every spatial point on the grid, for every time step, for each model or source location. For the case considered, this resulted in the wave field being calculated at 40 billion time points (4.0×10^{10}). To place a perspective upon such a number, even adding 40 billion floating point numbers on a high-speed computer requires about 5 hours. It is obvious then that a task of this magnitude absorbs considerable CPU time. This high demand motivated our use of a dedicated minicomputer system to make it cost effective for routine use. For less frequent use, the decision to use a dedicated system is not necessarily a good one since the algorithm is amenable for implementation on "super computers" such as the Cray I with much less programming effort.

It should be apparent that there is a massive amount of output information available to the modeler from this modeling system. In addition to the CS seismograms, snapshots displaying the wavefront as it exists and progresses through the model as a function of time are available. VSPs (recording the seismogram at vertical locations down through the model as a function of time) also exist. There are several ways in which to approach the interpretation of such a data set.

If the geometry of the model had varied more radically in the horizontal direction, such as would be the case for a complex thrust fault model or a growth fault model, it would have been advantageous to gather complete interpretation data, i.e., VSPs and snapshots in addition to the CS seismograms, at locations representative of the geologic variation from region to region.

REFERENCES

Alford, R. M., Kelly, K. R., and Boore, D. M., 1974, Accuracy of finite-difference modeling of the acoustic wave equation: Geophysics, v. 39, p. 834–842.

Alford, R. M., Kelly, K. R., and Whitmore, N. D., 1977, Simulation of seismic acquisition and processing of field data by finite-difference modeling: Presented at 30th Annual SEG Midwestern Meeting, April 6, Oklahoma City.

Angona, F. A., 1960, Two-dimensional modeling and its application to seismic problems: Geophysics, v. 25, p. 468–482.

Bamberger, A., Chavent, G., and Lailly, P., 1980, Etude de schémas numériques pour les equations de l'elastodynamique linéaire: Inst. National de Recherche en Inform. et en Automatique, raports de recherche, no. 41.

Brekhovskikh, L. M., 1960, Waves in layered media: New York, Academic Press.

Cagniard, L., 1962, Reflection and refraction of progressive seismic waves: New York, McGraw-Hill.

Červený, V., Molotkov, I. A., and Psenčik, I., 1977, Ray method in seismology: Praha, Univ. Karlova.

Clayton, R., and Engquist, B., 1977, Absorbing boundary conditions for acoustic and elastic wave equations: SSA Bull., v. 67, p. 1529–1540.

French, W. S., 1974, Two-dimensional and three-dimensional migration of model-experiment reflection profiles: Geophysics, v. 39, p. 265–277.

Gardner, G., Gardner, L., and Gregory, A., 1974, Formation velocity and density—The diagnostic basis for stratigraphic traps: Geophysics, v. 39, p. 770–780.

Hilterman, F. J., 1970, Three-dimensional seismic modeling: Geophysics, v. 35, p. 1020–1037.

———, 1975, Amplitudes of seismic waves—A quick look: Geophysics, v. 40, p. 745–762.

Hubral, P., and Krey, Th., 1980, Interval velocities from seismic reflection time measurements: SEG special publ., Tulsa.

Keller, J. B., 1962, Geometrical theory of diffraction: J. Opt. Soc. Am., v. 52, p. 116–130.

Kelly, K. R., Alford, R. M., Treitel, S., and Ward, R. W., 1975, Application of finite-difference methods to exploration seismology, *in* Numerical analysis II: John J. H. Miller, Ed., New York, Academic Press.

Kelly, K. R., Ward, R. W., Treitel, S., and Alford, R. M., 1976, Synthetic seismograms: A finite-difference approach: Geophysics, v. 41, p. 2–27.

Knopoff, L., and Gangi, A. F., 1959, Seismic reciprocity: Geophysics, v. 24, p. 681–691.

Lysmer, J., and Kuhlemeyer, R. L., 1969, Finite dynamic model for infinite media: J. Eng. Mech. Div., Proc. Am. Soc. Civil Eng. v. 95, p. 859–877.

Mitchell, A. R., 1969, Computational methods in partial differential equations: New York, John Wiley and Sons.

Peacock, K. L., and Treitel, S., 1969, Predictive deconvolution: Theory and practice: Geophysics, v. 34, p. 155–159.

Perkeris, C., Alterman, Z., Abramovici, F., and Jarosch, H., 1965, Propagation of a compressional pulse in a layered solid: Rev. Geophys., v. 3, p. 25–47.

Reynolds, A. C., 1978, Boundary conditions for the numerical solution of wave propagation problems: Geophysics, v. 43, p. 1099–1110.

Riley, W. F., and Dally, J. W., 1966, A photoelastic analysis of stress wave propagation in a layered media: Geophysics, v. 31, p. 881–899.

Robinson, E. A., and Treitel, S., 1980, Geophysical signal analysis: Englewood Cliffs, New Jersey, Prentice-Hall.

Smith, W. D., 1974, A non-reflecting boundary for wave propagation problems: J. Comp. Phys., v. 15, p. 492–503.

Stolt, R. H., 1978, Migration by Fourier transform: Geophysics, v. 43, p. 23–48.

Strang, G., and Fix, G. J., 1973, An analysis of the finite-element method: Englewood Cliffs, New Jersey, Prentice-Hall.

Trorey, A. W., 1970, A simple theory for seismic diffractions: Geophysics, v. 35, p. 762–784.

Tucker, P. M., and Yorston, H. J., 1973, Pitfalls in seismic interpretation: SEG monograph 2, Tulsa.

Chapter 5

MIGRATION—THE INVERSE METHOD

Seismic Exploration

Objectives

The purpose of seismic exploration in the petroleum industry is to produce models of subsurface geology that can be evaluated for the presence of hydrocarbons. The basic data consist of reflected elastic waves recorded at the surface or in boreholes. Other chapters address various details of seismic exploration, from acquisition to interpretation. In this chapter key aspects of transforming seismic data into images of the subsurface are discussed both as it is done presently and as we hope to do it in the future. In particular, we discuss

(1) concepts and associated computations necessary to produce an accurate picture of subsurface geology, a process called seismic migration;
(2) examples which demonstrate the rapidly advancing migration technology;
(3) estimates of data quantity and computer requirements of present and future migration and imaging technology.

Data acquisition

Figure 1 illustrates the seismic method. A vibratory or impulsive source directs elastic waves into the subsurface where they are reflected and scattered by subsurface variations in physical properties. The waves returning to the surface are detected by geophones which record time histories of ground motion over a few seconds. The amplitudes, frequencies, and phases of these trace recordings are governed by such physical properties of the subsurface as elastic constants, geometry, dimensions, anelasticity, and anisotropy. Ideally we would like to extract quantitative estimates of these properties from seismic data, but presently only the acoustic velocities and structural geometries can be quantified with much success.

Seismic surveys can be conducted with sources and receivers placed along a line to produce a two-dimensional (2-D) cross-section of the subsurface or with sources and receivers distributed areally to produce a three-dimensional (3-D) picture. Until recently nearly all seismic reflection surveys were conducted along single lines because of the cost, logistics, and data processing requirements of three dimensions. Also, nearly all surveys utilized motion detectors (geophones) sensitive only to vertical ground motion at each detector location; as such, they have recorded mostly longitudinal or P-wave reflections. In recent years, surveys have been conducted

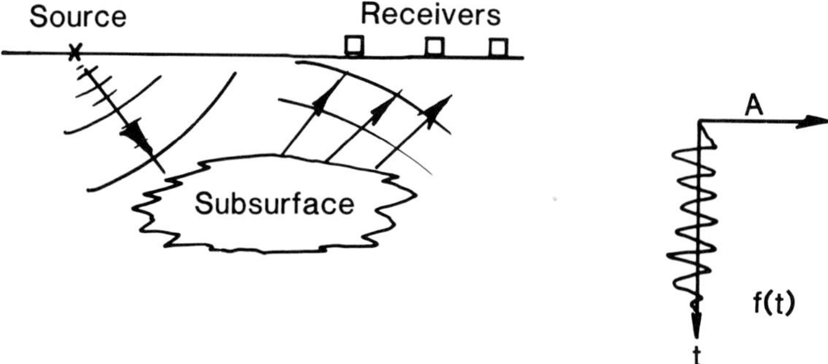

FIG. 1. The seismic concept.

with polarized (shear wave) sources and detectors sensitive to horizontal motion. A significant future development will be the recording of all three components of ground motion at each surface location as well as in boreholes at selected locations in a survey area.

Migration and Acoustic Imaging

Introduction

The technology of seismic migration has undergone remarkable advances in the last 10 years. As "deconvolution" was a buzzword of seismic data processing of the 1960s, "wave-equation migration" became a catch phrase of the 1970s. The purpose of migration, like deconvolution, is to transform seismic waveforms into an accurate picture of subsurface geology. Seismic data are migrated because subsurface reflecting points do not necessarily lie vertically beneath surface observation points. Figure 2 geometrically illustrates the "migration problem" for a single dipping reflector. The apparent positions of the end points are C and D, vertically beneath source-receiver positions. Migration moves points C and D to points C' and D', the true normal incidence positions. Some simple geometric relationships exist between true and apparent dip for this constant-velocity example:

$$\alpha_M > \alpha_{\text{app}}$$

$$\sin \alpha_M = \tan \alpha_{\text{app}}$$

$$\overline{CD} > \overline{C'D'},$$

where α_M, α_{app}, CD and $C'D'$ are indicated in Figure 2. Figure 3 shows a more complicated example of the difference between signal geometry and true structural geometry for normal incidence data recorded at different heights above a model. The figure indicates that the closer the sources and receivers are to the reflecting surface (T_0 = 263 msec versus T_0 = 1433 msec), the closer the reflection image resembles

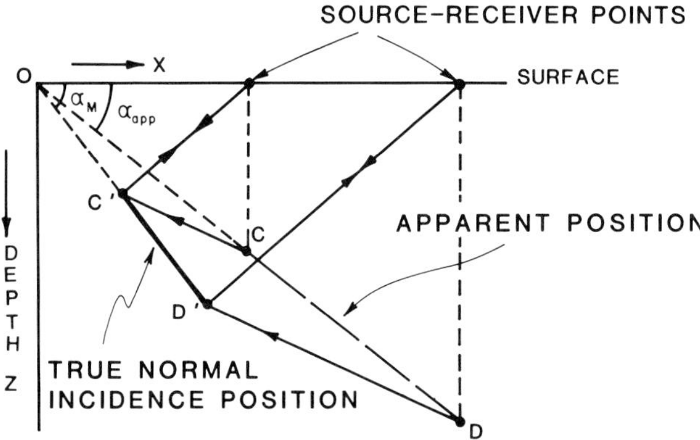

FIG. 2. Some elementary principles of reflector migration using straight raypaths.

the actual geometry. In modern computer migration the observation surface is moved progressively closer to the reflecting surface by simulation techniques. This process, called downward continuation, can be thought of as a focusing operation.

An operational definition of computer migration is: A space and time variant filtering process which maps observed seismic space-time amplitude data into either time or depth with correct amplitudes at true spatial positions. This shifting of digital information is dependent upon the rules of wave propagation, the time and time dip of the data, and the velocity distribution in the subsurface. Most of the discussion in this section concerns 2-D migration after common-depth-point (CDP) stack; however, 3-D and before stack processes are becoming increasingly significant in industry exploration efforts. Summarizing, the main reasons for migration are (1) correct structural placement of dipping events, (2) focusing of diffractions caused by point scattering centers and subsurface fault terminations, (3) sorting out of crossing events like those in Figure 3, (4) correction of amplitudes for geometric focusing effects and spatial smearing, and (5) improvement in resolution.

Early approaches to migration

The need for migration was recognized in the first reflection seismic survey by J. C. Karcher in 1921 (Sheriff, 1978). The evolutionary history of migration techniques is traced in Figure 4. Sherwood and Schultz (1980) presented a similar discussion. The earliest graphical methods used measurements of times and time shifts of reflection events picked by hand on field seismograms. True dips and lateral displacements were determined by plotting the times using ''common tangent'' procedures. These straight-ray methods usually employed constant or average velocity assumptions. Dobrin (1976) and Sheriff (1978) discussed these methods. Some of the rules governing this straight-ray migration are illustrated in Figure 2.

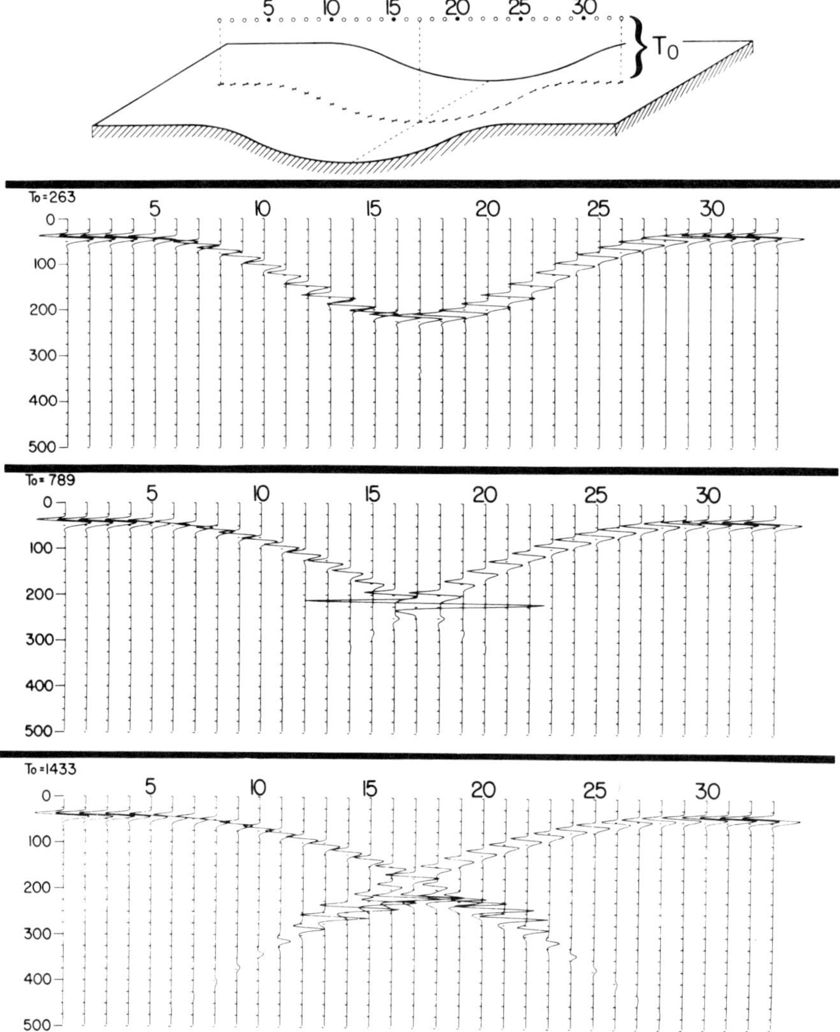

FIG. 3. For a physical model of synclinal structure, time sections recorded at various depths from the surface, as indicated by the time T_0. The task of migration is to map signal geometry, like the lower section, into true structural geometry. This is accomplished by algorithms that simulate seismic data that would be recorded successively closer to the interface of interest (after Hilterman, 1970).

A refinement of these early approaches used wavefront and raypath charts and overlays to represent more accurately velocity variation with depth. Families of orthogonal wavefronts and rays were usually constructed to fit a known linear increase of velocity with depth as shown in Figure 5. Time shifts between traces due to dip over some set distance determined which normal incidence raypath corresponded to a particular dipping event. Reflector segments were then plotted on a transparent overlay as shown in Figure 6 [see Musgrave, (1961)].

So far only reflections from a plane or nearly plane surface have been considered.

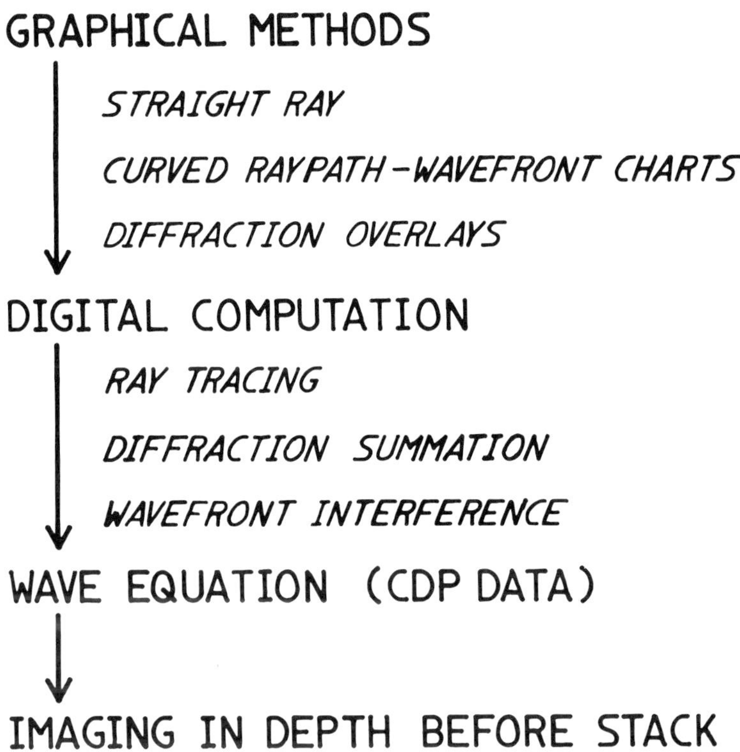

FIG. 4. Historical development of migration technology.

We now consider diffractions usually associated with a sharp edge or corner. A diffracting point accepts energy from any angle and radiates it along a spherical wavefront as if the diffracting point were a new energy source (Figure 7). This concept, known as Huygen's principle, provides an alternative approach to raypath migration called the diffraction or maximum convexity method. The method was introduced by Hagedoorn (1954) but was not generally implemented until the introduction of digital computers.

The concept of diffracting points can be extended to true reflections by considering that the reflection wavefront is the result of superposition and constructive/destructive interference of spherical waves emanating from elemental diffractors (treated as small reflecting spheres) which make up the reflecting surface (Figure 7). A useful presentation of this point of view was given by Trorey (1970). French (1975) provided a mathematical argument for this using the Kirchhoff integral solution to the scalar wave equation.

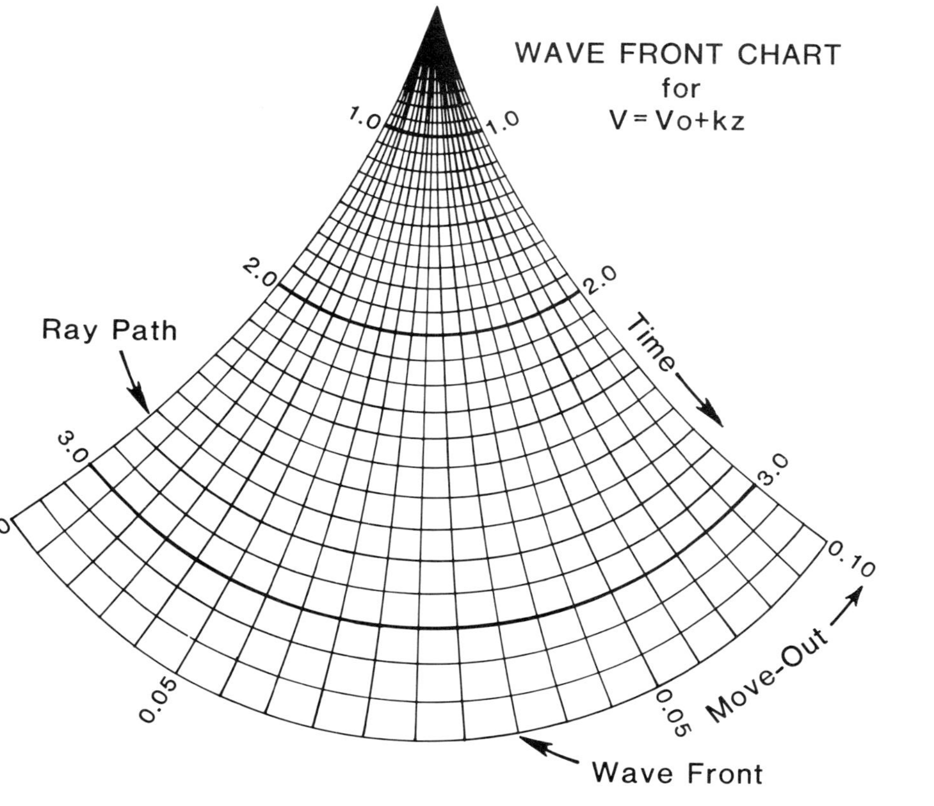

FIG. 5. Wavefront—raypath chart for linear increase of velocity with depth (after Dobrin, 1976).

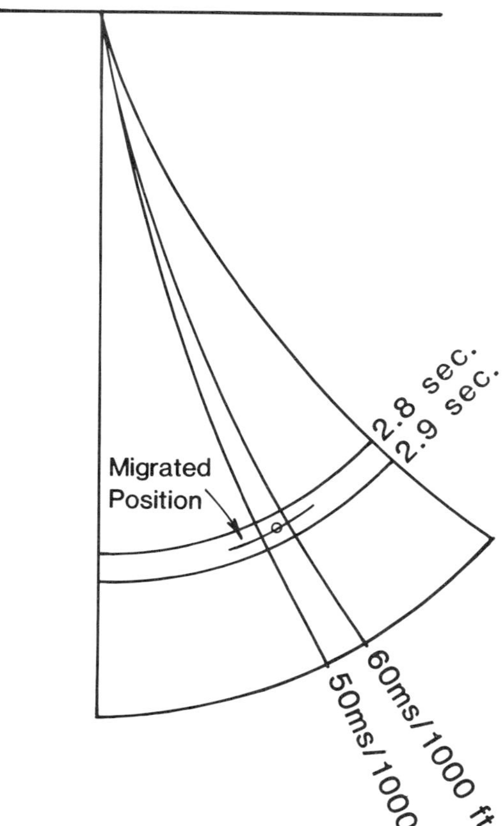

FIG. 6. Migration of reflection at 2.835 sec and time dip (moveout) of 57 msec/1000 ft using a wavefront—raypath chart (after Dobrin, 1976).

DIFFRACTION FROM POINT SOURCE

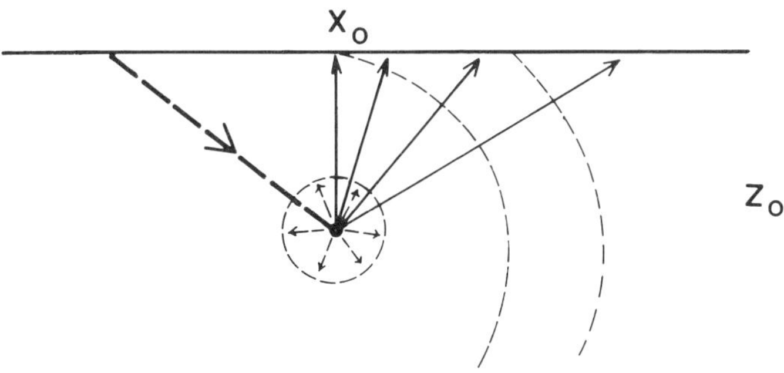

REFLECTION AS SUPERPOSITION OF DIFFRACTIONS

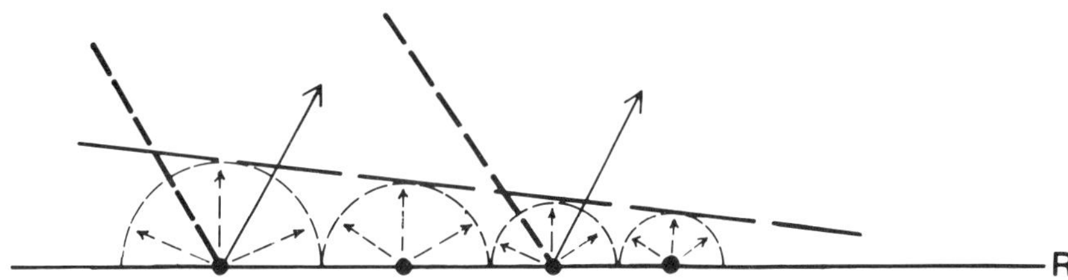

FIG. 7. Diffraction superposition according to Huygen's Principle, which states that every point on an advancing wavefront becomes the source of a new wavefront.

Each diffracting point generates a "diffraction curve" or "curve of maximum convexity" on a seismic time section. The true location of the diffracting point on the time section is at the apex of this diffraction curve. Migration using this method consists of moving arrivals along the diffraction curve to its apex. The power of the diffraction method arises in part because it is capable of handling plane reflecting boundaries, point or edge diffracting elements, or combinations of these without knowing beforehand the nature of the event. Migration proceeds point by point as shown in Figure 8.

Before computers, this method was implemented graphically using wavefront charts and diffraction overlays as shown in Figure 9. First, the event is located on the wavefront chart. Then we slide a diffraction curve laterally until a line-up is found tangent to the diffraction curve. The event is then moved to the diffraction apex and drawn tangent to the wavefront at that point. Often suites of curves are generated from known velocity functions for varyng arrival time and velocity.

Automatic migration by computer was developed in the mid-1960s. The first computer methods were simply automatic picking, shifting, and plotting routines that used raypath geometries. However, wavefront and diffraction procedures were also easily programmed since they could be formulated as statistical summing or stacking of sampled data according to well-defined mathemetical rules. To understand implementation of these as well as wave-equation techniques, a seismic section needs

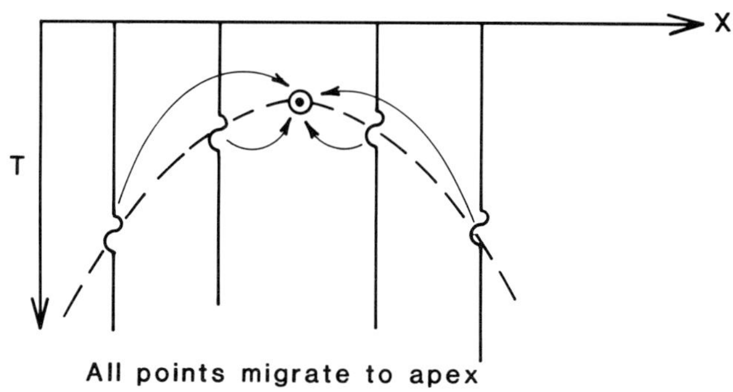

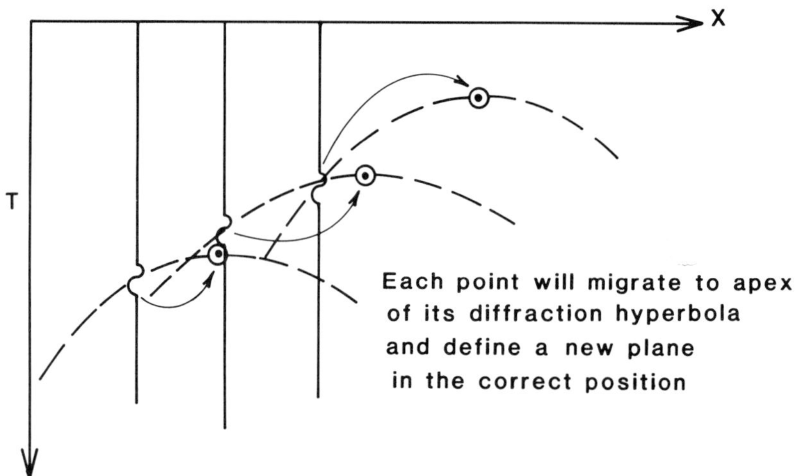

FIG. 8. Migration using the diffraction method.

to be defined as the computer sees it. The section is considered a recording of waves at the surface ($Z = 0$) sampled at discrete ground locations and arrival times. An amplitude sample is stored in the computer memory at each X_i, t_j coordinate location. The entire section is represented as an $M \times N$ amplitude matrix, where M is the number of traces and N the number of time samples per trace.

Digital migration can then be conceptually defined as nonlinear coordinate transformation that shifts amplitude samples to new locations.

$$\begin{array}{ccc} \text{Unmigrated} & & \text{Migrated} \\ (X_i,t_j) & \rightarrow & (X_k,t_l) \end{array}$$

where X_i is a horizontal spatial coordinate and t_j is a time sample value.

This matrix transformation is based on the nature of wave propagation in the earth

DIFFRACTION MIGRATION

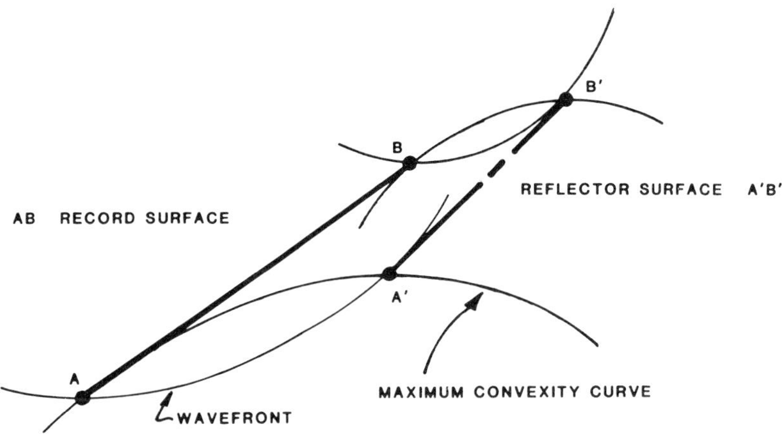

FIG. 9. Graphical implementation of the diffraction or maximum convexity method.

and the relationship between the true and apparent travel paths. A mapping of amplitude matrix A_{ij} is established so that

$$A_{ij} \xrightarrow{T} A_{kl}.$$

The transformation T is accomplished via a family of functional relationships of the general form

$$t_\ell = F[t_j, X_k - X_i, V(x,t)],$$
$$x = x_k,$$

where $V(x,t)$ is the media velocity field.

Diffraction summation and wavefront interference listed in Figure 4 are equivalent statistical migration methods. The first is based on Hagedoorn (1954); it treats every (x,t) sample location on a section as a diffraction apex and generates a maximum convexity trajectory for each sample based on an input velocity function. This is illustrated in Figure 10 (after Dobrin, 1976) for one sample location. The algorithm adds all amplitudes intersected by this trajectory and stores the sum at the apex location. The process is repeated for all samples down and across the section. Where a maximum convexity curve coincides with an actual zone of high energy (for example, an actual diffraction pattern on the section), a large signal is output at the apex. This "many to one" mapping results in constructive interference of true reflected data at migrated positions. Programs of this type often have some kind of coherency editing and summation aperture control options.

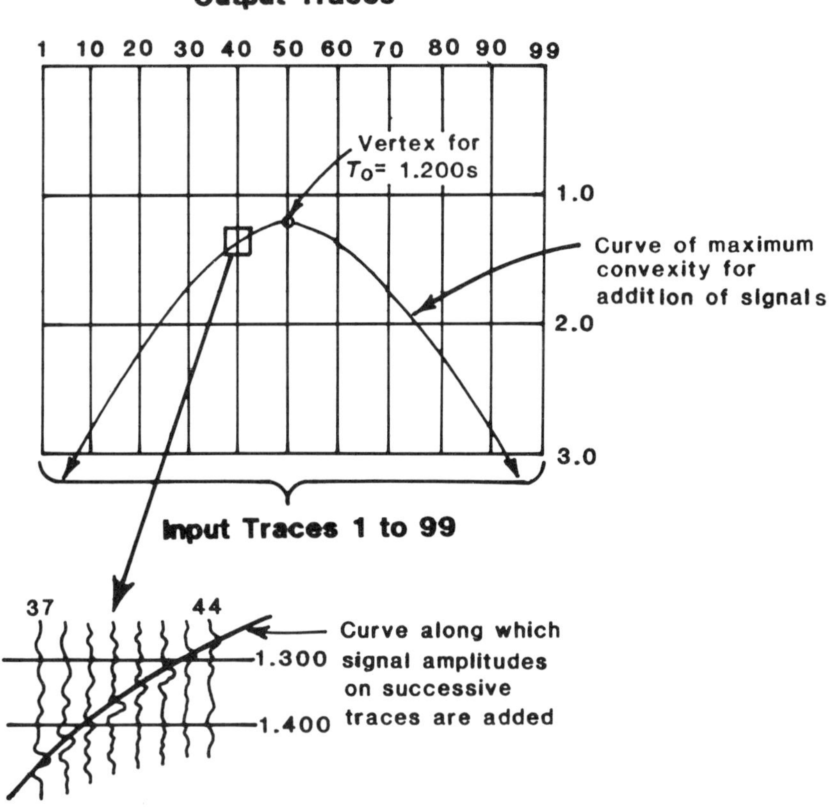

FIG. 10. Computer migration using curve of maximum convexity. Signals are summed along a maximum convexity trajectory associated with an apex time and rms velocity. Any lineup of strong energy will result in a high amplitude at the apex (after Dobrin, 1976).

The wavefront method (Rockwell, 1971), by contrast, is a "one to many" mapping based on common tangent principles. Individual amplitude samples are "swung out" and summed in the computer along wavefront trajectories to create interference patterns such that coherent reflections add in phase at their properly migrated positions and incoherent events and random noise tend to cancel. This method proved readily adaptable to migration before stack.

These two statistical methods had notable success in migrating complex seismic data in the late 1960s and early 1970s (see Schneider, 1971). However, they also have certain undesirable features: (1) mixing and destruction of event character and amplitude information because of summation process and presence of noise; (2) loss of resolution because summation is a low-pass filter operation; (3) strong, isolated high-amplitude events produce edge effects along wavefront trajectories; (4) no consideration of ray bending other than rms velocity compensation; and (5) detailed velocity models were difficult to apply. CDP wave-equation migration techniques, which are described in the following section, alleviate some of these difficulties.

The last item in Figure 4 represents present research and development of techniques to migrate common shot data into depth, followed by a stacking process to improve signal-to-noise (S/N) ratio.

Wave-equation migration concepts

Migration procedures discussed so far were based on ray theory concepts. An inverse problem such as migration requires a wave theory approach that maintains the amplitude and phase resolution contained in the input data. Such an approach was pioneered by Claerbout and his coworkers at Stanford in the early 1970s. They developed finite-difference solutions to the acoustic wave equation to simulate wave propagation, and by applying a new concept of reflector mapping they used these solutions to migrate seismic data.

The wave-equation migration process for 2-D data is conceptually illustrated in Figure 11 (after Alford, 1975, personal communication). The process of moving the waves into the earth, called downward continuation, is equivalent to pushing the recording surface into the earth closer to the seismic reflectors. A migrated time or depth section is obtained by evaluating the wave field at each depth for zero travel time. This process can be implemented by a recursive-iterative algorithm (finite-difference), or in a single transformation (*F-K*), or in combination (Kirchhoff). One hypothetical model suggested for this approach is the so-called "explosive reflector" model of Lowenthal et al (1974). It illustrates how three key concepts—wave fields, downward continuation, and imaging—fit together.

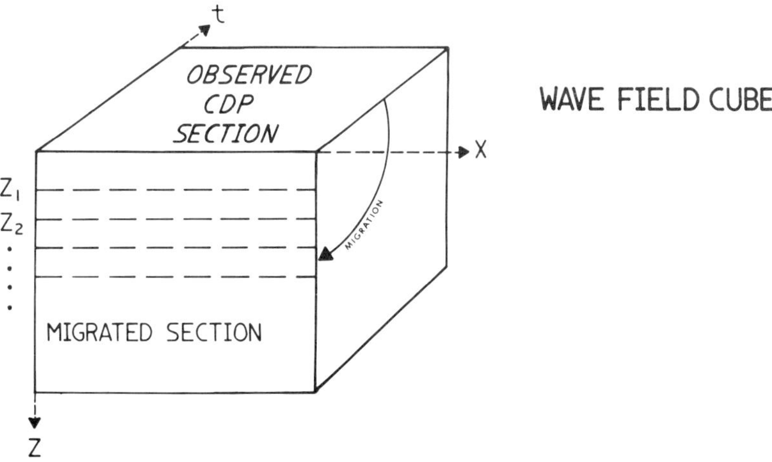

FIG. 11. A conceptual picture of seismic migration. The wave equation moves seismic data backward in time and down in depth to generate desired reflector configuration.

A CDP stacked seismic section is assumed to be equivalent to the data that would be recorded by a coincident source and receiver moving progressively across the surface. Multiple reflection energy is assumed to be attenuated by stacking. Thus, the data to be migrated correspond to two-way transmission of acoustic pulses along raypaths which are normal to the reflecting surfaces (see Figure 12a). By considering only one-way propagation along these normal rays, the stacked seismic section can be viewed as a record of a propagating wave field $P(x,0,t)$ recorded at $z = 0$, where $P(x,z,t)$ is the pressure amplitude of the traveling acoustic disturbance at (x,z,t). The variation in pressure amplitude is governed by the 2-D scalar wave equation:

$$P_{xx} + P_{zz} = \frac{1}{V^2} P_{tt}.$$

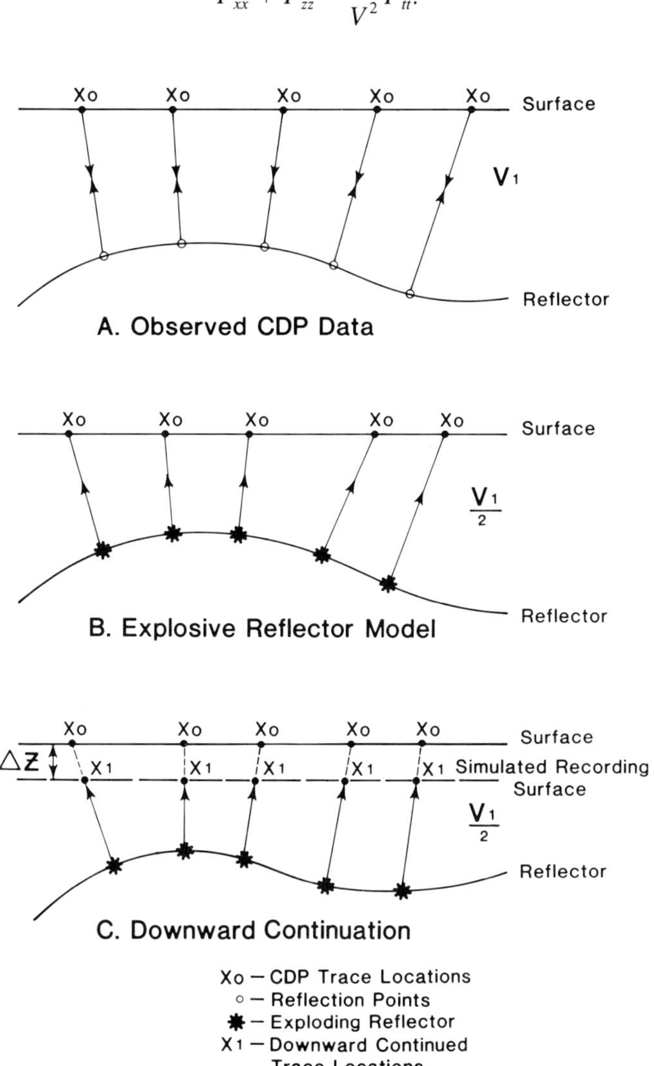

FIG. 12. (a) The CDP stacked section simulates data corresponding to two-way transmission along paths normal to reflecting surface. (b) The exploding reflector model for wave-equation migration. (c) Downward continuation of receivers to depth ΔZ.

The subsurface is modeled as a distribution of small explosives whose strength is proportional to the reflection coefficient at each point (Figure 12b). These charges are fired at $t = 0$ and the upgoing waves are recorded at the surface $[P(x,0,t)]$. Since in actual CDP data upgoing and downgoing energies are assumed to follow the same path, this model would be equivalent to our actual experiment if true earth velocities are divided by two. The data are migrated using the solution to the wave equation, obtained by any of the available techniques. $P(x,0,t)$ is an initial value used to calculate the wave field at successively greater depths $P(x,z,t)$. This is hypothetically eqivalent to lowering the geophones toward the reflector "sources" (Figure 12c). At each depth the wave field is evaluated for $t = 0$ since at zero traveltime, waves and sources coincide. This is called the imaging or reflector mapping principle (Claerbout, 1976). In another way of looking at this, as the coincident source-receiver pairs are pushed downward into the earth, the structures just beneath the buried locations of source and receiver come into focus. This downward continuation and imaging is done in successive steps of Δz until the complete migrated section is synthesized. In principle, a variable velocity model can be specified for geologic structure, and all propagation effects can be included during downward continuation of the waves through this model structure. Figure 3 in a way illustrates how this downward continuation procedure can successively unravel a buried focus reflection pattern as t_0 changes from 1433 to 263 msec.

Present day algorithms use solutions to the 2-D scalar wave equation as the basis for downward continuing seismic data. It is important to recognize the assumptions listed below.

(1) The data consist only of primary, upgoing *P*-waves and no undesirable waves such as sideswipes, multiply reflected waves, shear waves, and critically refracted energy. Computers will treat these coherent noise waves as if they were all *P*-waves and migrate them accordingly.
(2) The model for wave-equation migration assumes no random noise; accurate migration programs often treat noise as steeply dipping energy and move it in semicircular arcs. This is particularly a problem in low S/N data which can take on a mixed or "wormy" appearance after migration.
(3) Spatial aliasing is a persistent problem where dips are steep. Migration of undersampled data results in wavelet dispersion and loss of signal. Aliasing becomes a particular problem in migration of common shot data before normal movement (NMO) correction because cost considerations often lead to inadequate spatial sampling of the subsurface.
(4) The CDP section is equivalent to data recorded with zero source-receiver offset. This assumption is perhaps weak, since the CDP section is not itself the result of a physical experiment but rather the result of processing offset data to simulate coincident source-receiver data.
(5) The correct subsurface velocity distribution is known. The necessity of correct velocities occurs with any migration procedure and often results in the paradox that one has to know correct velocities before accurately migrating the structure, but to find correct velocities the migrated structure must be known. A further problem concerns lateral velocity variations; first generation programs do no better than the statistical approaches in that they ignore Snell's law for horizontal variations and can produce distorted,

misplaced images where overlying velocities vary laterally. This is commonly called the image ray problem (Hubral, 1977), and it was discussed by Hatton et al (1979) and Judson et al (1980).

Alternative mathematical operators

Computer implementation of wave-equation migration is based on numerical solutions to the wave equation. There are several approaches to algorithms. Three principal categories, all of which are in wide use today, are listed in Table 1.

Further modifications of these techniques were discussed by Berkhout and Palthe (1979) and by Hu et al (1980). Comparisons of various operators from both a practical and theoretical standpoint have been done by French (1975), Larner and Hatton (1976), Johnson (1979), and Berkhout (1979).

Finite-difference operators were pioneered by Claerbout (1976). His first-order method was limited to waves propagating within about 15 degrees to the vertical, but better approximations soon extended validity to steeper dip migration. Higher order approximations have been suggested (Stolt, 1978), but these have mostly proven uneconomic. Also, various constraint schemes are available to combat operator-generated phase error. Frequency domain-wavenumber methods seem to fall into two subcategories. The first, introduced by Stolt (1978), migrates in a wavenumber domain after transformation to an apparent or pseudo-depth domain. The second involves frequency-wavenumber shifts closely related to plane wave expansion—holographic methods of acoustics and optics. Gazdag (1978) and Bolondi et al (1978), discussed these. Kirchhoff integration, perhaps the oldest and most elegant technique, was discussed by French (1975), Kuhn and Alhilali (1976), and Schneider (1978). For areally recorded data, it entails integration over a spatial aperture; for 2-D applications, Schneider showed that it requires a time-dependent temporal convolution followed by a weighted spatial integration.

The complete mathematics of the three classes of operators will not be presented here, but instead their relative merits in practical use are discussed. Finite-difference methods use numerical derivative approximations so that a modified acoustic equation can be formulated in a way that allows recursive solution. In Claerbout's method (Claerbout and Johnson, 1971), a transformation is made to a moving coordinate system in which τ (see Figure 13) is constant for a vertically traveling wavefront. The second derivative term P_{zz} in this frame can be dropped or approximated so that the resulting equations have stable, recursive solutions for $P(x,\Delta z,\tau)$. The solutions can be cast as recursive filter operators that accomplish downward continuation and imaging as shown in the latter part of Figure 13.

Figure 14 is a diagramatic illustration of how one convolutional filtering scheme works. A spatial filter operates up each trace of the surface CDP data, sample by sample, to reconstruct traces at $z + \Delta z$. The downward continued data at $z + \Delta z$ are then input for the next step to $z + 2\Delta z$ in a similar fashion. In principle a filter operator can be built which depends upon the actual interval velocity between each Δz step and honors lateral changes in that velocity. The relative merits of finite-difference approaches are summarized below.

Table 1. Wave-equation migration operators.

Finite-difference—Recursive
- 15°
- 45°
- Higher order
- Dispersion minimization

Frequency domain—Wavenumber
- Stolt (K_x,K_z)
- Phase modification (K_x,ω)

Kirchhoff integral
- 3-D spatial integration
- 2-D convolutional

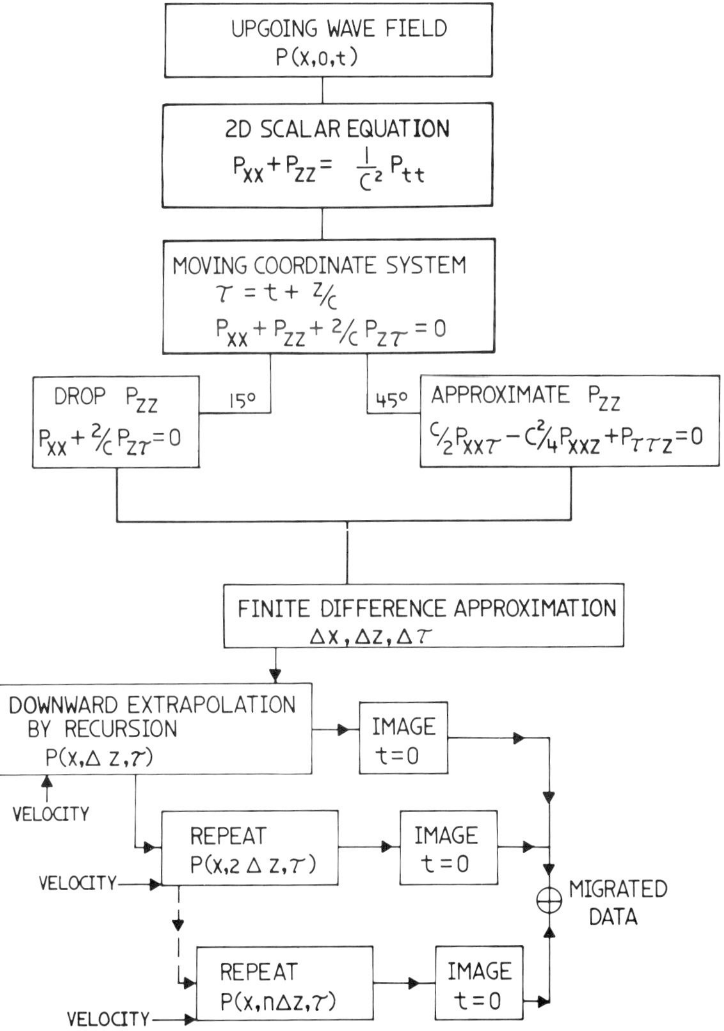

FIG. 13. A flow chart for finite-difference migration (expanded after Larner and Hatton, 1976).

A PICTORIAL FINITE DIFFERENCE SCHEME

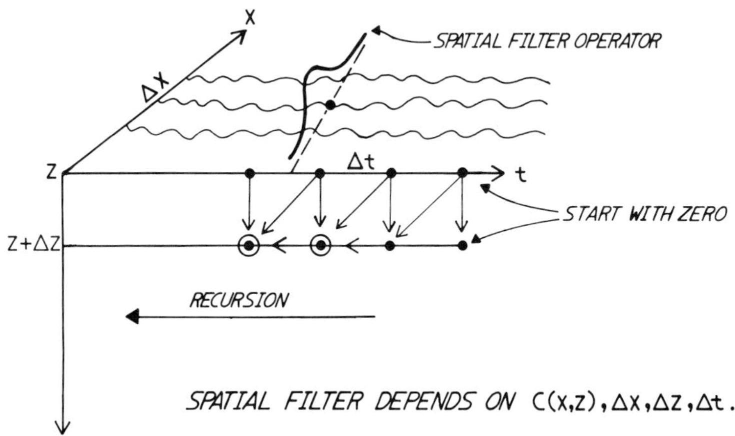

FIG. 14. Finite-difference migration formulated as recursive spatial filtering. Arrows indicate sequential reconstruction of a downward continued trace, sample by sample, at $Z = \Delta Z$ as filter moves up a representative trace at $Z = 0$.

Finite-difference approaches

Advantages

(1) In principle, raypath bending due to vertical and horizontal variations in input velocity models can be honored, resulting in true depth migration.
(2) Dip limitation avoids smoothing of noise data so that sections before and after migration appear cosmetically similar.

Disadvantages

(1) Dip is limited, to the extent that approximate equations are used.
(2) Dispersion is due to approximations to derivatives.
(3) It is relatively expensive because of data transfer operations in the recursive algorithm.

Frequency domain approaches

The frequency-domain approach discussed next is the widely used Stolt (1978) formulation. He showed that if seismic data could be transformed to a domain in which the wave equation and its boundary conditions were independent of velocity, simple geometric manipulations in spatial frequency could accomplish migration quickly and accurately. This transformation to an apparent or pseudodepth domain is a velocity-dependent, nonlinear scaling of the time axis. It compensates for squeezing of the seismic time data due to increasing velocity with depth. The data are in a sense made to approximate a constant-velocity earth so straight-ray migration rules (Figure 2) can be used. Table 2 lists the steps in this migration scheme.

Figure 15 (after Chun and Jacewitz, 1978) compares straight-ray migration in depth with the equivalent process in the frequency domain for one $K_x - K_z$ pair of the amplitude spectrum. The scaling factor in Table 2 occurs because migration shortens bounded reflector segments and results in lowered temporal frequency. Additional wavenumber scaling and scaling in the pseudodepth conversion can be done to combat migration noise. Chun and Jacewitz (1978) presented an excellent geometric discussion of this migration approach.

Advantages

(1) It makes no approximations to the acoustic equation and in theory can migrate true dips up to 90 degrees.
(2) There is no dispersion out to Nyquist frequencies.
(3) It is inexpensive compared to other methods.

Disadvantages

(1) It cannot accommodate multivalued velocity points on section.
(2) Distortion occurs where time-velocity functions vary rapidly.
(3) It has great difficulty honoring lateral changes in velocity.
(4) Noisy sections appear mixed or wormy.

Table 2. $K_x - K_z$ frequency domain migration.

Convert section to pseudo-depth
Transform data to K_x-K_z domain
Shift and scale wavenumber components
Inverse Fourier transform
Convert migrated pseudo-depth to time or depth

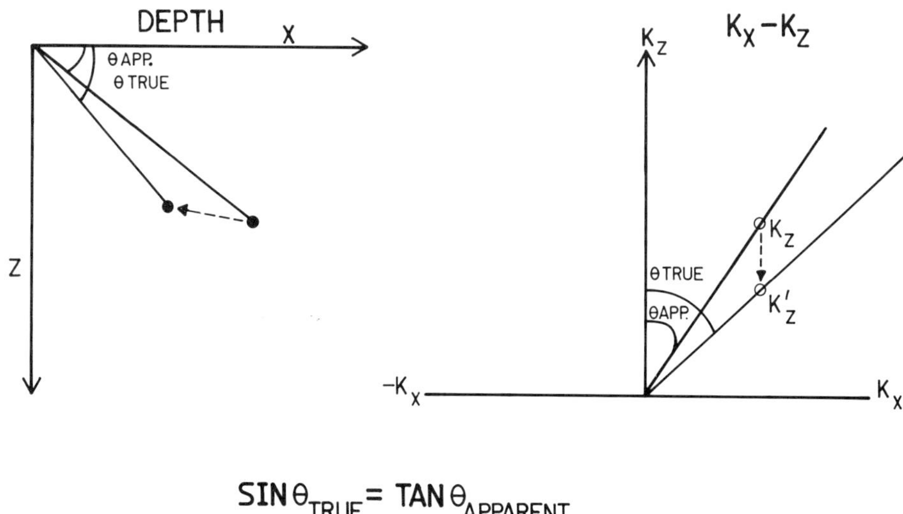

FIG. 15. Comparison of wavenumber and time-space shifts according to straight-ray rules (after Chun and Jacewitz, 1978).

Kirchhoff approaches

Kirchhoff migration is a close relative of wavefront or diffraction methods, but it puts summation operations in a weighted form that honors the wave equation. The method is based on Kirchhoff's theorem for a point source solution to the wave equation. This theorem is nothing more than a quantitative statement of Huygen's principle and formulates exactly the inverse scattering problem of acoustic physics. It states that if the value of a wave field is known on a closed surface along with the normal and time derivative on that surface, the wave field at any point interior to the surface can be calculated by integration of the known data over the surface.

As shown in Figure 16 (after Schneider, 1978), the downward continued wave field at Z for 3-D data can be determined point by point by spatial, time-retarded integration over the known areal wave field at (x,y,z_0). Setting $t = 0$ in the lower equation results in a 3-D migration equation.

Extension from 3-D to 2-D profile migration is not as simple as just dropping y dependence, but rather it requires either integration of y dependence or reformulation of the Kirchhoff integral for cylindrical rather than point sources. Figure 17 shows Schneider's 2-D migration formula which requires time-domain convolution of data beneath hyperbolic apertures followed first by spatial summation and then by a differentiation (phase correction). The media velocity distribution determines convexity of hyperbolic summation curves just as with the older diffraction summing techniques.

Advantages

(1) It offers full wave equation, limited only by finite aperture size.
(2) The method performs well with steep dip. Weighting schemes can easily be applied to combat noise. It is adaptable to migration before stack.

Disadvantages

(1) There is difficulty with lateral velocity variation.
(2) It is expensive and can enhance and mix noise.

Table 3 is a subjective summary that indicates the trade-offs in the three techniques; 1 represents best performance, 3 represents worst performance. The varying

Table 3. Subjective summary.

Space	*F-D* (t,x,z) (t,K_x,z)	"*F-K*" (K_x,K_z)	Kirchhoff (t,x,z)
Cost	2	1	3
Velocity variation	1	3	2
Steep dip	3	2	1
Dispersion	3	1	2
Treatment of noise	1	2	3
Cosmetic appearance	1	3	2
Before stack	3	2	1

3D KIRCHHOFF DOWNWARD CONTINUATION

$$P(r,t)=\frac{1}{2\pi}\int dA_0 \frac{\cos\theta}{RC}\left[P'(r_0,t_0)+\frac{C}{R}P(r_0,t_0)\right]_{t_0=t-R/C}$$

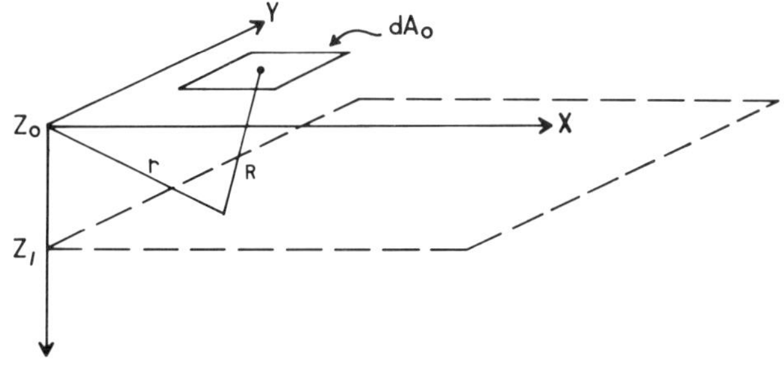

SIMPLIFIED AFTER SCHNEIDER (1978)

$$P(x,y,z,t)=-\frac{1}{2\pi}\,\partial/\partial_z\iint d_x\,d_y\,\frac{P(x,y,0,t-R/C)}{R}$$

FIG. 16. Reconstruction of areal wave field at *Z*, by spatial integration over $P(x,y,z=0)$.

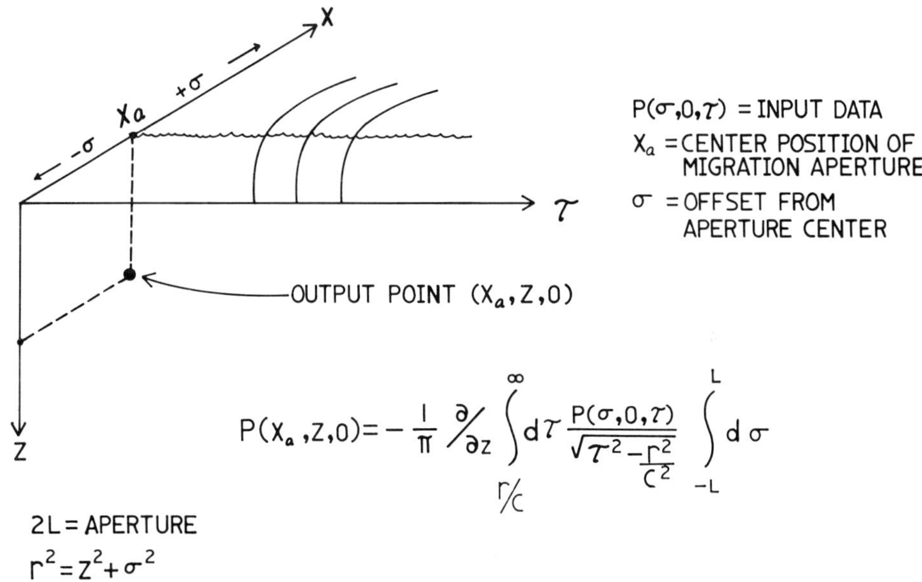

FIG. 17. Implementation of a 2-D Kirchhoff operator as a combination of temporal convolution and spatial integration over maximum convexity apertures (after Schneider, 1978).

numbers indicate that decisions on what operator to use in a given area are data dependent.

Future developments

Table 4 lists new and future developments in migration and subsurface imaging technology.

Until recently lateral velocity changes (image ray problem) were not handled properly in routine migration programs. Many exploration companies and contractors now have so-called depth migration programs that properly honor wave refraction if a depth-interval velocity model can be specified beforehand. Depth migration before stack is still mostly in a research stage (see Schultz and Sherwood, 1980).

Table 4. New and future migration developments.

Imaging in presence of lateral velocity inhomogeneities
Imaging of midpoints in depth before stack
Imaging of velocity during migration process
Complete elastic wave-equation processing

As indicated earlier, depth migration approaches require an accurate geologic model. Ideally we would like to have a full inversion process, one that simultaneously images velocities and migrates data to correct subsurface locations. This probably necessitates a before stack process. One approach is a direct acoustic inversion (Cohen and Bleistein, 1979). Another is an interpretative combination of modeling and migration. A preliminary model is specified, the data are depth mi-

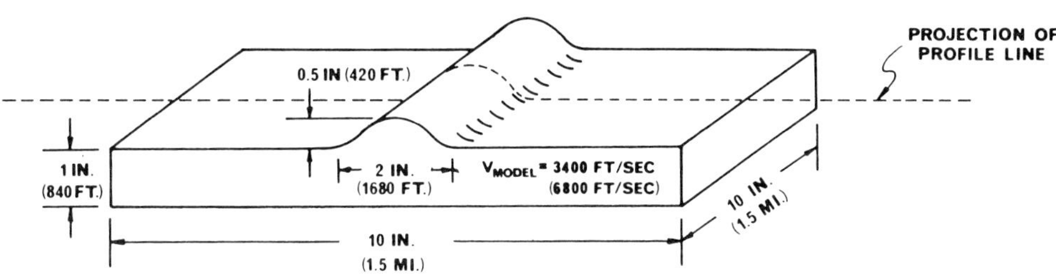

FIG. 18. High curvature ridge model.

grated through the model, inconsistencies between model geometry and migrated wave field are noted, and the model is successively changed until the interpreter judges the agreement satisfactory. A third approach is to image the subsurface recursively step by step from the surface downward using a combination of migration, velocity analysis, and downward continuation procedures. This approach will be referred to as the "unified solution." Rockwell et al (1979), Taner (1979), and Chun and Jacewitz (1979) discussed certain aspects of this kind of technology.

Further advances in migration imaging will require complete elastic wave theory including so-called undesirable waves such as source-generated shear waves, multiples, and converted waves in the imaging process. These waves contain subsurface information which, if recorded with three-component acquisition systems in true vector amplitude, could be inverted to subsurface elastic parameters. Unfortunately, a theory for this inversion does not yet exist, although several researchers seem to be pointed in this direction (see Marfurt, 1978; Graves and Schneider, 1979; Crampin and Radovich, 1979).

Migration Processing

2-D examples

Some examples of physical model, computer model, and actual data are shown to illustrate present migration technology. A physical model is shown in Figure 18. The rubber model was immersed in a water tank, and the data were gathered with an ultrasonic transducer-receiver system (French, 1974). The actual dimensions are as shown and the equivalent field dimensions are given in parentheses. Since the sound speed in the model is less than that in water, the model represents a structure with velocity less than that of the overburden.

A reflection profile was taken over this model with the model 8.9 inches below the recording line. This scales to a depth of 7440 ft. Since the data were recorded

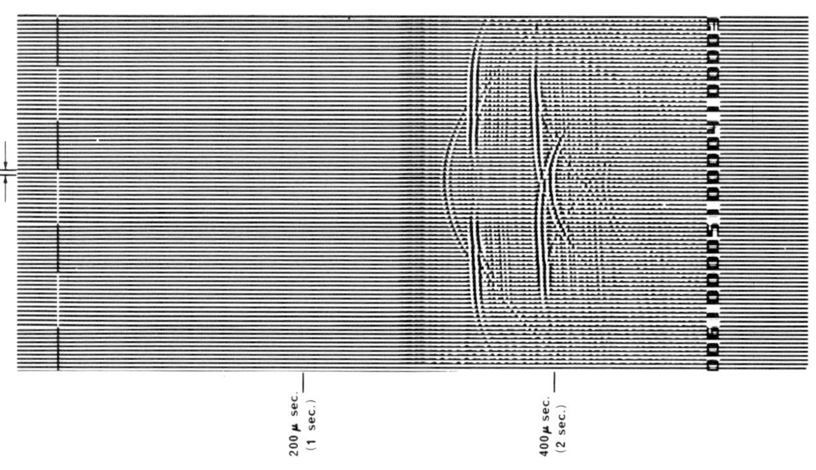

FIG. 19. Reflection profile over ridge model.

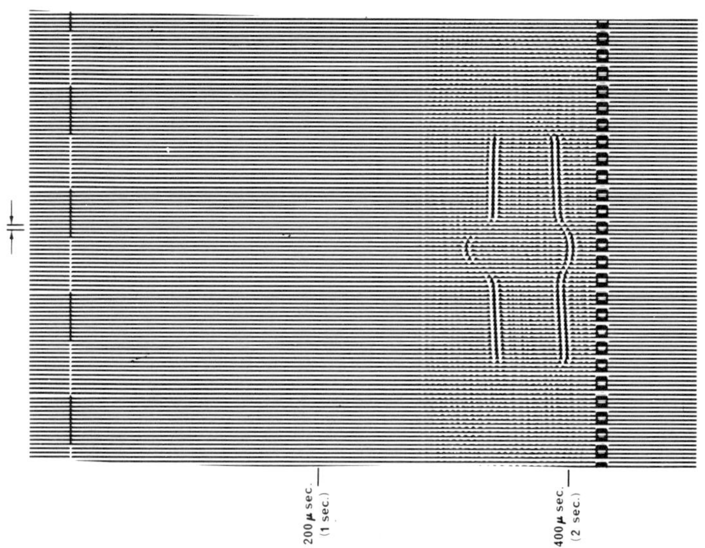

FIG. 20. Migrated time section.

in the same format as field data, the results could be output to a seismic plotter in wiggle/variable area format as shown in Figure 19. The following effects are apparent on the profile:

(1) The ridge has been spread out laterally far beyond its actual dimensions.
(2) Diffractions are apparent from the edges where the model terminates.
(3) Although the bottom of the model was flat, the reflection profile shows it to have a reflection crossover pattern suggestive of a syncline.
(4) A long legged diffraction emanating inward from the edge of the model bottom can be seen.

The results of migration by a Kirchhoff type algorithm are shown in Figure 20.

(1) The ridge in the upper surface has been imaged to its proper geometry.
(2) The edges of the upper surface of the model are sharply in focus (diffractions collapsed).
(3) The false syncline in the lower surface due to lateral velocity variation in the overburden is in focus.

Figure 21 shows an anticline model, and Figure 22 shows a computed normal incidence synthetic section. Highest temporal frequencies are spatially aliased on the steep limbs of the synthetic where actual dips approach 45 degrees. *F-K* migration (Figure 23) is superior to the finite-difference case (Figure 24) where dispersion is more pronounced. The *F-K* approach also used only one-fourth the computer time. The migration noise generated by *F-K* migration is due to aliasing and termination points in the depth model.

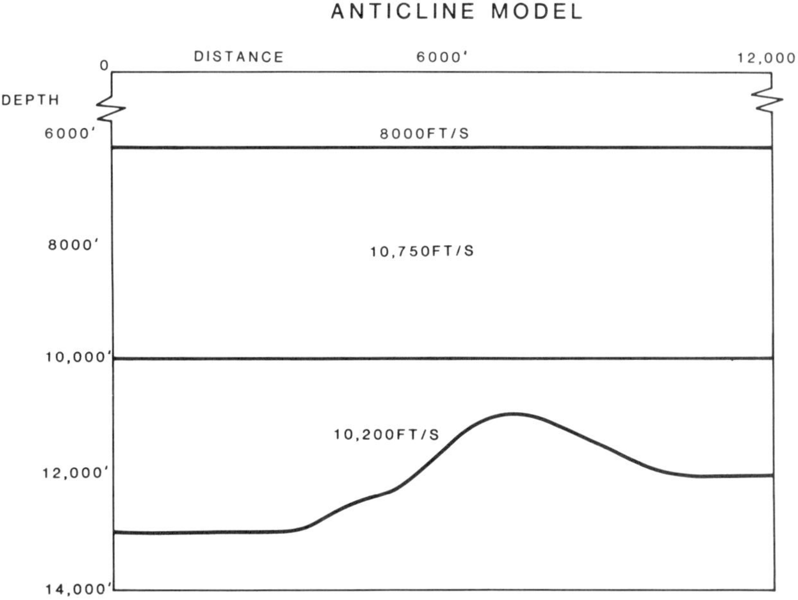

FIG. 21. Depth-velocity model input to normal incidence modeling program.

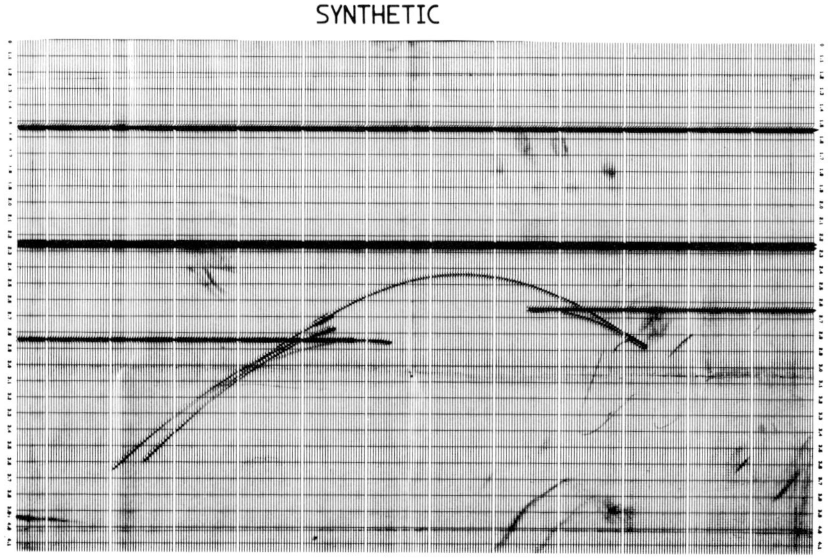

SYNTHETIC

FIG. 22. Normal incidence synthetic section. Wavelet is 35-Hz Ricker, trace spacing is 100 ft.

F–K

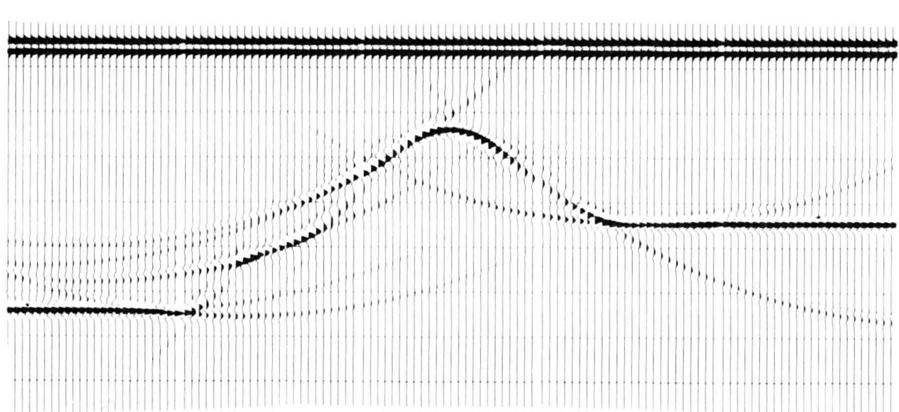

FIG. 23. *F-K* migration. Steeper dips are poorly imaged because of spatial aliasing.

Figures 25–27 illustrate the problem some *F-K* schemes have with multivalued velocities. The unmigrated data show steeply dipping reflections from a salt flank. Migrations were done using a stacking velocity function taken from the flat layered area on the right side of the section. The pseudodepth stretching of the dipping reflections uses velocities at the apparent time position of the events that are much higher than the actual raypath velocities above the true position of the events. The result is overmigration and smearing of the salt face (Figure 26). This problem is handled correctly by finite-difference recursive methods (Figure 27), and the result is a clearer image of the interface.

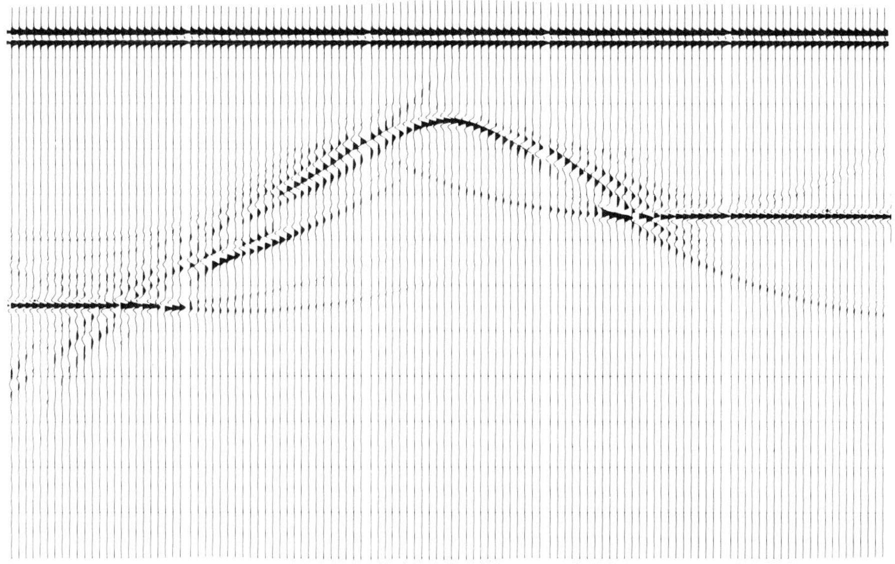

FIG. 24. 45 degree finite-difference migration. Dispersion is due to operator error and aliasing.

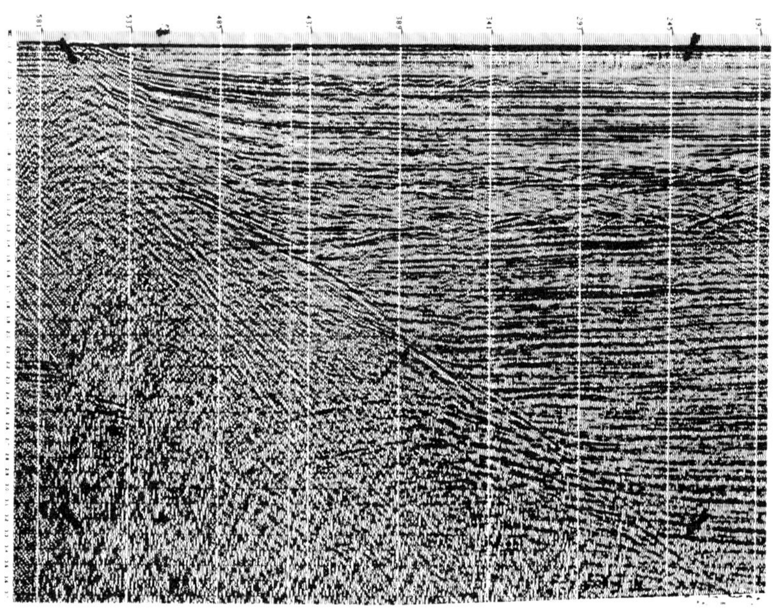

FIG. 25. CDP stacked section across a Gulf Coast salt flank.

Figures 28–32 (courtesy of K. Larner, Western Geophysical) compare finite-difference (15-degree operator) with Kirchhoff migration. Figure 28 shows migration of a constant-velocity depth model with dips up to 60 degrees. The Kirchhoff method exhibits superior steep dip performance, but it generates considerably more migration noise, even in flat dip portions of the model. Figure 29 further illustrates these points with the addition of a noise trace to a model. Kirchhoff migration swings the noise data out into syncline-like arcs but accurately positions the signal. The finite-difference operator used here only partly swings out the noise, but it disperses the steep dip parts of the signal, creating a different kind of migration noise. It should be

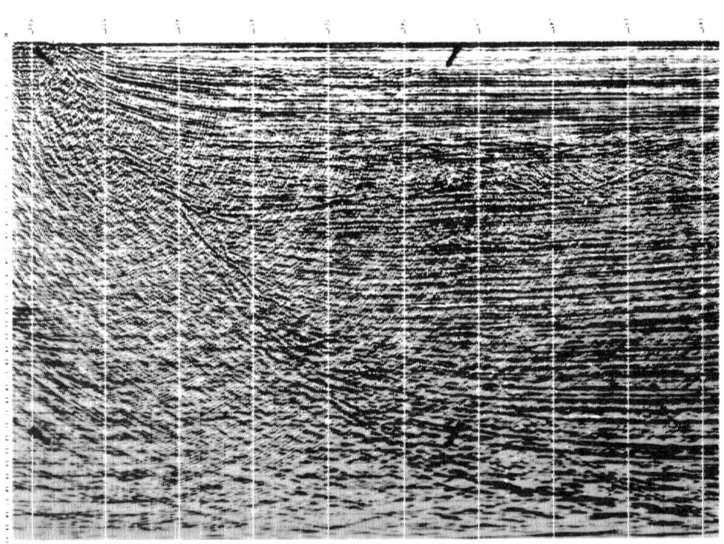

FIG. 26. *F-K* approach overmigrates salt flank because of multivalued velocity problem.

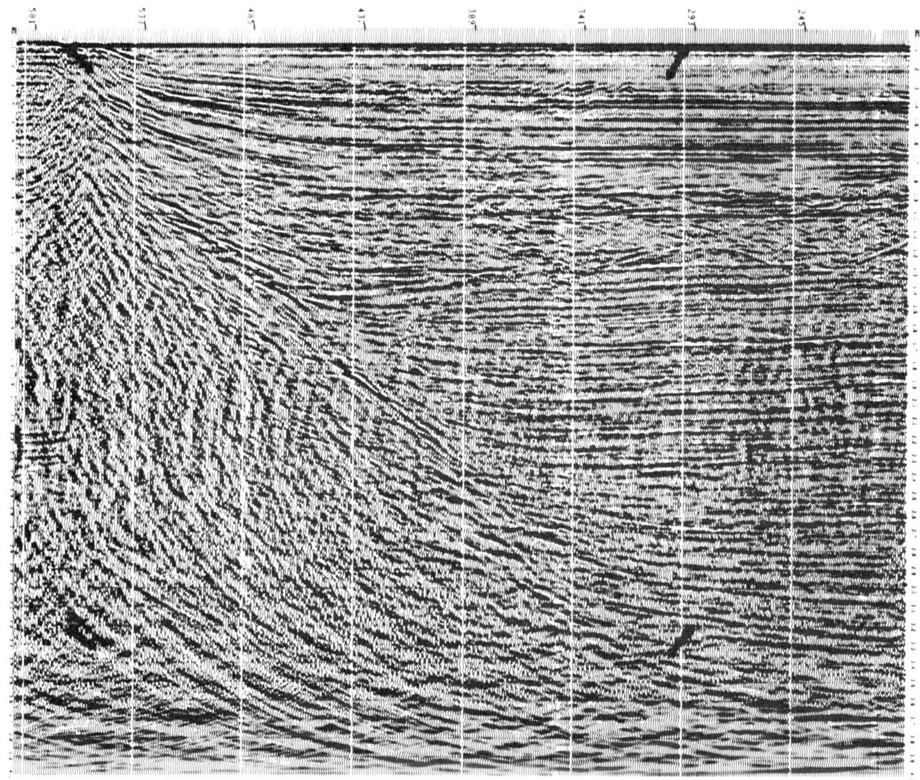

FIG. 27. Finite-difference recursive approach produces a more accurate migration for this data case.

pointed out that this is a first generation finite-difference operator. New differencing schemes have significantly improved steep dip performance, but usually at the expense of more mixing of noisy data.

Figure 30 is a stacked section from the Gulf of Mexico. Both migration approaches (Figures 31 and 32) successfully unscramble the faulted data (in the lower left), and the resulting migrations are markedly similar. The Kirchhoff operator performs slightly better with the steep dip data (lower left). It is our experience that the present state-of-the-art programs using any of the three contrasting approaches on high S/N ratio data produce near identical migrated sections. Differences become pronounced when velocities change rapidly or data are very noisy.

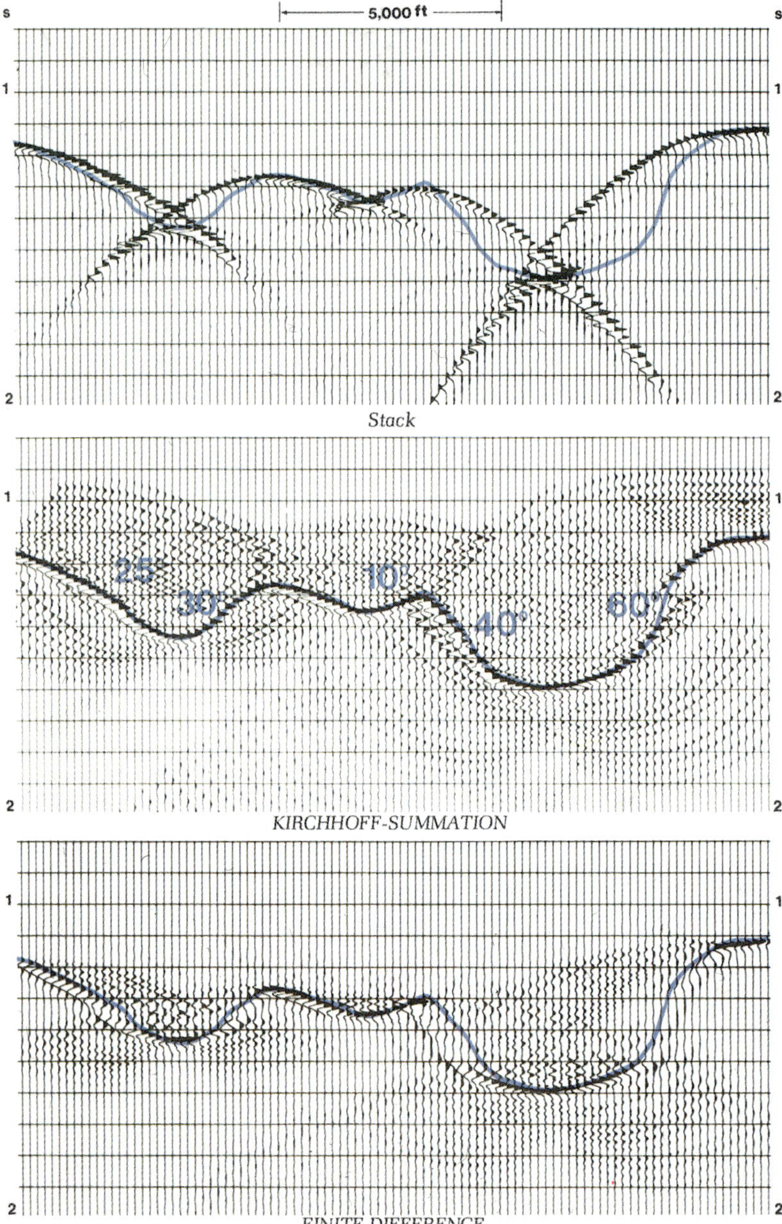

FIG. 28. Comparison of Kirchhoff and finite-difference migrations for constant velocity model. The Kirchhoff output has better steep dip performance, but overall signal to noise characteristics are worse.

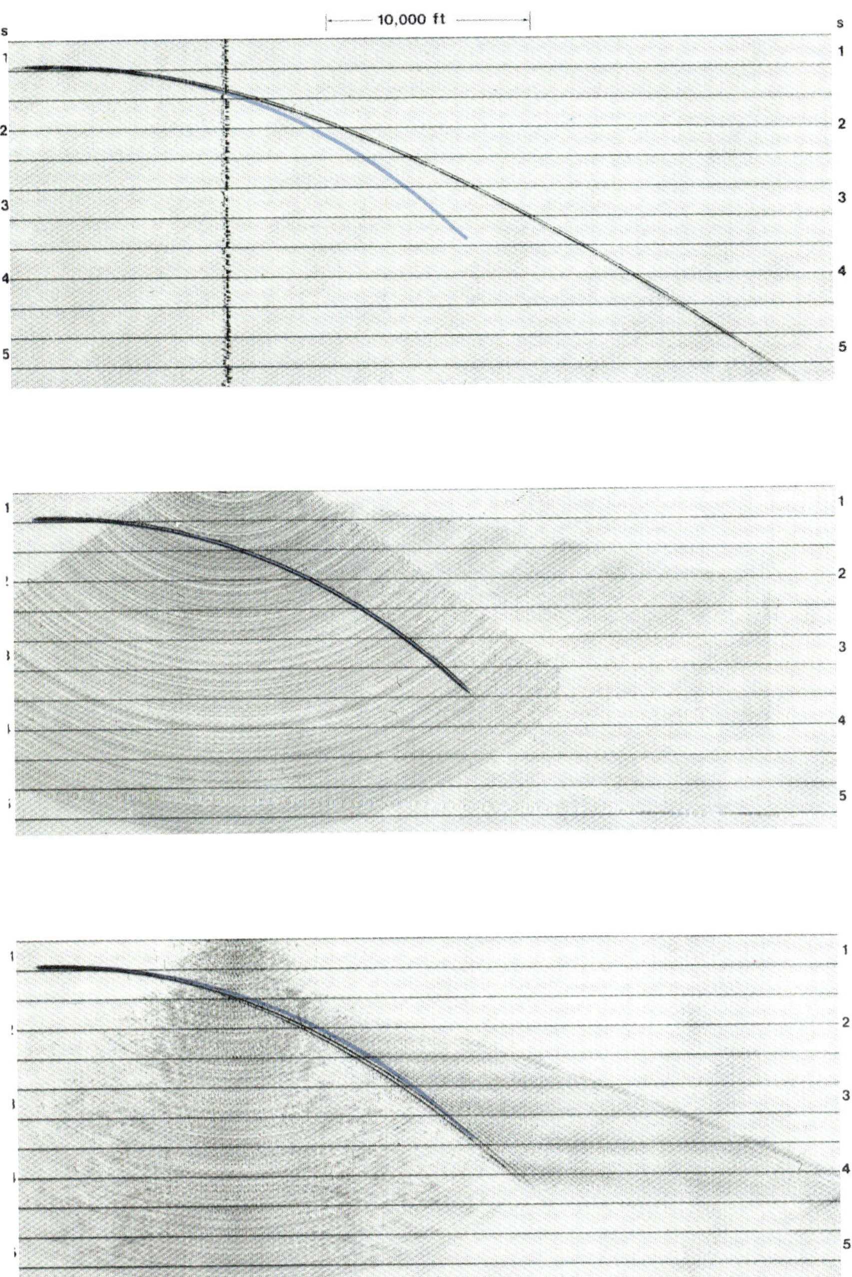

FIG. 29 (a) A synthetic section for a depth model which grades by three degree increments from horizontal to a maximum dip of 45 degrees. Band-limited random noise has been added to one trace. (b) Kirchhoff migration treats all dips well. Noise, however, has been organized into arc-like events that can obscure deeper reflections. This is an inevitable consequence of a full acoustic wave approach. (c) This finite-difference approach disperses reflections for dips greater than about 20 degrees. A fortuitous consequence is that algorithm approximations have spatially restricted the migration noise.

FIG. 30. Gulf of Mexico stacked section (horizontal scale is in miles). Note particularly the complex region of crossing events between 4 and 5 sec.

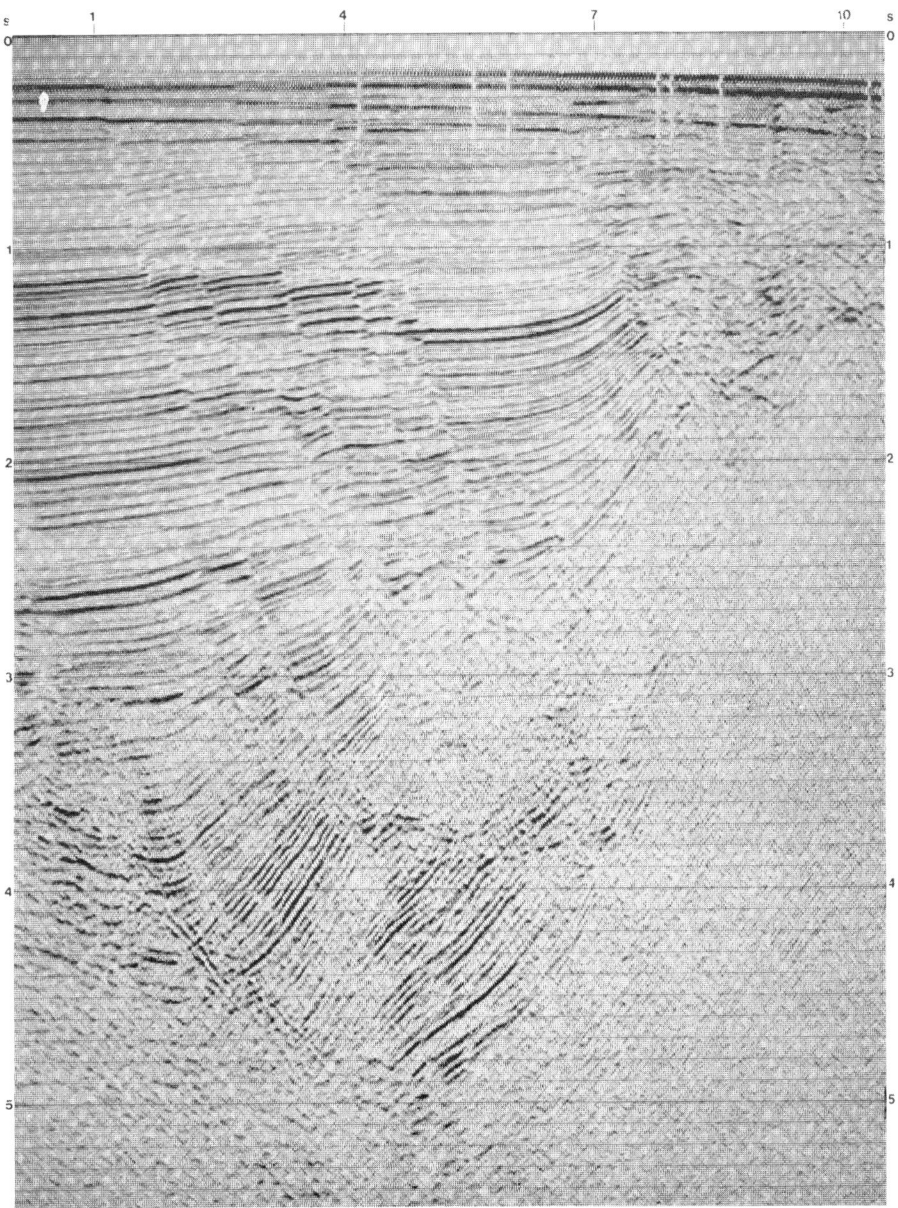

FIG. 31. Migration by the Kirchhoff method sorts out the crossing events and clarifies faults by collapsing diffractions.

FIG. 32. Finite-difference migration produces a near identical result with slightly poorer results for steeply dipping data.

3-D examples

Acquisition and migration of seismic data in three dimensions are becoming more widespread in the exploration industry; more than 100 surveys have been completed as of 1980, some with dramatic improvements in data quality and interpretability. In fact, 3-D methods will be required if we are to find future hydrocarbon reserves in complex stratigraphic traps, such as meandering channel sand deposits, and in deep structural traps, such as those associated with complex faulting.

These methods in particular are needed when poor data quality is due to structure and not reflectivity or where offline reflections (sideswipe) interfere with target data zones. 3-D migration algorithms are natural extensions of any of the three wave-equation methods already discussed.

The effects of 2-D and 3-D migrations can be seen by looking again at a physical model example. A photograph of the model used is shown in Figure 33. There are two domal features and an oblique fault in the upper surface of the model. In terms of field dimensions, the model is roughly two miles across, the domal features are roughly 3/4 mile in diameter and about 550 ft high (about two wavelengths of the dominant frequency of the 30–60 Hz broadband source).

Figure 34 is a plan view of the 3-D model. The dotted line represents the cross-section labeled no. 6 on the previous photograph. Figure 35 shows a conventional 2-D reflection profile recorded above this cross-section. When the shape of the reflection is known, most of the arrivals present on the profile can be accounted for. However, it would be impossible to reconstruct the model from the measured data alone.

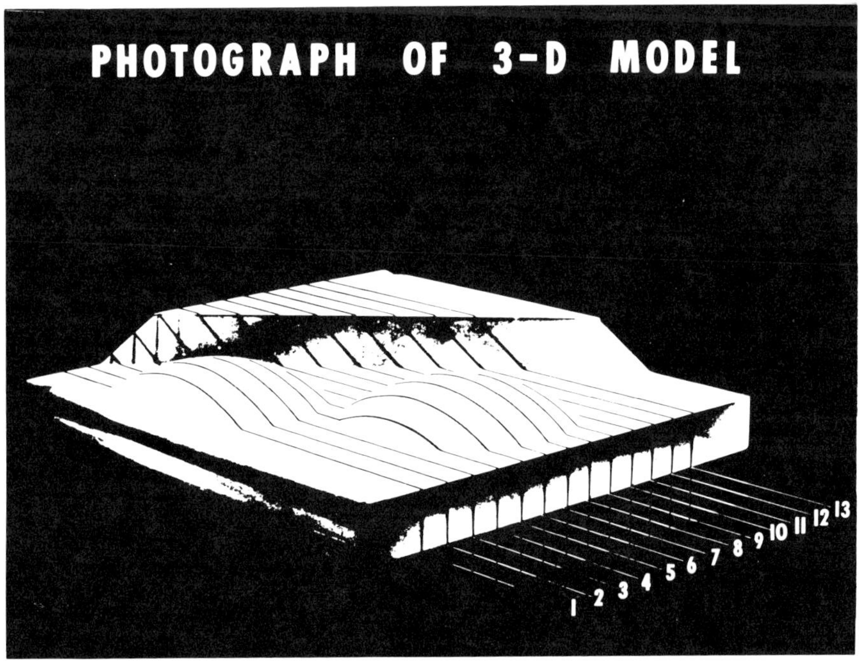

FIG. 33. Photograph of 3-D physical model.

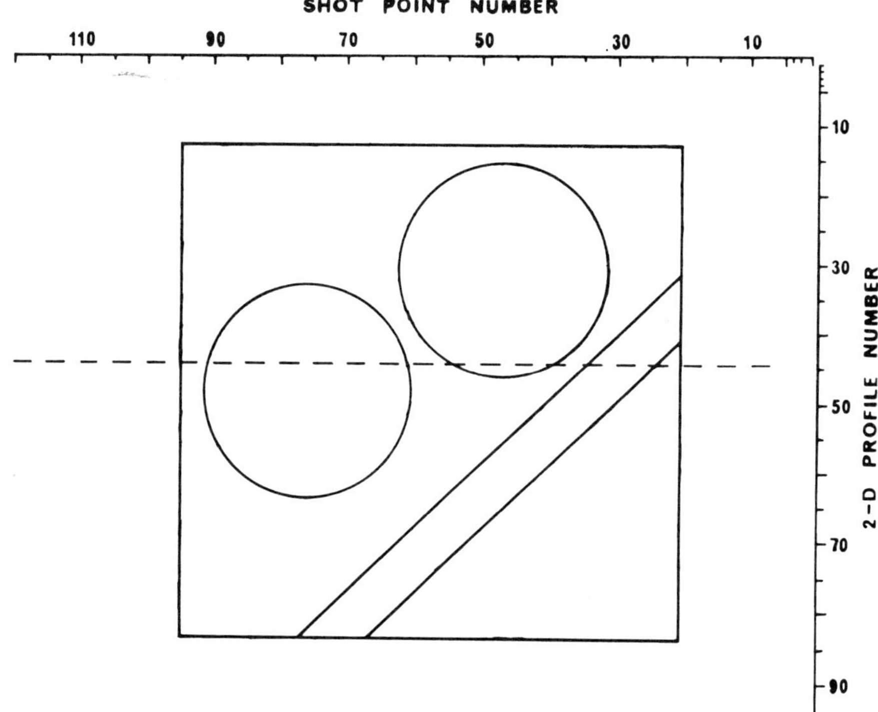

FIG. 34. Plan view of 3-D model.

A 2-D migrated time section of this profile is shown in Figure 36. This migrated section looks more like the actual cross-section, but there are still serious problems.

(1) Sideswipes from the central dome and the fault are not eliminated but are falsely focused. The image in the center of the upper surface could be erroneously interpreted as fluid-gas contact under an anticlinal trap.

(2) The fault plane beneath the profile line is a blind structure, and how to correlate over this blind zone is not known. This correlation problem would be compounded if additional layers were present.

Return now to the plan view of the data collection scheme shown in Figure 34. In this figure (1) the top numbers are shotpoint locations for the first data profile (110 ft apart). (2) The right-hand numbers are starting locations for the 96 parallel profiles shot over the model. The profiles were separated by 110 ft providing a square grid of zero-offset reflection data. (3) The dotted line represents the first 3-D migrated cross-section to be displayed. 24 data profiles to each side of this line were migrated into this cross-section (48-trace aperture).

The 3-D migration (Kirchhoff) is shown in Figure 37. It is seen that the structural geometry of this cross-section is precisely reproduced by 3-D migration. There is no ambiguity in picking and following the entire upper surface. The following comments can be made: (1) Sideswipes and blind structures have been eliminated; (2) the background noise generated is comparable to that in the 2-D case and can likely be eliminated by a direct extension of more sophisticated 2-D data processing algorithms.

Figure 38 compares unmigrated, 2-D migrated, and 3-D migrated field data. 3-D migration produces a more coherent, higher S/N ratio section that has a significantly different interpretation than the 2-D case. In view of physical model results, the 3-

A: RAW DATA

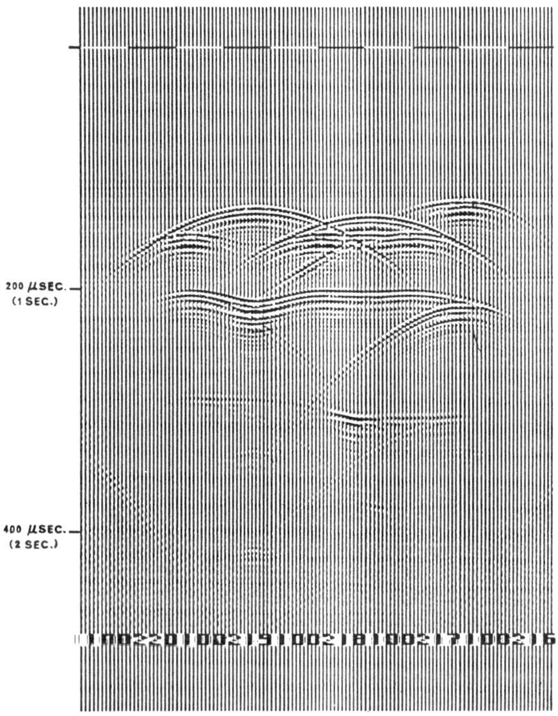

FIG. 35. 2-D profile over cross-section no. 6.

B: 2-D MIGRATION

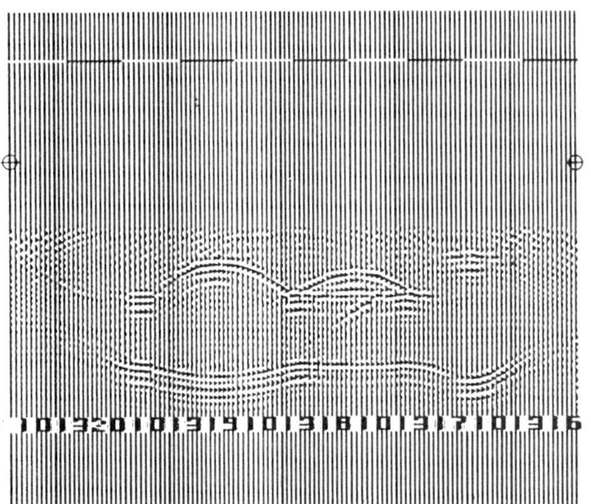

FIG. 36. 2-D migration produces an inaccurate image because some of the data are not in the plane of the profile.

C: 3-D MIGRATION

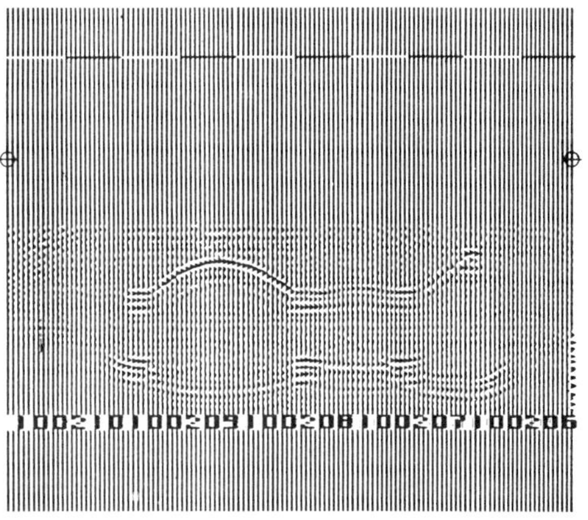

FIG. 37. 3-D migration correctly images cross-section.

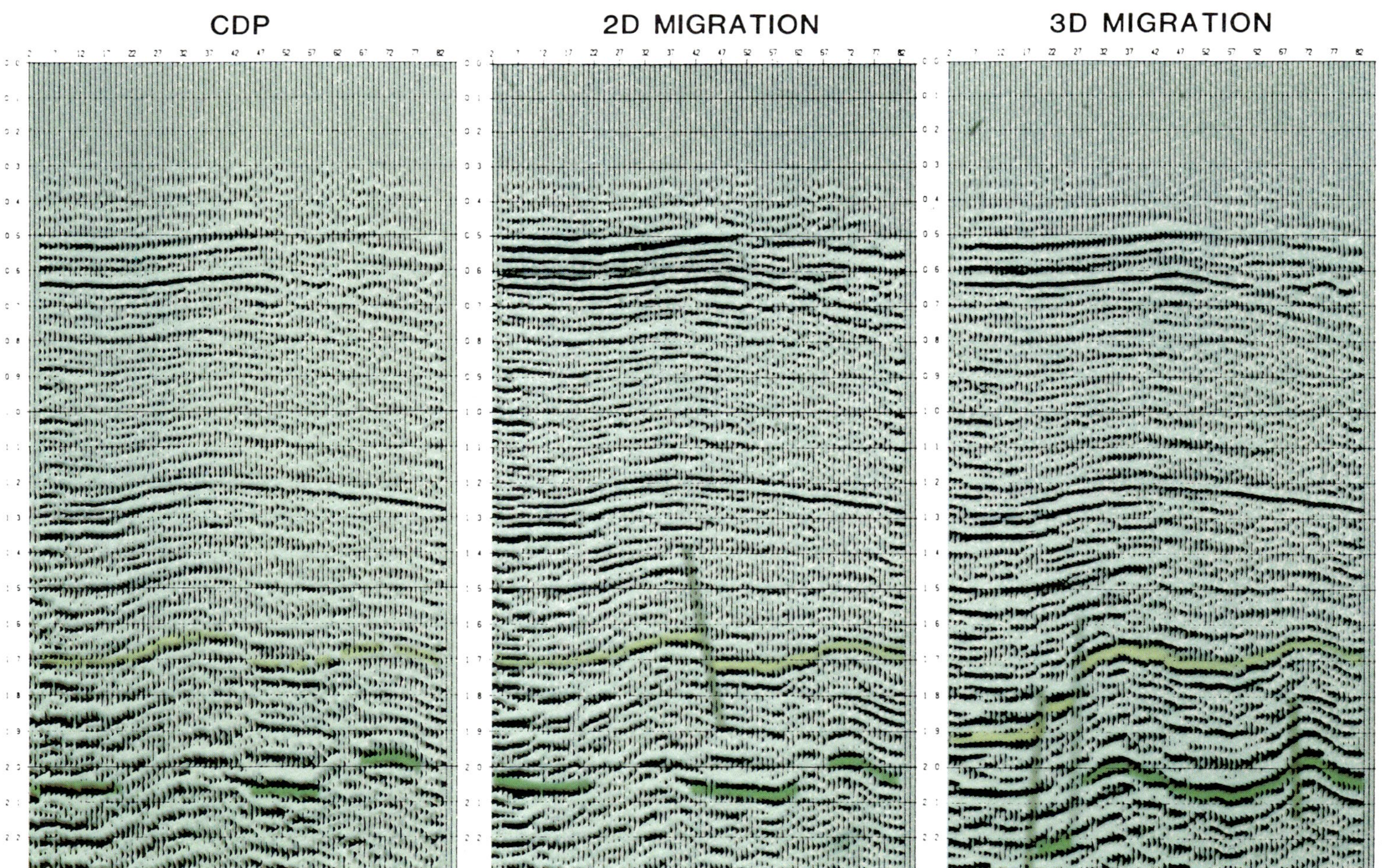

FIG. 38. 3-D migration of a field example significantly changes the subsurface picture.

D migration shown here produces a more reliable picture upon which to base exploration decisions.

Data Quantity and Computer Requirements

Figure 39 summarizes estimates of past, present, and future data volumes for typical seismic surveys (10 km linear or 100 km^2 areal coverage) in units of 16 bit words per survey. In the 1940s and 1950s, 12 to 24 trace analog recording was used with wide (100 m or so) receiver spacing and single fold (100 percent coverage of subsurface points) shooting techniques. Six seconds of recording time and a 15–55 Hz bandwidth were typical recording parameters. These numbers in digital terms translate to 10^4–10^5 words per 10-km line.

Digital recording and processing as well as multifold shooting began in the early 1960s. 48-trace, 12-fold data acquisition with 50 to 75 m group intervals resulted in an order of magnitude increase in data per survey. Bandwidth increased to typically 8–75 Hz.

Increased fold, shortened group intervals, and increased bandwidth resulted in another order of magnitude increase to tens of megawords per survey in the 1970s. The most dramatic increase, to approximately 10^{14} words, occurred with the advent of 3-D multifold recording in the late 1970s. Increases in acquisition requirements in the 1980s should result from increased receiver density (25 m group interval), trace length to 12 sec, and sampling rate of 1 msec, but most importantly from recording three components of motion at each detector position. This will be necessary if all elastic wave arrivals—*P*, *S*, and converted waves—are to be used in imaging true subsurface properties.

In the 1990s, three-component areal surveys will be extended into boreholes which will allow direct observation of upgoing waves. Although the amount of additional data is not significant, the corresponding processing is.

It is worth pointing out that some forecasters predict the use of group intervals one-tenth as large as we have used. Thus, the number of data collected in a 2-D survey could be an order of magnitude larger, and in a 3-D survey could be two orders of magnitude larger, than the above estimates.

What may be done with all these data in the 1980s and 1990s is shown in Figure 40. The term ''omnibus seismic data'' means a set of data adequate to solve the complete problem. It may be that broadband sources at every point on some 3-D area have to be used above the structures of interest. Much finer spatial sampling may be required than estimated above. Thus, presently it is not known how many data are actually necessary to solve the inverse problem for the elastic displacement equation shown. Indeed, it is really not known whether a unique solution can be found.

Figure 41 indicates what is actually done now. The diagram is overly simplified; shown are essentially two processing streams, one for a structural solution and the other for the velocity (lithology) picture. These processing streams interact with each other along the way when necessary. Ideally the future approaches will have some sort of stream incorporating the generalized equation given in Figure 40. This will require massive data acquisition as discussed above, an inverse theory that does not yet exist, and computer processing capabilities (estimated below) that are enormous.

Estimates of computer time required for migration processes are described here. These requirements are then extrapolated for algorithms and data bases 5 to 10 years

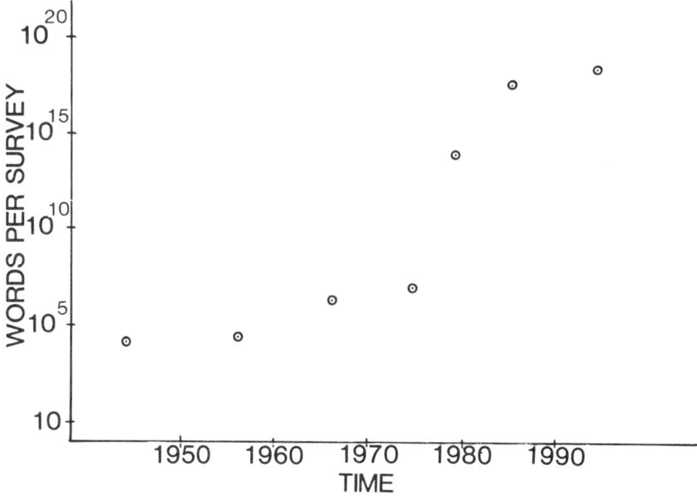

FIG. 39. Data volume estimates for typical seismic acquisition. Major jumps represent extension to 3-D and then to three-component elastic data.

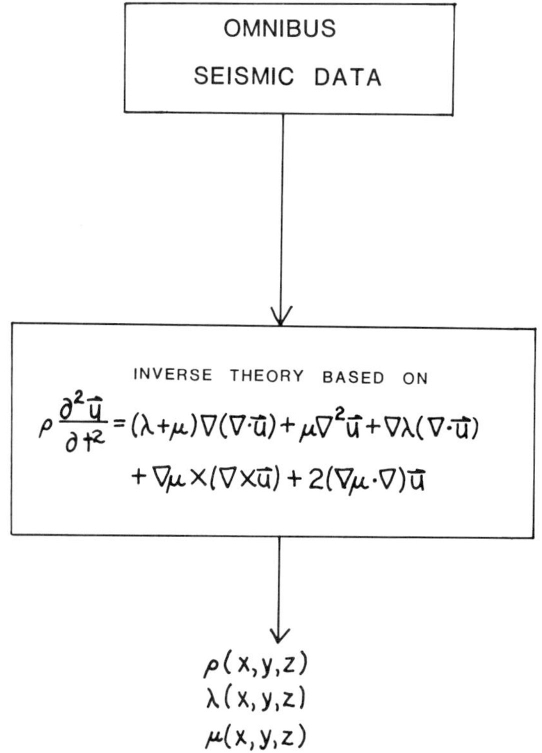

FIG. 40. Desired elastic wave data processing scheme.

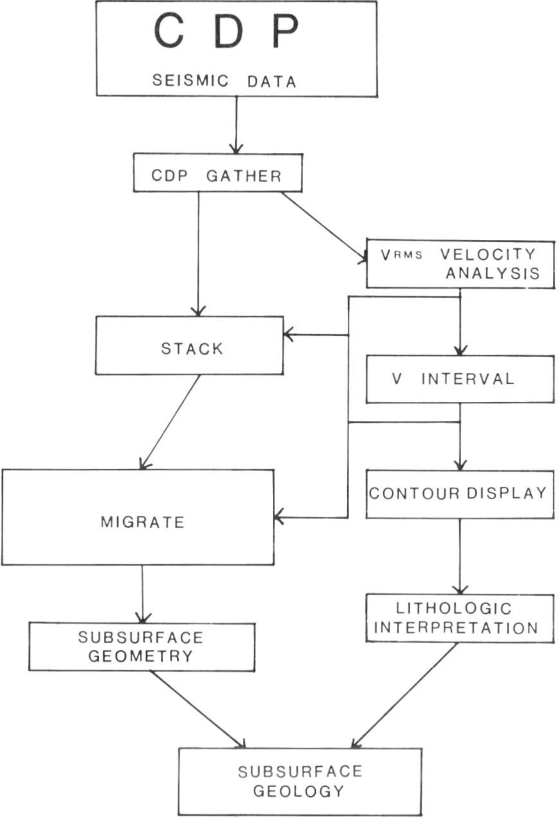

FIG. 41. Actual seismic data processing scheme now in use.

hence to solve the subsurface inverse problem more completely. In order to relate these estimates, a baseline called a benchmark data unit (BDU) is defined. One BDU is defined as the computer effort to resample in time 24 traces of data from 3000 samples per trace to 1500 samples per trace. This operation requires antialias filter and decimation routines. One BDU on an IBM 3033 central processing unit takes approximately .2 sec. The relationship of this unit to migration algorithms depends, of course, on particular hardware available and on the formulation of a specific algorithm. These estimates are based on a combination of direct computer comparisons and the author's experience. Table 5 gives estimates for present migration approaches. They are based on 12 fold data with a 50-m group interval (25-m CDP interval). Computer time estimates are for a 10-km profile or 100 km^2 areal survey, as were the acquisition estimates given earlier.

What was called the "unified" solution attempts to image geologic pictures like Figure 42 where structure and velocities vary in all dimensions. To accomplish this, an iterative imaging of structure and velocity step by step from the surface downward is envisioned. This is nothing more than an acoustic wave theory approach to the

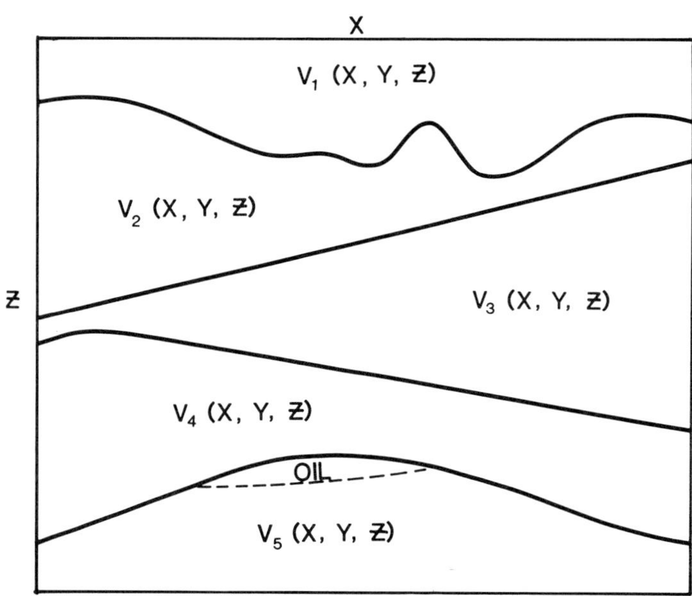

FIG. 42. In the real earth, interface geometry and interval velocities vary in three dimensions. The unified solution iteratively images velocity and structure layer by layer.

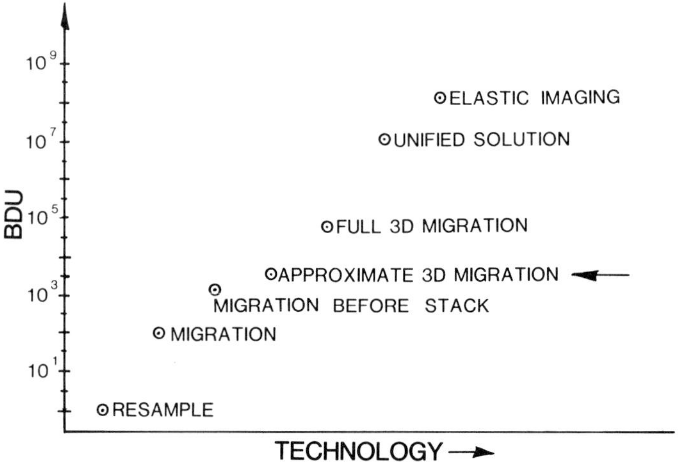

FIG. 43. Summary of computer estimates in BDU per 24 trace output.

old idea of seismic stripping. It uses existing seismic technology but requires data sufficiently sampled in space and probably an interactive processing system. The first step in this technique (V_1 layer) solves the statics problem. Table 6 summarizes estimates for such computations. Again, these are based on a 50-m group interval and 12-fold data.

Extending these estimates to the 1990s, when full elastic inverse theory may be available, is highly speculative. However, assuming three-component recording of a vector wave field with 25-m group spacing, the estimates in Table 7 can be obtained by extending the unified approach to elastic waves.

Table 5. Estimation for present migration algorithms.

Method	BDU/24 trace output	Computer time
Finite-difference	100	6.0 minutes
Frequency domain	20	1.0 minutes
Kirchhoff	150	8.5 minutes
Before stack[1]	1,500	85.0 minutes
Approximate 3-D[2]	4,800	76.0 hours
Full 3-D[3]	76,800	51.0 days

[1]Before stack migration assumes a 100 BDU process, migration of 12 common offset sorts, and subsequent resorting and stacking.
[2]The approximate 3-D estimate is based on a two-step migration for each midpoint (Gibson et al, 1979), first in an x-direction and then in a y-direction.
[3]Full 3-D uses a complete x-y aperture. BDU per output trace can range from 12 times the approximate 3-D value for a 24-trace aperture radius to much larger numbers for bigger apertures. A 32-trace aperture radius was chosen for these calculations.

Table 6. Unified solution: Migration—velocity—statics.

Method	BDU/24 Trace Output	Computer time
Migration before stack	1,500	
Velocity analysis	1,000	
Downward continuation of common shot records	1,200	
Total per step	2,700	2 hours
10 iterations	27,000	1 day
Approximate 3-D	10^6	2 years
Full 3-D	20×10^6	30 years

Table 7. Elastic imaging—1990.

Method	BDU	Computer time
2-D	.2 Mega	16 days
Full 3-D	180 Mega	500 years
3-D + Downhole	200 + Mega	??

Although the numbers in Table 7 seem ridiculous, they are in many ways based on minimum assumptions of data volume and algorithm complexity. They also consider none of the processing and massaging necessary to prepare data for imaging. Figure 43 summarizes the analysis of present and future technology requirements. The arrow indicates what is possible now. These estimates of future data processing needs are so enormous that major breakthroughs will be needed in computer technology. Improved algorithms, specialized digital hardware, and hybrid computer concepts will be required in concert to further subsurface imaging capabilities. The earth can transmit, reflect, and scatter elastic waves for a few seconds and keep our present digital processing systems busy for years imaging the results.

REFERENCES

Berkhout, A. J., 1980, Seismic migration: Amsterdam, Elsevier.

Berkhout, A. J., and Van Wulften Palthe, D. W., 1979, Migration in terms of spatial deconvolution: Geophys. Prosp., v. 27, p. 261.

Bolondi, G., Rocca, F., and Savelli, S., 1978, A frequency domain approach to two-dimensional migration: Geophys. Prosp., v. 26, p. 750.

Chun, J. H., and Jacewitz, C. A., 1978, Fundamentals of frequency domain migration: Seismograph Service Corporation (preprint).

——— 1979, Application of partial NMO and partial stack to prestack migration: Presented at the 49th Annual International SEG Meeting, November 6, in New Orleans.

Claerbout, J. F., 1976, Fundamentals of geophysical data processing: New York, McGraw-Hill Book Co., Inc.

Claerbout, J. F., and Johnson, A. G., 1971, Extrapolation of time dependent waveforms along their path of propagation: Geophys. J. Roy. Astr. Soc., v. 26, p. 285.

Cohen, J. K., and Bleistein, J., 1979, Velocity inversion for acoustic waves: Geophysics, v. 44, p. 1077.

Crampin, S., and Radovich, B. J., 1979, Synthetic CDP gathers for one-layer anisotropic models: Estimation of elastic constants from calculated *P* and *SH* curves: Presented at the 49th Annual International SEG Meeting, November 7, in New Orleans.

Dobrin, M. B., 1976, Introduction to geophysical prospecting: New York, McGraw-Hill Book Co., Inc.

French, W. S., 1974, Two and three dimensional migration of model experiment reflection profiles: Geophysics, v. 39, p. 265.

——— 1975, Computer migration of oblique seismic reflection profiles: Geophysics, v. 40, p. 961.

Gazdag, J., 1978, Wave equation migration with the phase shift method: Geophysics, v. 43, p. 1342.

Gibson, B., Larner, K., Solanki, J., and Ng, A., 1979, Efficient 3-D migration in two steps: Presented at the 49th Annual International SEG Meeting, November 7, in New Orleans.

Graves, J. E., and Schneider, W. A., 1979, Investigation of converted wave (*P-SV*) reflection prospecting systems: Presented at the 49th Annual International SEG Meeting, November 8, in New Orleans.

Hagedoorn, J. G., 1954, A process of seismic reflection interpretation: Geophys. Prosp., v. 2, p. 85.

Hatton, L., Larner, K., and Gibson, B., 1979, Migration of seismic data from inhomogeneous media, Migration and Modeling: Symp., Denver Geophysical Society, March 15.

Hilterman, F. J., 1970, Three-dimensional seismic modeling: Geophysics, v. 35, p. 1020.

Hu, T., Wang, K., and Hilterman, F. J., 1980, Comparison of some seismic imaging techniques: Presented at the 33rd Midwestern SEG Meeting, March 25, in Tulsa.

Hubral, P., 1977, Time migration—Some ray theoretical aspects: Geophys. Prosp., v. 25, p. 738.

Johnson, J. D., 1979, Comparison of finite difference, *F-K*, and Kirchhoff approaches to wave equation migration. Migration and modeling: Symp., Denver Geophysical Society, March 15.

Judson, D. R., Lin, J., Schultz, P. S., and Sherwood, J. W. C., 1980, Depth migration after stack: Geophysics, v. 45, p. 361.

Kuhn, M. J., and Alhilali, K. A., 1976, Weighting factors in the construction and reconstruction of acoustical wave fields: Geophysics, v. 42, p. 1183.

Larner, K., and Hatton, L., 1976, Wave equation migration: Two approaches: Presented at the 46th Annual International SEG Meeting, October 24, in Houston.

Lowenthal, D., Lu, L., Robinson, R., and Sherwood, J. W. C., 1974, The wave equation applied to migration: Geophys. Prosp., v. 24, p. 380.

Marfurt, K. J., 1978, Elastic wave equation migration-inversion: Ph.D. thesis, Columbia Univ.

Musgrave, A. W., 1961, Wavefront charts and three-dimensional migration: Geophysics, v. 26, p 738.

Rockwell, D., 1971, Migration stack aids interpretation: Oil and Gas J., April 19, p. 202.

Rockwell, D. W., Berkhout, A. J., and Larson, D. E., 1979, Lambda-space, the bridge between complex *X-Z* models and frequency domain migration processing: Presented at the 49th Annual International SEG Meeting, November 7, in New Orleans.

Schultz, P. S., and Sherwood, J. W. C., 1980, Depth migration before stack: Geophysics, v. 45, p. 376.

Schneider, W. A., 1971, Developments in seismic data processing and analysis: 1968–1970: Geophysics, v. 36, p. 1043.

——— W. A., 1978, Integral formulation for migration in two dimensions: Geophysics, v. 43, p. 49.

Sheriff, R. E., 1978, History and overview of seismic migration methods: Wave equation migration: Symp., Geophysical Society of Houston, April 3, 1978.

Sherwood, J. W. C., and Shultz, P. S., 1980, Migration in the 1970s: presented at the 33rd Midwestern SEG Meeting, March 25, in Tulsa.

Stolt, R. H., 1978, Migration by Fourier transform: Geophysics, v. 43, p. 23.

Taner, M. T., 1979, Common image point stacking system: Presented at the 49th Annual International SEG Meeting, November 6, in New Orleans.

Trorey, A. W., 1970, A simple theory of seismic diffractions: Geophysics, v. 35, p. 762–784.

WELL LOGS

Chapter 6

PHYSICAL BASIS OF WELL LOGGING

Introduction

The hydrocarbon potential of an area is often judged by the existing production and reserve estimates of the area. For a new exploration program, all available geophysical, geologic, and engineering data must be interpreted. The results of the new wells must be integrated for subsequent development operations. Such exploration studies normally include an understanding of the regional geology, geologic and/or geophysical maps, show maps, cores, tests, mud logs, and wire line logs.

Cores and wire line logs are among the most important data to a formation evaluationist, whether he is a geologist or an engineer. Petrophysicists integrate these data to derive indirectly many basic reservoir parameters, such as porosity and fluid saturation, and to determine volumetrically the reservoir oil in place. There are many current logging techniques which can be combined for this purpose.

A well log is a graphical presentation of a physico-chemical characteristic of the geologic formations measured in a borehole as a function of depth. The logs determine lithologic characteristics of the sedimentary rocks and pinpoint hydrocarbon-bearing strata. Before the invention of well logs, such geologic variables were ascertained by (1) inspection and analysis of drill cuttings and cores and (2) formation tests. Many of these operations are being used less today because the necessary subsurface information can be obtained from well logs at lower costs and with sufficient resolution.

The first well log was designed and recorded in 1927 by the Schlumberger brothers in Pechelbronn, Alsace, France (Allaud and Martin, 1976). Figure 1 shows part of this first resistivity log, which is the prototype of modern resistivity well logs. In 1931 they added the spontaneous potential (SP) curve. This historic log, shown in Figure 2, identified a conglomerate porous bed. Years later the two logs were recorded simultaneously by improved logging equipment; since then they have been used extensively for qualitative and quantitative formation evaluation.

During the fifty some years since the first well log was run, logging techniques and their interpretation methods have notably improved. It is now possible to determine many rock characteristics and parameters for comprehensive applications in such areas as geology, geophysics, engineering, and economics. Rock porosity and fluid saturation determination are common quantitative uses of the logs. Figure 3 is an example of a subsurface cross-section made through log correlations (Gomez-Rivero, 1975).

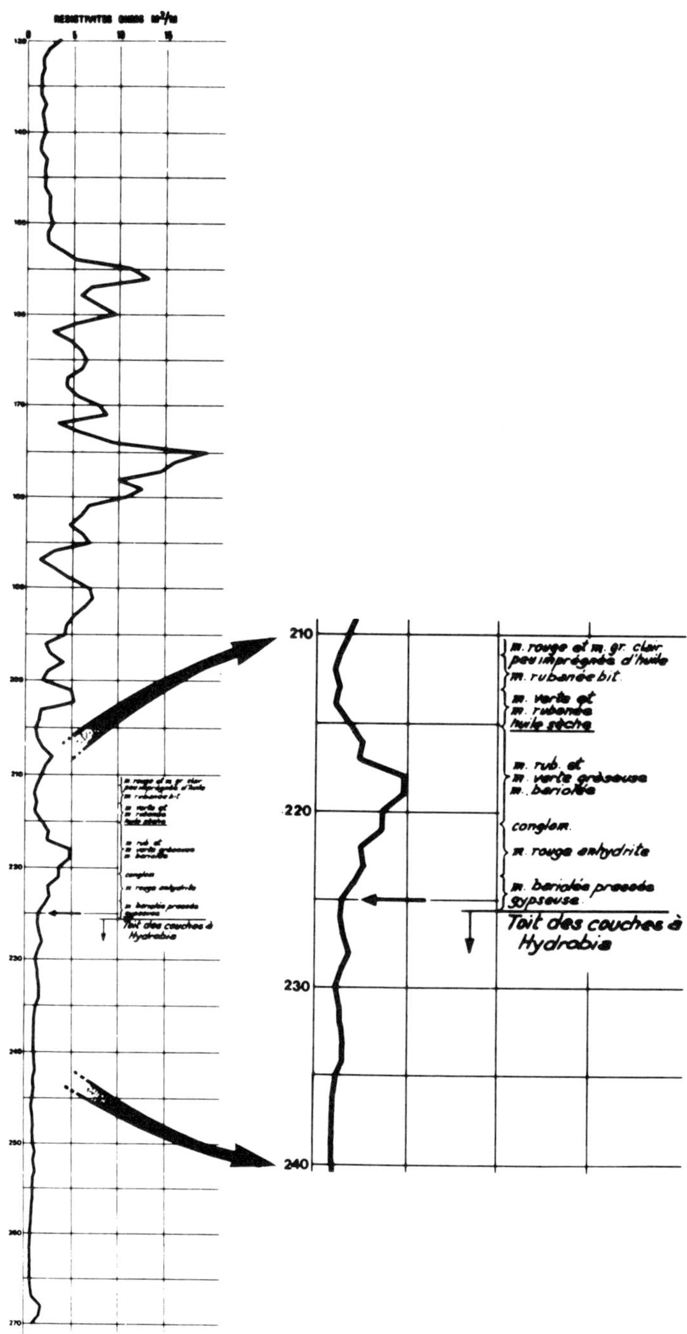

FIG. 1. Partial diagram of the first resistivity log run in Pechelbronn, Alsace, France in Sept., 1927.

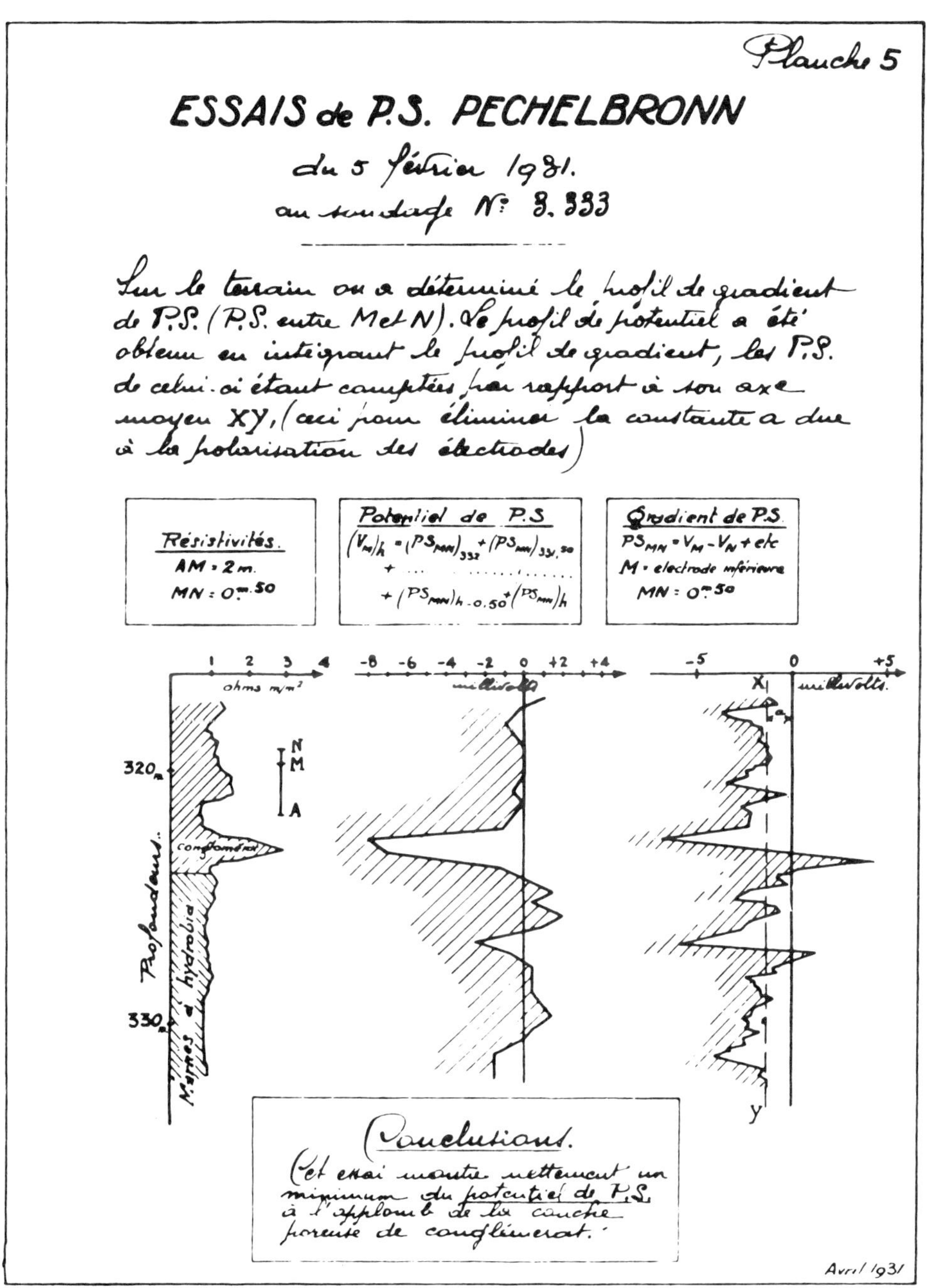

FIG. 2. The first spontaneous potential (SP) log run in Feb., 1931, which identified a congolmerate porous bed.

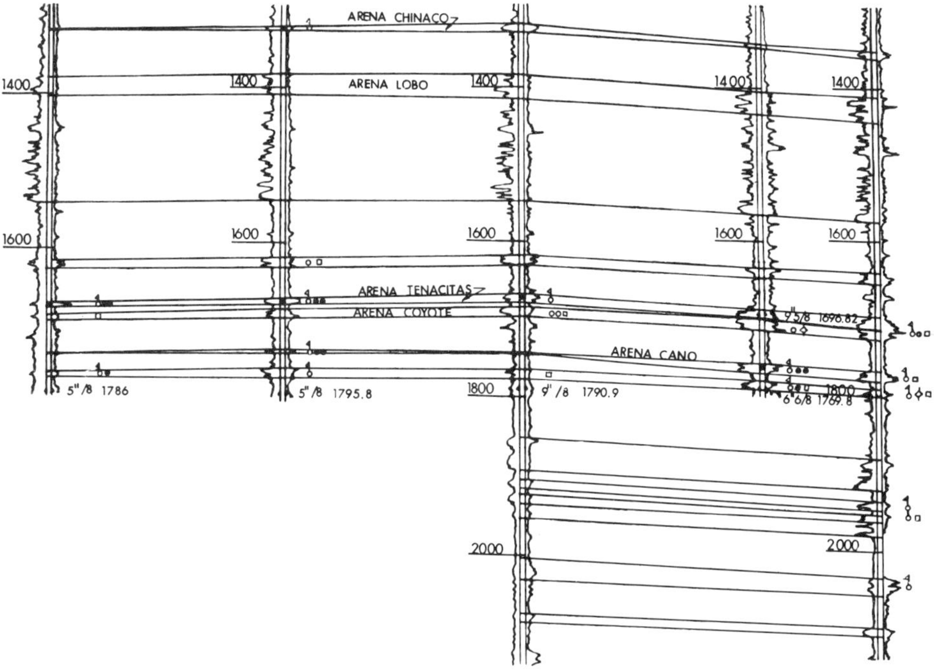

FIG. 3. The resistivity logs used for well to well correlation.

This chapter presents an overview of some of the logging techniques used in exploration and their relationships to rocks and fluids. The principles of the important logging techniques, from electrical to nuclear, are reviewed and the common approaches for deriving the reservoir parameters are outlined.

Exploration Applications of Logs

Well logs are used in a variety of earth science applications, from petroleum to mining. In the petroleum industry they are used in exploration, field development, and exploitation. The following is a partial list of the qualitative and quantitative information derived from well logs:

Qualitative

Identification of rock types
Detection of hydrocarbons
Detection of permeable beds
Reservoir limits
Stratigraphic correlation
Hydrocarbon-water and gas-oil levels
Well completions
Selection of new drilling locations

Quantitative

Rock types
Porosity
Fluid saturation
Permeability index
Amount of oil and gas in place
Recoverable reserve
Selection of completion intervals

An ever increasing variety of well logs are now available, which can be classified in different ways depending upon the nature of the rock characteristic being measured. Two main groups are: (1) those measuring properties which exist naturally in rocks (the SP and gamma-ray curves belong to this group); (2) those having as a common denominator the transmission of a certain physical signal through the formation, whose energy level, whether original or transformed, is measured and recorded after having traveled a certain distance through the formation. Examples of this group are resistivity, acoustic, and nuclear/radioactivity logs.

Some logs may be run in both cased and open holes, but others are limited to open holes. Similarly, some logs can be run in empty holes, while others require a hole filled with drilling fluid.

Rock Characteristics

In order to understand and interpret well log response properly, a knowledge of the rock types penetrated by the well is necessary. [The reader is referred to the chapter on exploration strategy for appropriate details on this subject.] A brief description is presented here in the context of the well log applications.

Common petroleum reservoir rocks are sandstones and carbonates. The rocks serving as seals which trap hydrocarbons are fine grained, compacted formations with zero or negligible permeability such as shales, salts, and anhydrites.

Quartz grains are the principal constituent of sandstones. The most common cementing material is silica or calcium carbonate. Sand grade may vary from very loose to compacted; in the latter case, the cementing material may be clay.

Carbonate rocks are made up of the remains of marine animals and plants such as corals, mollusks, and algae. The texture of carbonate rocks may vary from unconsolidated bodies of shells to compacted crystalline rocks. Rocks containing certain proportions of calcium and magnesium carbonates are called dolomites. Limestones are mainly calcium carbonate. Calcareous clays or mixtures of clay and particles of calcite or dolomite are called marls, which are impervious rocks.

Shales are fine clastic sediments indurated to reach the grade of a rock. Generally speaking, they are the noncommercial part of an oil field. They normally act as seal rocks and prevent the natural migration of hydrocarbons. The organic-rich shales are source rocks, which are intimately related to the origin of hydrocarbons. In some areas, the oil-bearing shales are now being exploited for hydrocarbons by mining methods.

Basic Principles in Log Interpretations

Well logs are affected by rock characteristics as well as by their fluid content. Some basic definitions and concepts are briefly reviewed here to make qualitative and quantitative well log interpretation more understandable.

Porosity and fluid saturation are two main rock properties which characterize a reservoir. *Porosity* provides a direct measure of the total void space available for the storage of fluids (Clark, 1960). Normally, it ranges between 5 to 50 percent in sandstones and 1 to 25 percent in carbonate rocks. Two types of porosities—effective and absolute—are defined, both of which can be determined from well logs (Figure 4).

Effective porosity is the ratio of volume V_{ip} associated with the interconnected pores to bulk volume V_t of rock, i.e.,

$$\phi = \frac{V_{ip}}{V_t}. \tag{1}$$

Absolute or total porosity (ϕ_A) is the ratio of total pore volume (V_p) to bulk volume (V_t) of rock; the V_p includes the isolated or noneffective pores and vugs within the rock

$$\phi_A = \frac{V_p}{V_t}. \tag{2}$$

Another classification can be made in terms of primary and secondary porosity. Primary porosity is based on the pores resulting from original deposits and further cementation of original sediments, whereas secondary porosity is the result of such diagenetic effects as percolation and diastrophism, which gives rise to vugs and solution channels. The fractures are also common forms of secondary porosity.

Fluid saturation is another important rock property. Rock pores are generally filled with water. The original water which existed when sediments were deposited is called connate or interstitial water. In hydrocarbon-bearing rocks, the residual con-

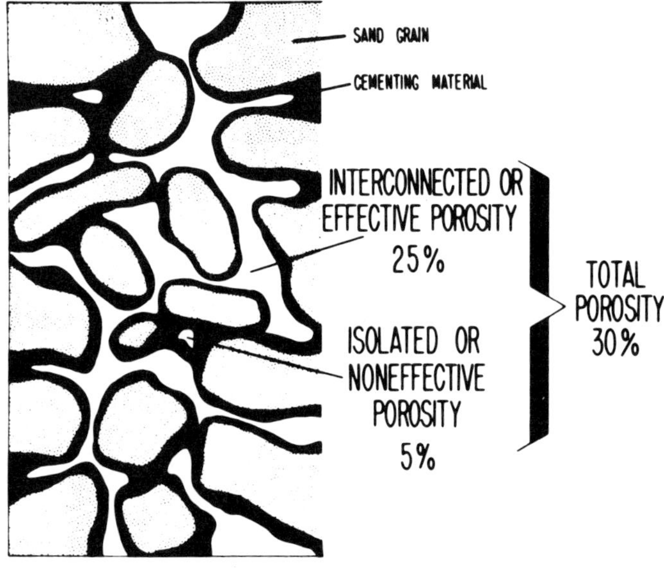

FIG. 4. Effective, noneffective, and total porosity in a rock matrix.

nate or interstitial water is normally located along the walls of pores, whereas hydrocarbons occupy the remaining pore volume.

Water saturation (S_w) is defined as the ratio of the pore volume occupied by water (V_w) to the total pore volume (V_p):

$$S_w = \frac{V_w}{V_p}. \tag{3}$$

Similarly, hydrocarbon saturation is the ratio of pore volume occupied by hydrocarbons (V_{hc}) to total pore volume:

$$S_0 = \frac{V_{hc}}{V_p} = (1 - S_w). \tag{4}$$

Water salinity is an important parameter for computing fluid saturation. The formation waters are generally saline, ranging from about 5000 to 300,000 ppm. Sodium chloride (NaCl) is the most common salt found in the formation waters. In solutions, the constituent molecules of the salt are dissociated in positive (Na^+) and negative (Cl^-) ions. Formation waters thus act as electrolytic solutions capable of conducting electric current. This chemical property is used to compute fluid saturation from electrical and porosity logs, as described below.

At a constant temperature water resistivity (R_w) decreases with increasing salinity, whereas at a constant salinity R_w decreases with increasing temperature. Figure 5 illustrates this relationship.

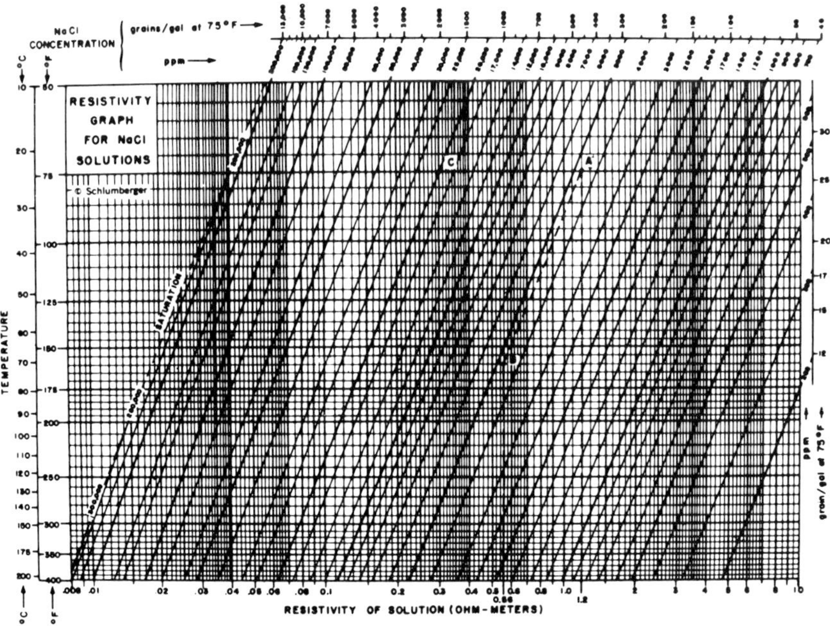

FIG. 5. Water resistivity and salinity as functions of temperatures (from Schlumberger charts, 1969).

The temperature increases with the subsurface depth; the temperature gradient is generally assumed to be constant with respect to depth. In log interpretation, the resistivity and temperature relationship (Arps, 1953) is given by

$$R_2 = R_1 \frac{T_1 + 7}{T_2 + 7}, \tag{5}$$

where R is resistivity and T is temperature in degrees Farenheit.

The *formation resistivity factor* (F) is the ratio of resistivity (R_0) of a rock saturated with brine to the resistivity of the brine (R_w)

$$F = \frac{R_0}{R_w}. \tag{6}$$

Archie (1942) found that a definite relationship exists between the resistivity factor and porosity. Figure 6 and 7 show the original graphical presentation of Archie's experimental results. The quantitative relationship is

$$F = \frac{1}{\phi^m}, \tag{7}$$

where m is cementation factor (also known as cementation or porosity exponent), which ranges between 1 to 3 depending upon the degree of rock cementation. The more cemented the rock, the higher should be the value of m. Some other factors modify m, such as fractures (Aguilera, 1976). Figure 8 is a graphical solution of equation (7) for different values of m.

A more general relationship between formation resistivity factor and porosity, known as the Humble relation (Winsauer et al, 1952), is

$$F = \frac{0.62}{\phi^{2.15}}. \tag{8}$$

Equation (8) has been used successfully for a long time. However, a detailed study of the relationship between F and ϕ seems to show that equations (7) and (8) are only special cases of a more general relationship of the form (Wyllie, 1953; Gomez-Rivero, 1976):

$$F = \frac{a}{\phi^m}, \tag{9}$$

where the coefficient a, in theory, may vary from zero to infinity and the exponent m, from plus infinity to minus infinity. Further, a and m are interrelated quantities according to the following relationships:

$$m = 1.8 - 1.29 \log a \tag{10}$$

for sands, and

$$m = 2.03 - 0.9 \log a \tag{11}$$

for carbonate rocks.

The *resistivity index* (I) of a formation is the ratio of true resistivity (R_t) of the formation to resistivity R_0 of the rock saturated only with brine.

Archie (1942) found that a relationship exists between the resistivity index and water saturation (S_w), which is of the form

$$S_w = \left(\frac{1}{I}\right)^{1/n} = \left[\frac{R_0}{R_t}\right]^{1/n}, \tag{12}$$

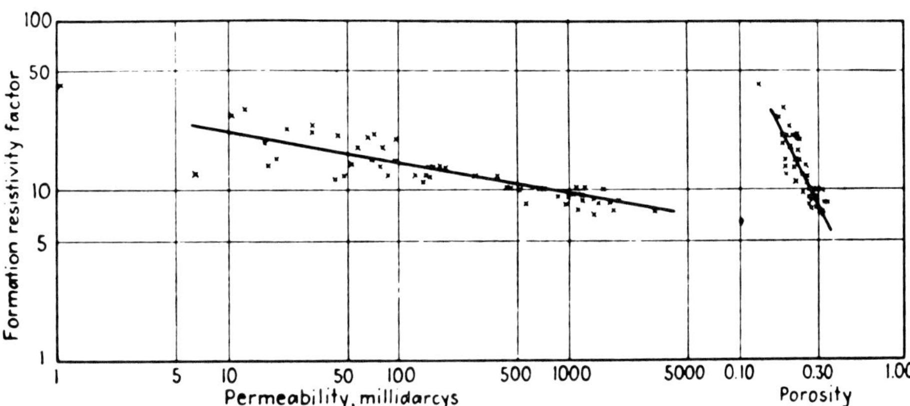

FIG. 6. Relation of porosity and permeability to formation resistivity factor for consolidated sandstone cores of the Gulf Coast.

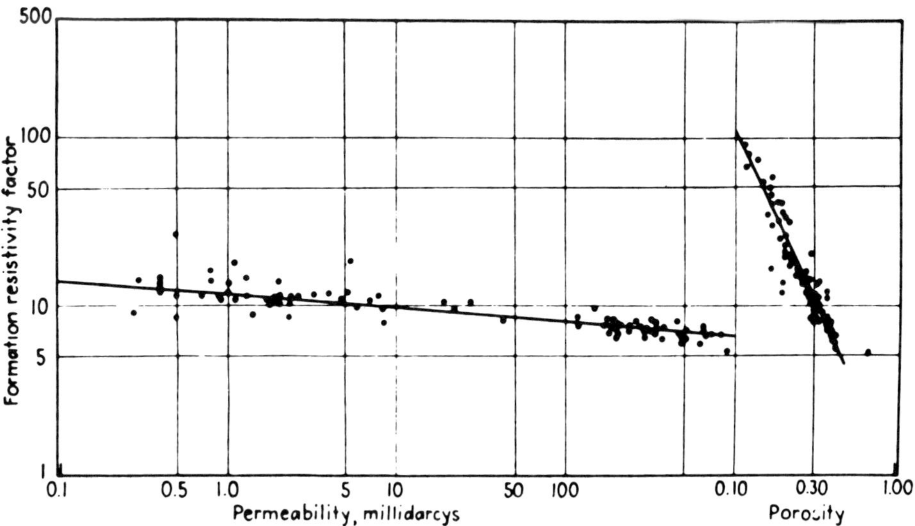

FIG. 7. Relation of porosity and permeability to formation resistivity factor for Nocatoch sand, Bellevue, Louisiana.

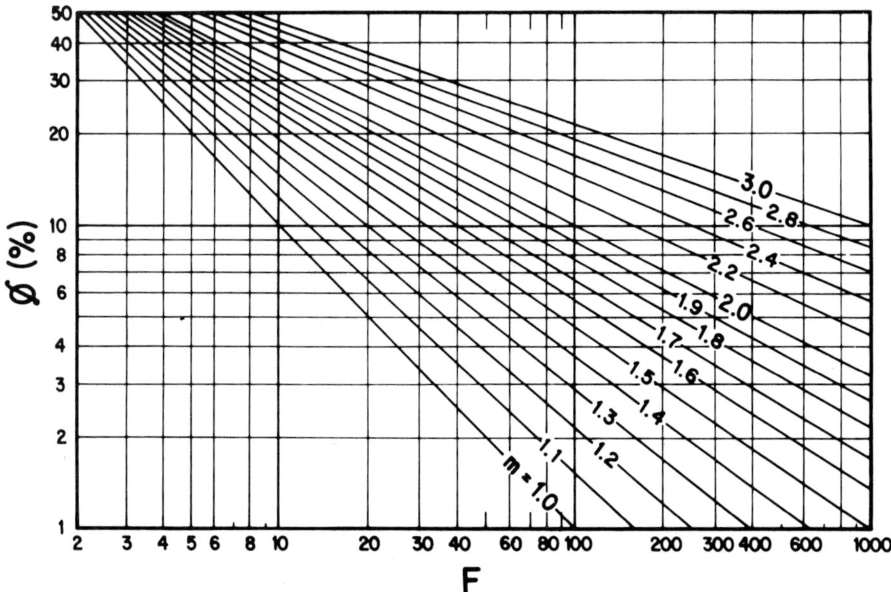

FIG. 8. Porosity (ϕ) and formation factor (F) as functions of the cementation exponent (m).

where the exponent n is called saturation exponent. According to some authors, its value is mainly dependent upon water resistivity and rock permeability (von Gonten and Osoba, 1969). Several relationships exist, some of which are

$$n = 0.904 - 0.515 \log R_w + 0.325 \log K, \tag{13}$$

and

$$n = 1.347 - 0.519 \log R_w, \tag{14}$$

where K is permeability in millidarcies. A simpler relationship (Gomez-Rivero, 1976) in terms of salinity instead of resistivity

$$n = 1.095 + 0.442 \log P, \tag{15}$$

where P is salinity in thousands ppm. A commonly used value for n is 2. Figures 9 and 10 are graphical representations of equations (13) and (15), respectively. Figure 10 also shows some actual data which indicate that the model for saturation exponent is rather complex.

Permeability is another important rock property for identifying a commercial reservoir. It may be defined as the ease with which fluids can move through the interconnected pore spaces of a rock. Permeability is measured in darcies. A rock is defined to have one darcy (1D) permeability when it allows a fluid of one centipoise viscosity to flow at a rate of one cubic centimeter per second through rock of 1 centimeter length and one square centimeter cross-section area under 1 atmosphere of differential pressure. The common permeability values found in reservoir rocks are much less than 1 D. Hence the normally used permeability unit is 1/1000 of a darcy, i.e., the millidarcy (mD).

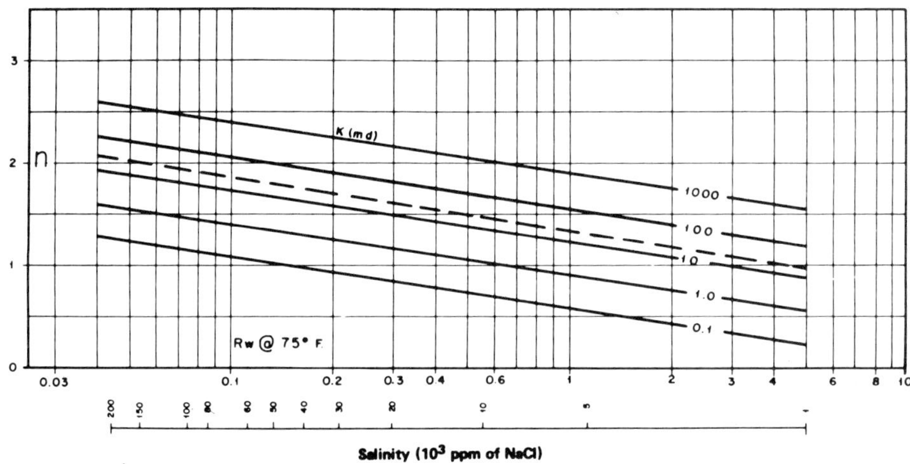

FIG. 9. Graphical representation of the saturation exponent [equation (13)].

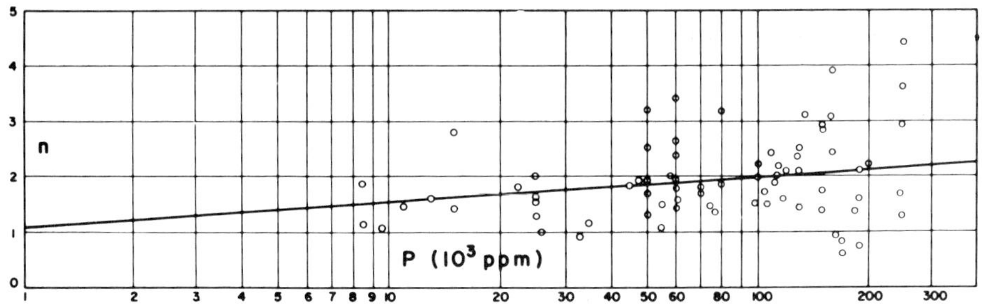

FIG. 10. Simplified graph for the saturation exponent n using equation (15), plotted against actual measurements. The data scatter shows that the saturation exponent model is quite complex.

Permeability is closely related to porosity. In intergranular porosity rocks, permeability generally increases with porosity. Reverse cases are principally found when dealing with fractured rocks.

The Concept of Formation Invasion

Well logs measure properties of the surrounding media within a certain distance of the borehole. This distance may vary from a few inches to several feet, depending upon the type of log. Normally, the hole is filled with drilling mud, which is conditioned so that the hydrostatic pressure it exerts on the hole wall exceeds the natural pressure of the formations. This causes some filtration of the aqueous phase of the mud to penetrate the permeable beds, while the solid mud particles form a ''mud cake'' on the exposed face of the bed. Knowledge of the characteristics of this mud filtrate invaded zone is very important in well log interpretation, because it affects the measurements of most logs; many logs measure the properties just within the invaded zone.

Figures 11 and 12 show schematically the radial and vertical resistivity characteristics of the invaded zone in a bed containing only water and a hydrocarbon-bearing bed. As can be observed, there is a gradual invasion of the mud filtrate from the wall of the hole to the uncontaminated zone. Close to the wall the invasion is maximum; this is called the flushed zone. In the case of a hydrocarbon-bearing formation there will also be a residual hydrocarbon saturation which increases from the wall to the uncontaminated zone. The resistivity of the formation in the noninvaded zone is designated R_t. When the formation is water-bearing, R_t equals R_0. These are the same variables as in equations (6) and (12), and they can normally be determined from resistivity logs. The general notation for the resistivity in the flushed zone is R_{x0}, which can also be determined from the resistivity logs.

Physical Description of Logs

There are far too many varieties of well logs to include in this chapter. The reader is referred to logging company documents and appropriate textbooks for this purpose. We will restrict the following discussion to the logs which are generally useful in exploration studies. Based on their principal attributes, these logs may be classified as follows:

Rock type identification: SP and gamma-ray logs
Fluid saturation measurement: resistivity logs
Porosity measurement: neutron, density and acoustic logs

SP log

The spontaneous potential of the formations in a well is defined as the electrical potential difference between a moveable electrode M in the borehole and a fixed electrode N at the surface, as schematically illustrated in Figure 13. These measurements are useful, among other things, for detecting permeable beds, for well-to-well correlation, for determining the formation water resistivity R_w, and for estimating the shaliness of some beds. The SP readings are made from the shale base line to the left (Figure 14).

SP curves may be recorded simultaneously with resistivity and acoustic logs, and they require that the hole be filled with a relatively conductive fluid of salinity lower than the formation water salinity.

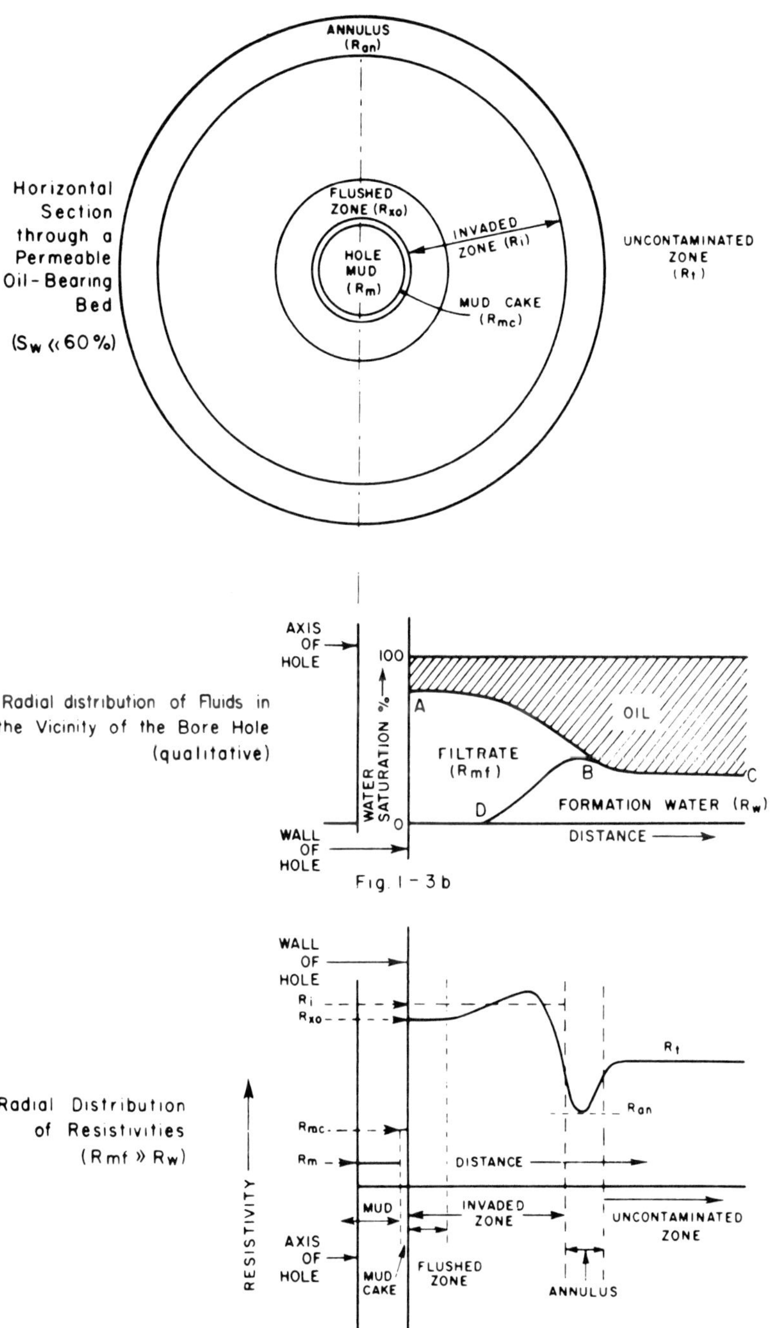

FIG. 11. Radial and vertical distribution of fluids in a permeable oil-bearing formation penetrated by a well.

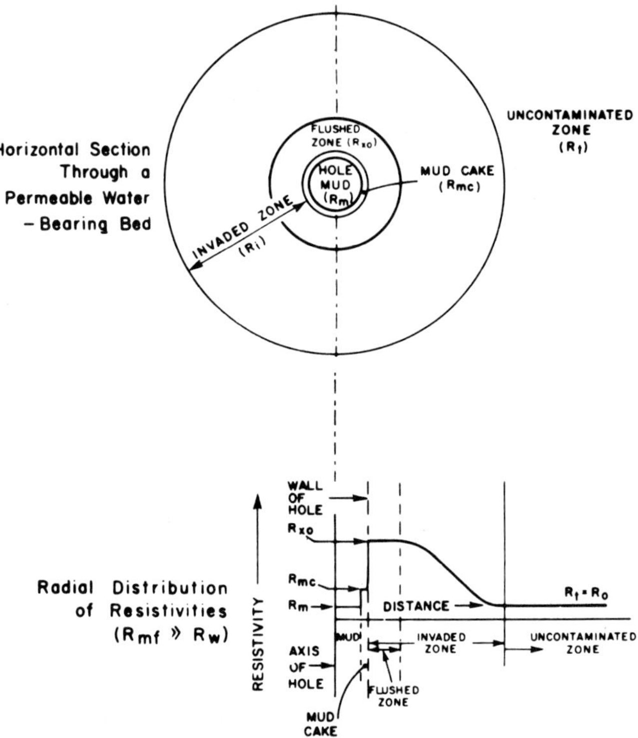

FIG. 12. Radial and vertical distribution of fluids in a water-bearing formation penetrated by a well.

The SP curve measures electromotive force of electrokinetic and electrochemical origins. The electrokinetic or electrofiltration potential originates in the filtration of mud through the mud cake and depends upon, among others, the mud resistivity and pressure differential across the mud cake. This component of SP is generally regarded as negligible in practice. Electrochemical potential may be subdivided into diffusion and membrane potentials.

Diffusion potential is the potential difference generated when two saline solutions come in contact through a permeable membrane. This is the case of mud filtrate and the water in a permeable formation in a well such as sands and limestones (Figure 15). It is originated by movement of ions from the more concentrated solution (formation water) to the less concentrated one (mud filtrate).

Membrane potential is generated when two saline solutions are separated by a cationic membrane which allows the cations Na^+ to pass from the more concentrated solution to the dilute one. A case of the potential circuit for the media with mud-shale-permeable beds is illustrated in Figure 15.

The main component of total electrochemical potential is the membrane potential, whereas the diffusion potential is another component. The general expression for spontaneous potential (*SP*) is

$$SP = -K \log \frac{R_{mf}}{R_w}, \tag{16}$$

where K is a function of temperature.

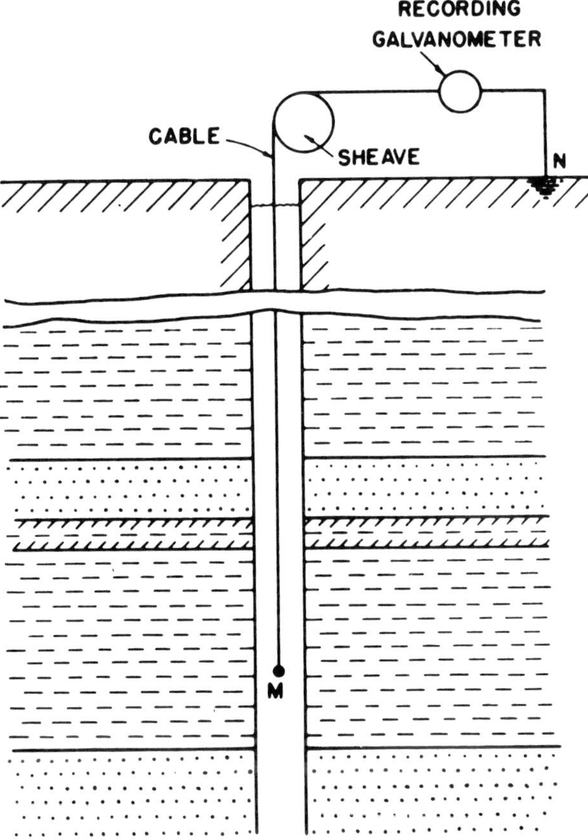

FIG. 13. Schematic circuit for SP logging. M and N are electrodes across which electrical potential difference is recorded.

Gamma-ray log

All geologic formations contain variable amounts of radioactive material. The natural gamma-ray log measures the natural radioactivity of rocks. Rocks are constantly emitting gamma rays as a result of natural disintegration of small amounts of radioactive elements. Shales normally contain more radioactive material than sands and carbonate rocks. A natural gamma-ray curve detects the difference in radioactivity between shales and normal reservoir rocks. Thus, it can be considered a lithology log similar to the SP log. Figure 16 illustrates schematically the typical responses of the SP, resistivity, and radioactivity curves, including a gamma-ray curve. The response characteristics of these curves in terms of specific interpretations are pointed out next to the curves.

Radioactivity can be defined as a spontaneous disintegration of atoms accompanied by emission of radiation. The simpler atoms have stable nuclei. There are some heavier and more complex atoms which are only partially stable, and they spontaneously transmute themselves to other, more stable isotopes with a change of mass. Potassium 40, uranium, and thorium are among the most common elements having radioactive characteristics.

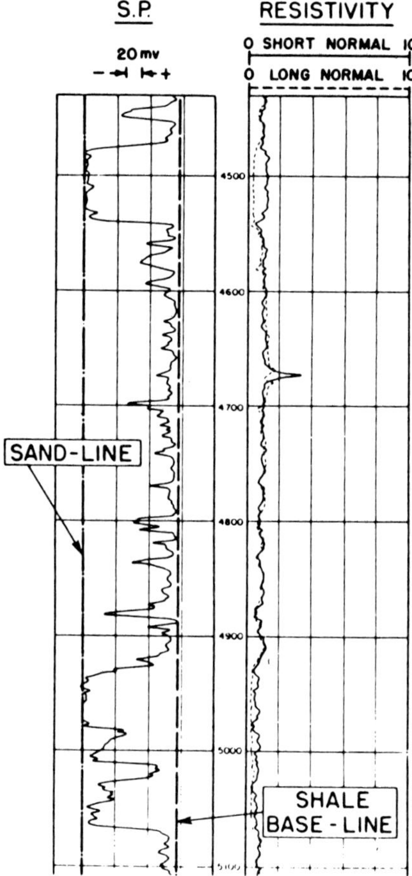

FIG. 14. Example of the SP log in a sand-shale sequence.

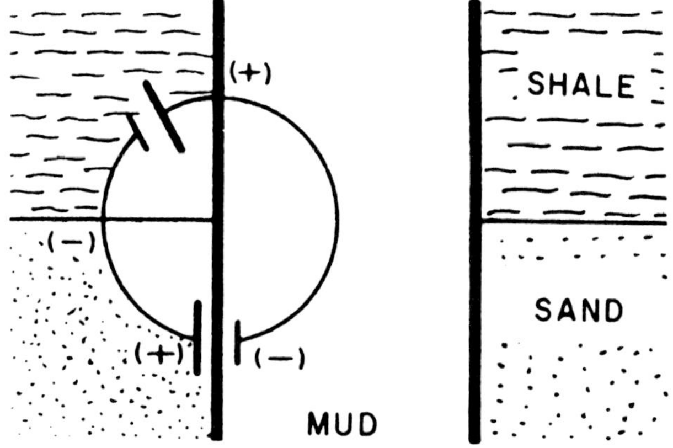

FIG. 15. Electromotive forces due to diffusion and membrane potentials.

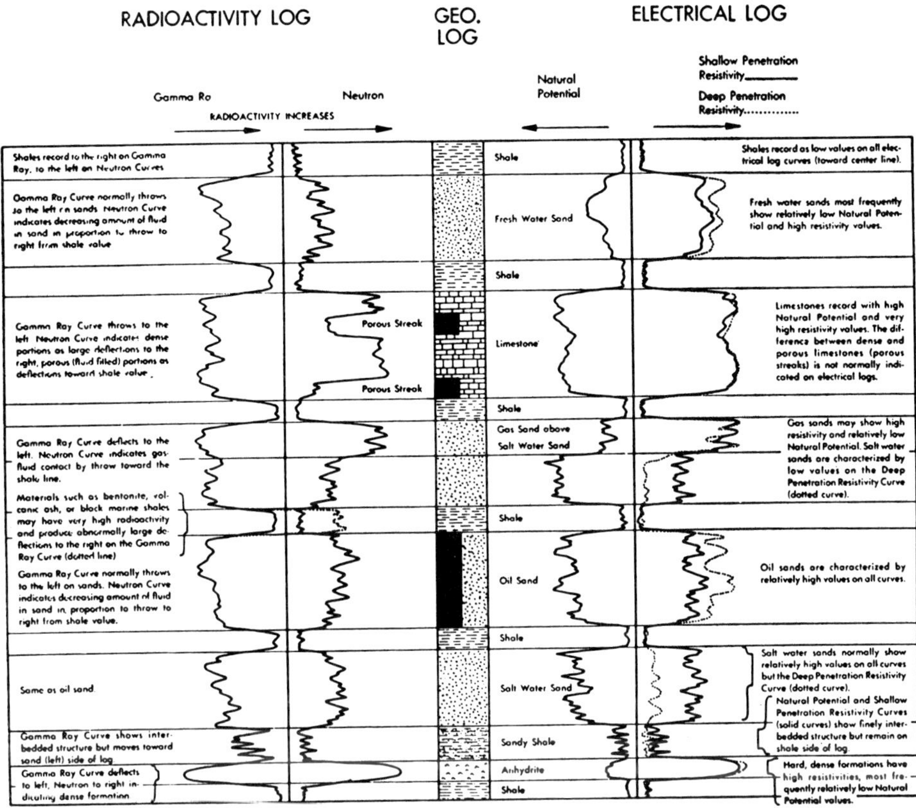

FIG. 16. Typical response of radioactivity and electrical logs along with the interpretations of the borehole media logged.

In disintegrating, an atom emits three kinds of radiation, commonly known as alpha, beta, and gamma rays. Alpha rays are higher energy particles but have a short penetration; a few sheets of paper can stop this kind of radiation. Beta rays are of lower energy than alpha rays but have a higher penetration power. Gamma rays have the highest penetration radiation and can penetrate even through the well casing strings.

Figure 17 shows the relative radioactivities of various sedimentary rocks as indicated by the intensity of gamma radiation. The vertical height of the lines increases with frequency of occurrence, and intensity of radioactivity increases toward the right.

Detection of gamma-ray logs may be accomplished by ionization chambers, Geiger-Mueller, or scintillator counters. An ionization chamber, schematically illustrated in Figure 18, is essentially formed by a metal cylinder containing an inert gas under high pressure and an inner metallic wire insulated from the cylinder. A potential difference of about 100 V is maintained between the inner wire and the cylinder. Energetic gamma rays entering the chamber react with the inert gas, producing free fast electrons. When enough charge has accumulated, it flows to the

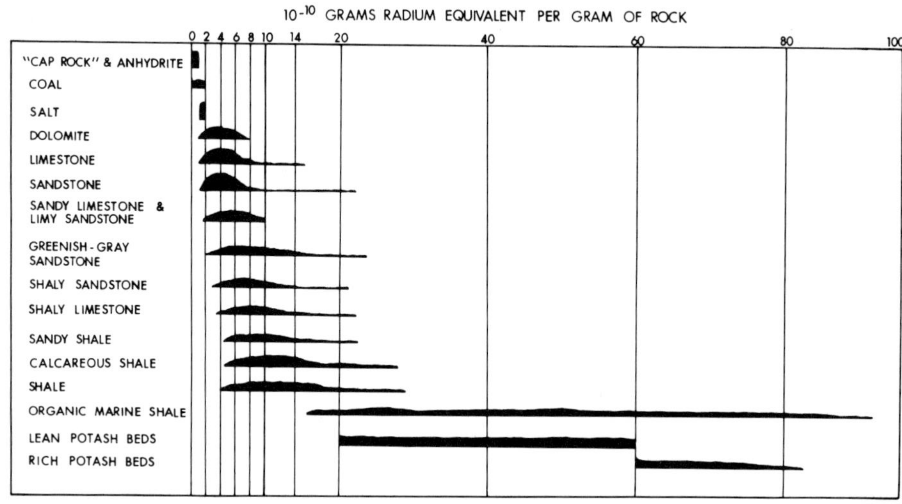

FIG. 17. Diagram showing relative radioactivities of various sedimentary rocks.

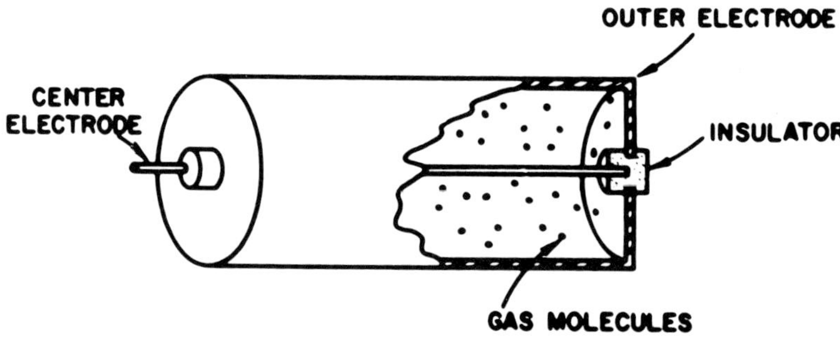

FIG. 18. Basic configuration of an ionization chamber.

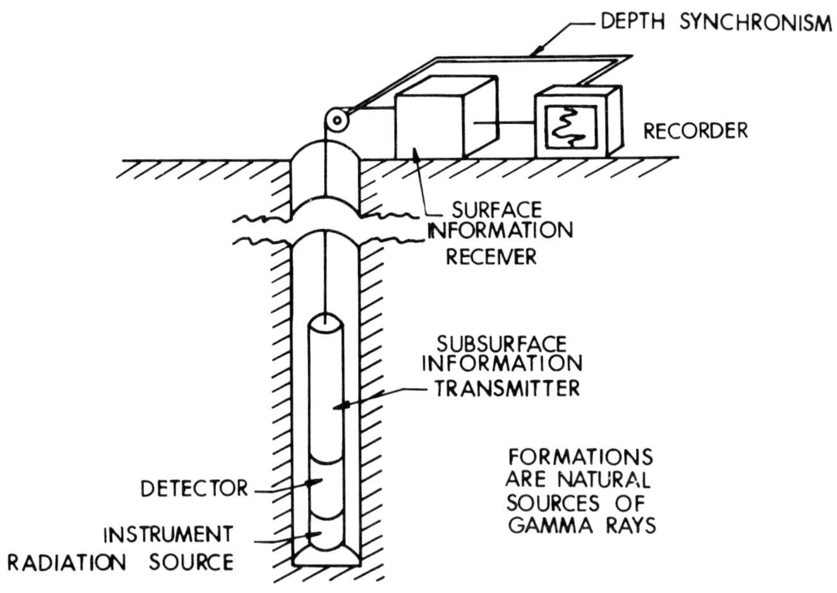

FIG. 19. Components of radioactivity logging devices.

positive pole, i.e., the inner wire. The subsurface amplifier transmitting system sends the signal to the surface where it is recorded.

In well logging practice, gamma rays are logged along with other logs such as induction and neutron logs. Figure 19 is a schematic display of the basic subsurface and surface equipment for obtaining neutron and gamma-ray logs. Gamma-ray logs can be obtained in open or cased holes, whether filled with mud or empty.

Gamma-ray logs are normally scaled in API units. An API unit is defined as 1/200 of the difference in log deflections between two formations of known radioactivity in an artificial calibration pit existing at the University of Houston (Belknap et al, 1959). The gamma-ray signal from the formation is mainly affected by bed thickness, casing, cement behind casing, and mud within the borehole. Service companies provide correction charts for compensating these effects.

Resistivity log

Since the first resistivity log was tested by the Schlumberger brothers in 1927, many resistivity devices with different physical characteristics have been developed and are in widespread use. We have grouped these logs into the following broad categories.

Conventional resistivity log.—These early resistivity logs have been replaced by improved devices. However, it is important to understand them because they are the only logs available for most of the older wells. All available subsurface information is valuable for exploration as well as for reservoir studies. These logs can be used for qualitative and quantitative determination of fluid saturation. As a general criterion, high resistivity indicates low permeability, low porosity, high hydrocarbon saturation, fresh water, or a combination of these conditions.

Figure 20 is a sketch of the sonde, which can simultaneously log the resistivity and spontaneous potential curves. A typical record is shown in Figure 21. Three

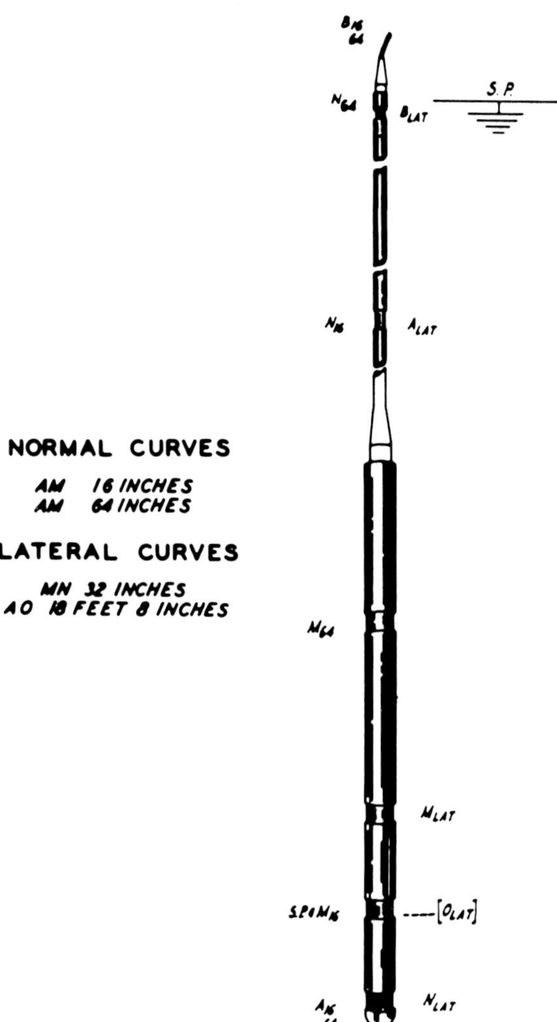

FIG. 20. A conventional resistivity logging sonde showing electrode spacing for normal and lateral curves.

resistivity curves shown are short normal, long normal, and the lateral. The electrode spacings (AM in Figure 20) are 16 and 64 inches for the normal and 18 ft 8 inches for the lateral (A_{LAT} O_{LAT} in Figure 20) curves. These curves are obtained by sending an alternating current through an electrode contained in the sonde. The potential drop between the current electrode and several reference electrodes is then measured and the corresponding resistivity is computed. The borehole must be filled with a conductive mud to establish the necessary electrical contact with the formation.

Figure 22a illustrates the principle of resistivity logging. Electric current leaving electrode C flows in all directions, creating equipotential surfaces which are spheres in a homogeneous isotropic medium. Electrodes S and L measure the potential drop between these points and the surface electrode D, respectively. Most of the potential

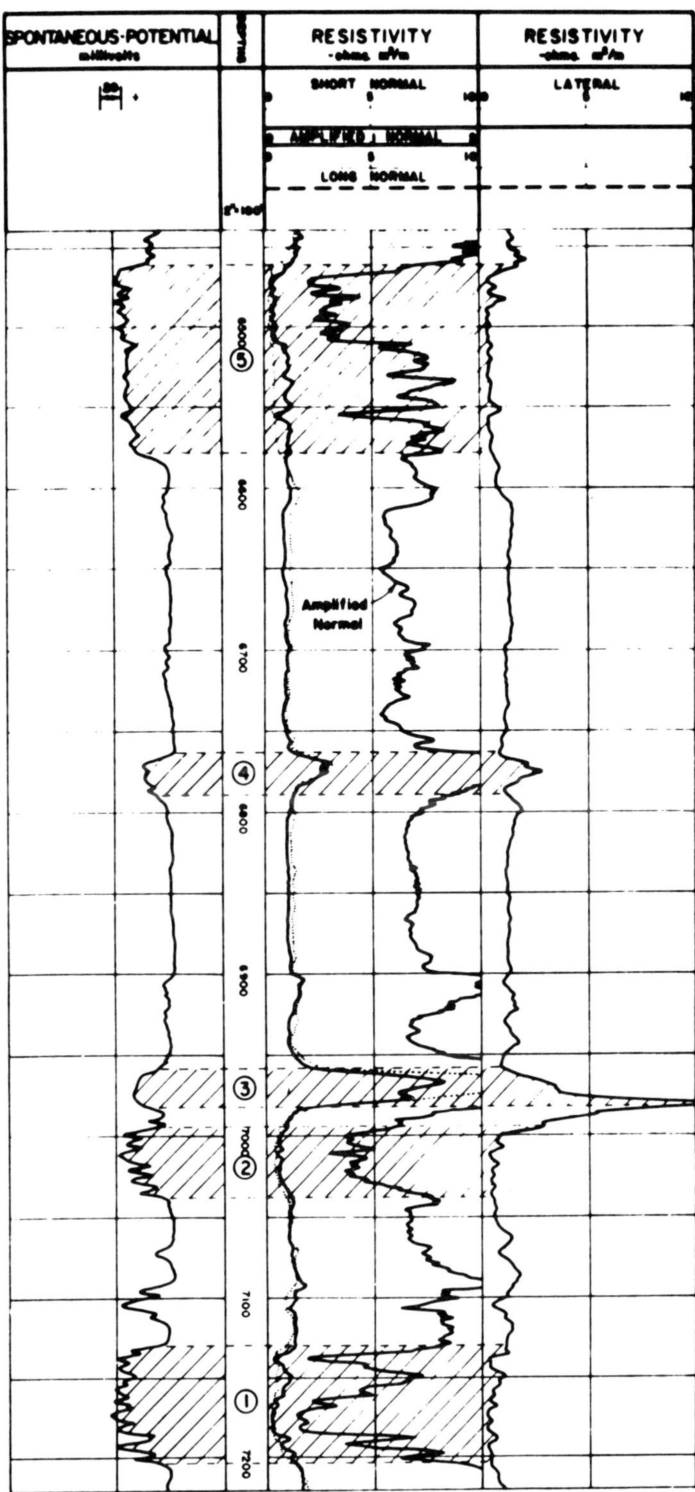

FIG. 21. A typical conventional resistivity log. The sandstones in the borehole are identified by the shaded areas.

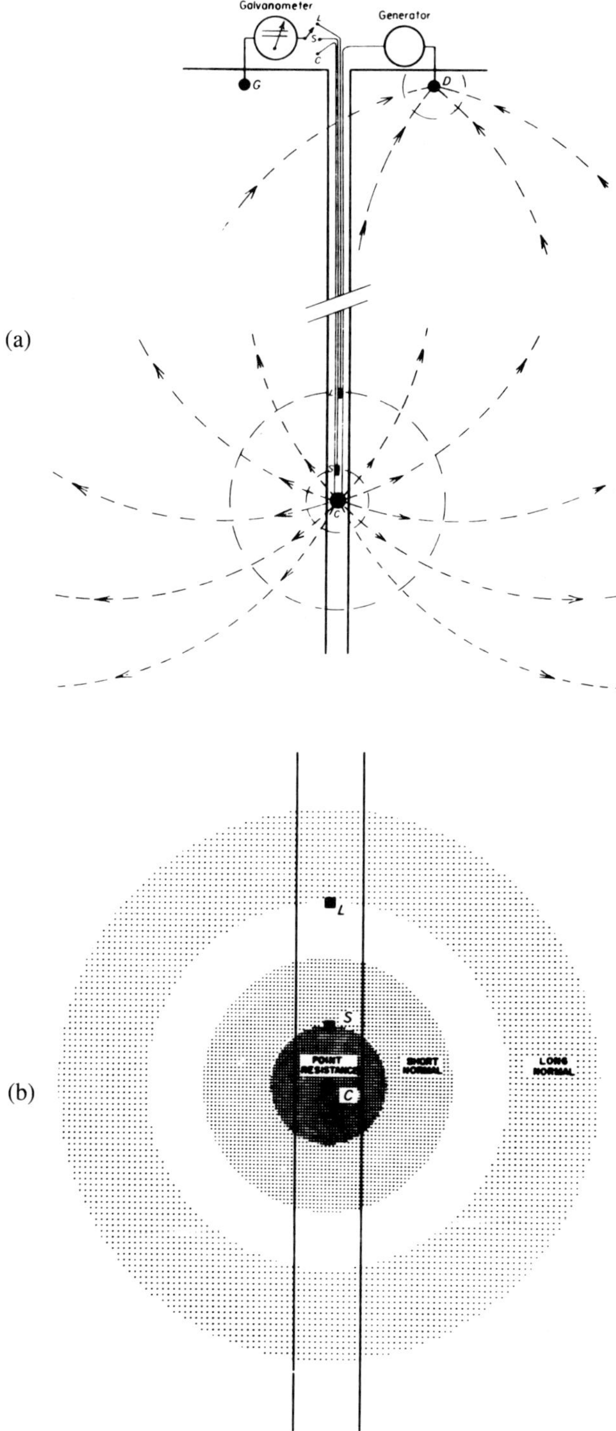

FIG. 22. (a) The distribution of current patterns when logging in a homogeneous medium (C = current electrode; CS = short normal; CL = long normal). (b) Zones of investigation of the short normal and long normal resistivity devices in a homogeneous medium.

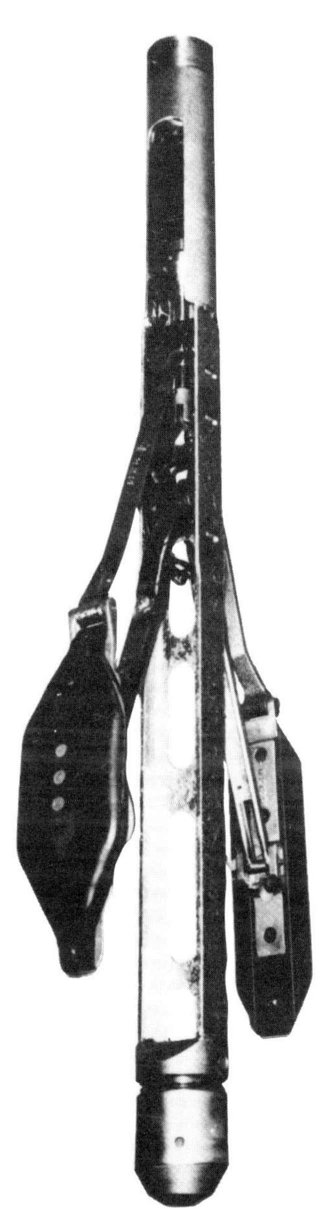

FIG. 23. One of the early sondes for microresistivity logging.

drop occurs within a short distance from current electrode C. Figure 22b shows the zones of investigation of the normal devices. The short normal is influenced by the mud filtrate invaded zone, whereas the long normal measures a value close to the true formation resistivity of the formation. Charts are available from service companies for borehole effect corrections.

In the lateral device the potential drop between two electrodes M and N is measured (Figure 20). The spacing AO is large compared with the distance MN. Similar to the normal resistivity, the lateral device also measures the resistivity of the formation material of thickness MN. However, it penetrates deeper into the formation than the long normal, and hence provides more accurate values of the formation resistivity.

Conventional microresistivity log.—The current and measurement electrodes are short spaced in this log and are mounted in an insulating material pad which is pressed against the hole wall during the logging operation (Figure 23). The electrode spaces AM_1 and M_1M_2 (Figure 24) are 1 inch; hence the zone of investigation is very small. An alternating current is sent through electrode A. When current flows through mud cake into the formation, it tends to spread out. Mud cake is normally formed across the permeable beds. Two microresistivity curves are micronormal and microinverse. Microinverse resistivity is proportional to the potential difference between M_1 and M_2; micronormal resistivity is proportional to the potential difference between M_2 and the surface electrode. A microcaliper curve is also obtained, which records the hole diameter variations throughout the hole.

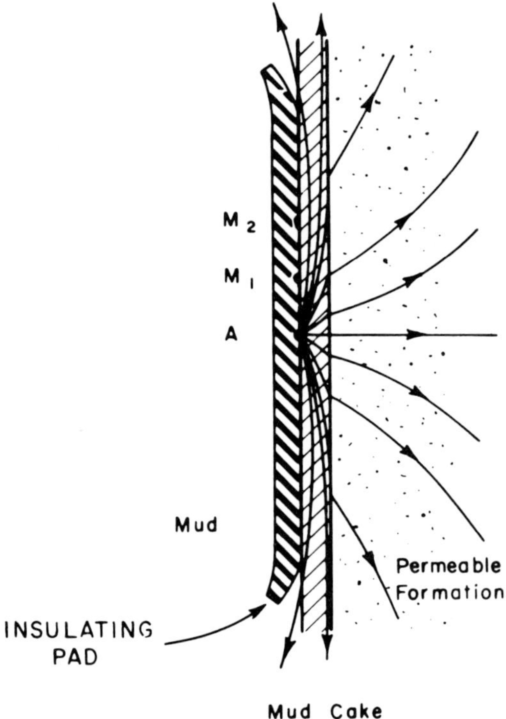

FIG. 24. Microresistivity device showing the current and measurement electrodes and the current path distribution.

Figure 25 is an example of an old type of microresistivity log. In modern logs the caliper is recorded simultaneously. Sandstone beds are identified by the crosshatched lines in the figure.

The difference between the two microcurves, called separation, is commonly used to interpret these logs qualitatively. When micronormal resistivity is higher than microinverse, the separation is called positive. The permeable beds are characterized by a positive separation and presence of mud cake. Nonpermeable hard formations show no mud cake, high resistivity, and no separation. These formations are often brittle, causing hole diameter enlargement as in the case of highly cemented carbonate rocks. Soft nonpermeable formations such as shales also have no separation and no mud cake, but the recorded resistivity is low; hole enlargement also occurs. The variations in mud cake thickness are indicated by the microcaliper curve.

Induction log.—Conventional resistivity logs provide useful measurements only when the muds are conductive. Often, the wells are drilled with air or nonconductive muds. Induction logs, originally designed for such situations, can also be used in wells drilled with normal water-base muds. Thus, they have also replaced the conventional resistivity devices where the formation resistivity is less than 200 Ω-m.

The basic subsurface equipment consists of a sonde which carries two main coils, namely, the transmitter and receiving coils (Figure 26). A high-frequency alternating current of constant intensity is sent through the transmitter coil, creating a magnetic

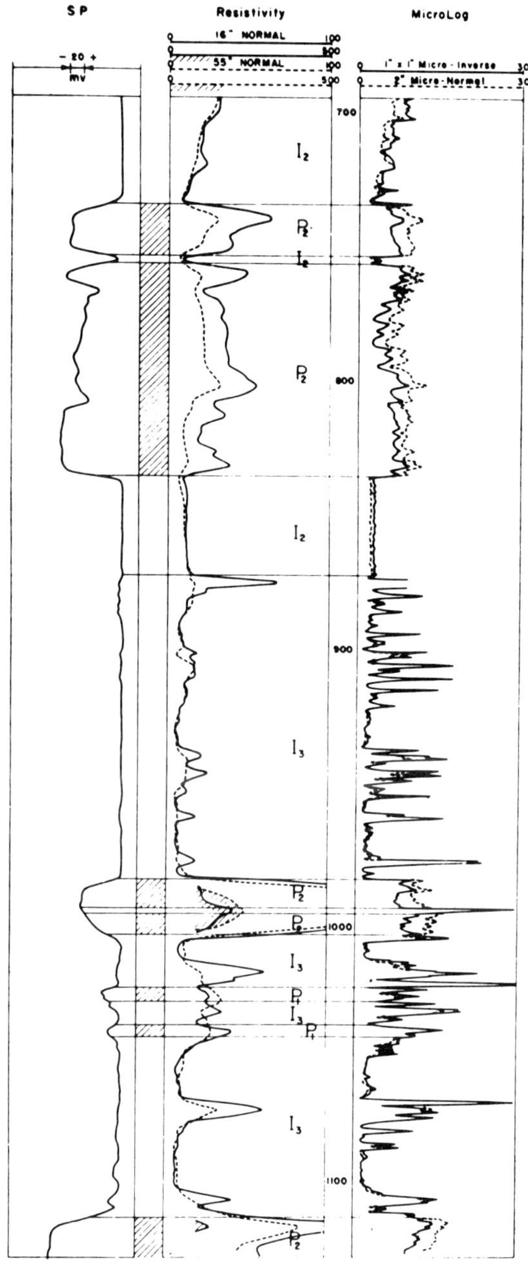

FIG. 25. Example of a microresistivity log showing the permeable beds as shaded zones.

field which induces secondary currents in the formation. These currents flow in circular ground-loop paths coaxial with the transmitter coil, creating fields which induce signals in the receiver coil. The receiver signals are then transformed into conductivity and recorded as apparent formation conductivity.

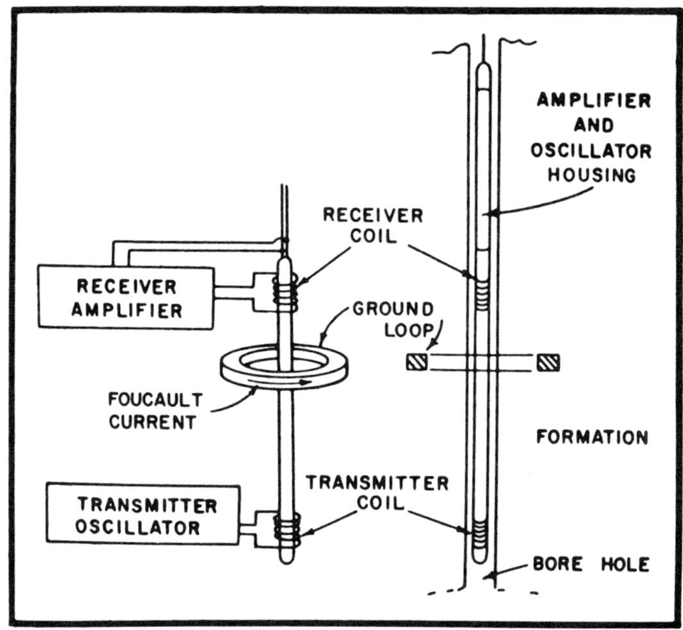

FIG. 26. The basic two-coil induction system.

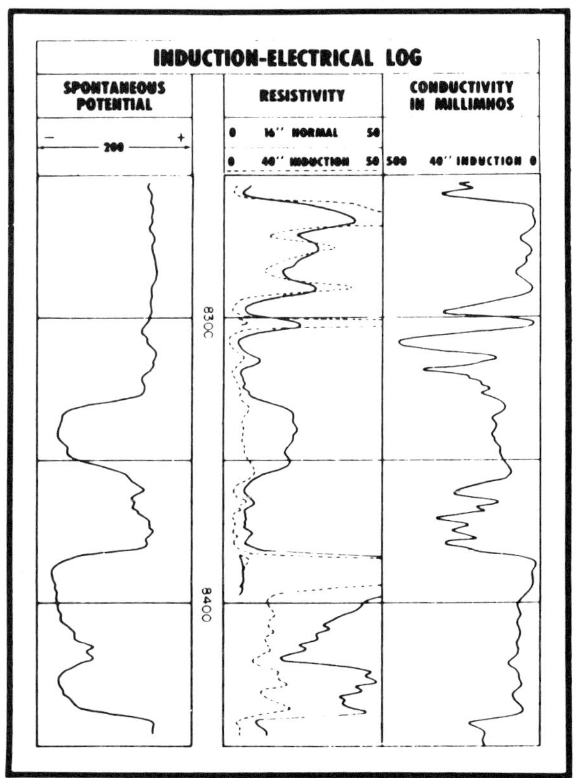

FIG. 27. A typical induction log.

Figure 27 shows an example of the induction log. These logs are interpreted similar to the conventional resistivity logs; accordingly, the variations in resistivity represent changes in rock characteristics and fluid content. Since there is no flow of current through the mud, induction logs are less affected by borehole conditions than conventional logs.

Focused current logs.—The focused current logs are useful in moderate to small thick formations, in highly resistive formations, and in the wells drilled with salt muds. Their interpretation is similar to those of the conventional and induction devices.

Figure 28a shows the electrode arrangement for obtaining one of the commonly used focused current logs. A constant intensity current is fed through electrode A_0, and adjustable currents of the same polarity are sent through electrodes A_1 and A_2. The intensity of these currents is automatically adjusted so that monitor electrodes M_1 and M_2 and M_1' and M_2' are always at the same potential. The potential drop between monitor electrodes and a far electrode will be proportional to formation resistivity. Figure 28b compares the lines of current flow for normal and focused devices, when logging across a formation with a high resistivity contrast $R_t >> R_s$. In a normal device, the current lines tend to disperse by following the path of less resistance, such as through mud and the adjacent formation resistivity R_s. Thus the measurements obtained differ from true formation resistivity. In a focused log the current is forced normal to the hole wall with practically no current dispersion.

Figure 29 shows an example of focused log. For a dual laterolog device, three resistivity curves obtained simultaneously with caliper and gamma-ray curve are shown. The logarithmic scales are used, since the focused devices are designed for high resistivities. R_{LLd} and R_{LLs} are the deep and shallow penetration devices, respectively. R_{MSFL} is the resistivity measured by a focused microresistivity device.

Focused microresistivity log.—This log determines more accurately the resistivity R_{x0} of the mud filtrate flushed zone as compared to the conventional microresistivity logs. Figure 30 illustrates the measurement principle of a typical log, which is essentially the same as for focused resistivity logs. The current electrodes A_0, A_1 and measurement electrodes M_1, M_2 are imbedded in an insulating pad, which is pressed against the wall of the borehole by means of a spring system. A constant current is sent through the central electrode A_0, and another current of the same polarity is fed through the outer electrode A_1. Intensity of current through A_1 is automatically adjusted to make the potential difference between M_1 and M_2 essentially zero. Potential at M_1 or M_2 is measured, which is proportional to the formation resistivity. Figure 30 shows, for both permeable and impervious formations, how the current emitted by electrode A_1 prevents the current emitted by A_0 from diverging in all directions, which occurs in conventional microresistivity devices. It is confined to a beam perpendicular to the hole wall.

Modern logging tools allow simultaneous recording of conventional microresistivity logs for determining permeable beds and microresistivity focused logs for deriving R_{x0} values. Figure 31 is a sketch of the sonde used showing the position of the pads for obtaining the two logs. Figure 32 illustrates a typical presentation for the microlaterolog, and also for the proximity log. This device is less accurate when the mud cake reaches certain thickness. Other logging tools have been designed to cope with this problem (Monti, 1969).

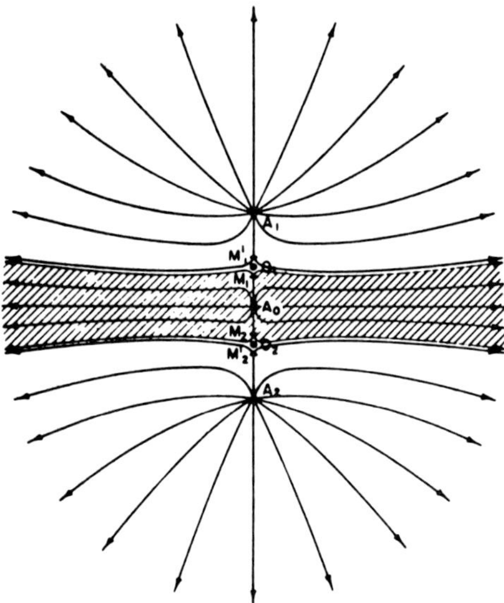

FIG. 28a. Electrode arrangement and current distribution in a typical focused log.

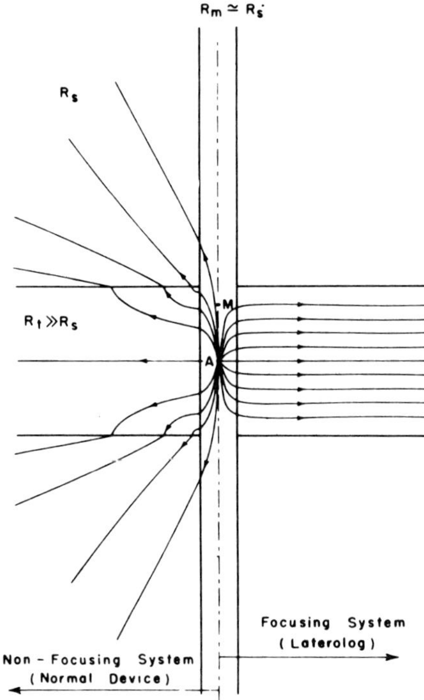

FIG. 28b. Comparative distribution of current lines for a normal device (left) and a focused current log (right).

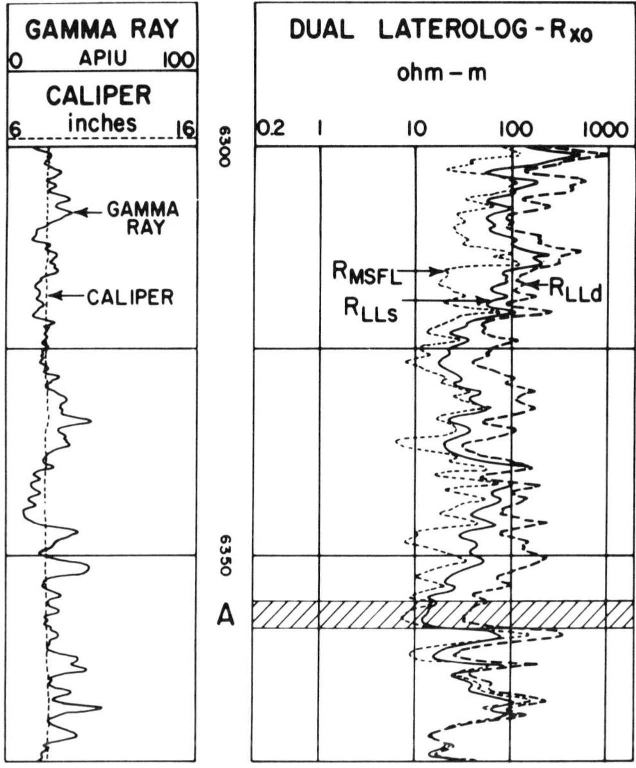

Fig. 29. Example of a dual laterolog—microresistivity log recorded in a dolomite formation.

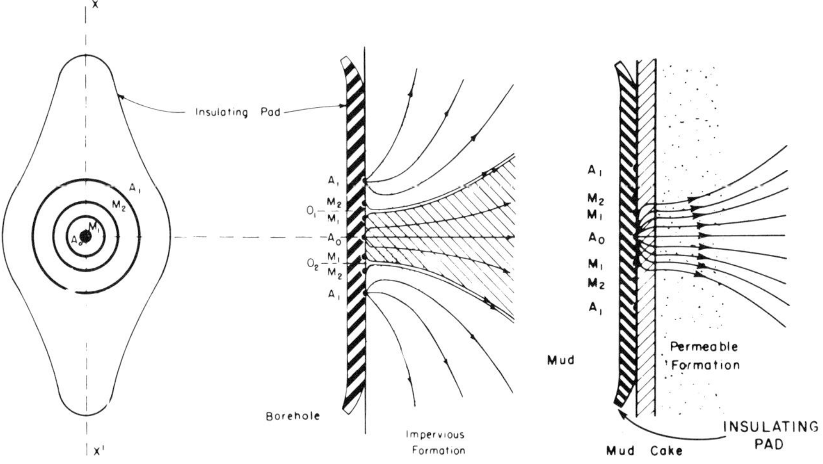

FIG. 30. A microresistivity pad showing the electrodes and schematic representation of current lines in an impervious (left) and permeable (right) formation.

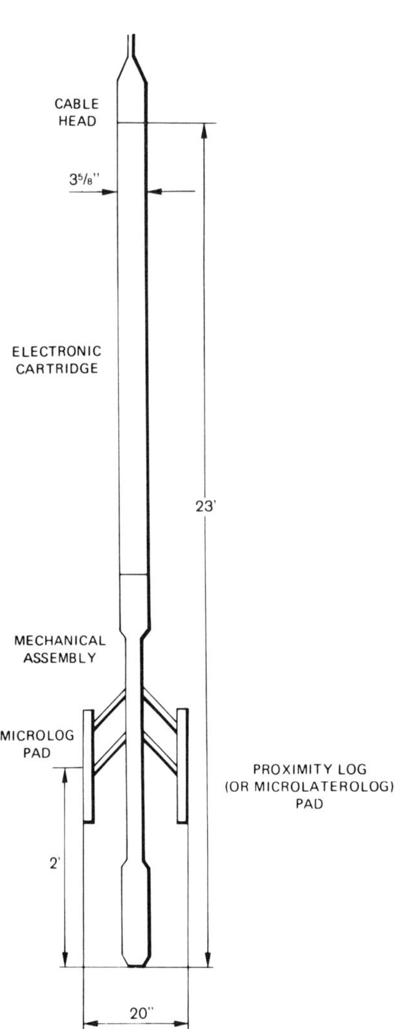

FIG. 31. Sketch showing a sonde for simultaneous recording of microlog and proximity or microlaterolog.

Neutron log

Some elements, such as berilium, contain in their nuclei more neutrons than others, and they are also loosely bound. These elements give up their neutrons when bombarded with alpha rays. The common neutron sources are mixes of berilium and an alpha-ray emitting radioactive element such as americium or berilium. These sources emit fast neutrons.

Upon leaving its source, a fast neutron suffers a series of collisions. If the neutron hits a heavy nucleus, it bounces with energy almost identical to the energy of emission (Figure 33); this is an elastic collision. But if the collision is centered against a nucleus of about the same weight, the energy of the fast neutron is practically transferred to the other nucleus, as in the case of two billiard balls. The effect is to slow down the fast neutron. A hydrogen nucleus has about the same weight as a neutron, and hydrogen is the most abundant element in formations; therefore, it is the major source for slowing down neutrons. Each time a neutron collides, it becomes slower until it finally reaches its minimum velocity at formation temperature and diffuses. It is at this stage in the reaction chain that the neutron (called thermal neutron) is absorbed by an atom, which then emits one or several gamma rays.

The concept of neutron logging is based on the bombardment of formation by neutrons and their detection in the formation. Different kinds of neutron logs are available depending upon the point in the nuclear reaction chain in which the neutrons are detected. The basic design of the logging system consists of a radiation source,

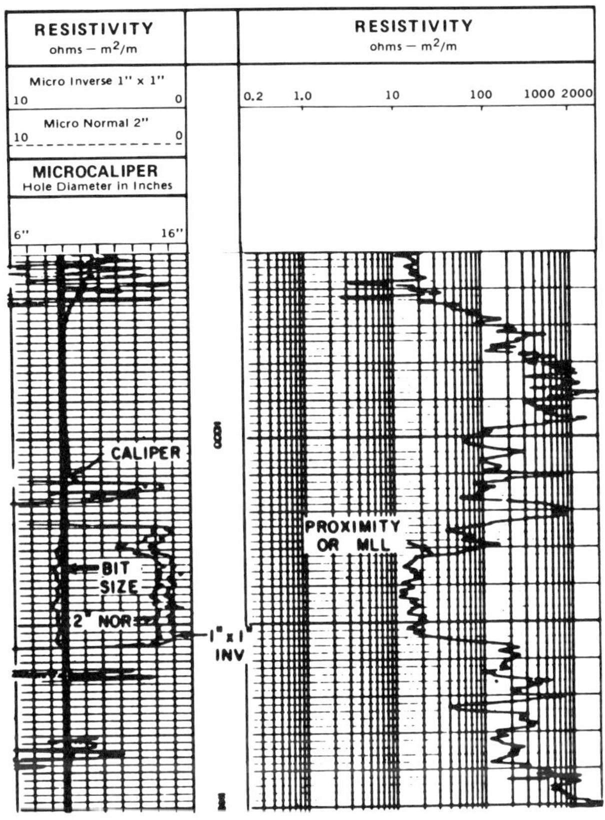

FIG. 32. Example of simultaneous recording of conventional and focused microresistivity logs.

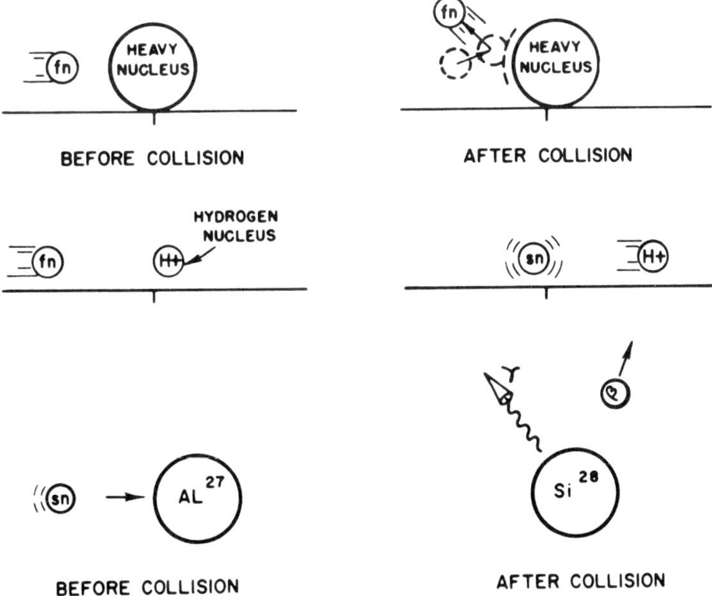

FIG. 33. Effect of fast neutron collisions with a heavy and a hydrogen nucleus, and capture of slow thermal neutrons accompanied by gamma-ray radiation.

a detector, and a recorder. The detecting instruments may be the same as those used for the natural gamma-ray logs.

The neutron-gamma log is one of the early logs in this category, and it is based on measurement of capture gamma rays. In interpretation of this log, it is assumed that capture gamma rays are proportional to the amount of hydrogen present in the formation. Since most of the hydrogen is contained in the water and hydrocarbons filling the rock pores, deflections of the neutron curve recorded should be proportional to formation porosity. The neutron-gamma logs are now scaled in API units. An API neutron unit is defined as 1/1000 of the difference in deflections observed when a sonde is introduced and passed through the artificial formations of a calibration pit at the University of Houston, with and without the neutron source (Belknap et al, 1959). The API units can be converted into porosity index units by using logging company charts. Besides porosity determinations, this log can be used for well-to-well correlations in the same manner as resistivity-SP logs.

Considerable advances have been made in the development of thermal neutron logs which eliminate the drawbacks of the neutron-gamma logs. The compensated neutron tool (Alger et al, 1971) is an improved device for thermal neutron logging. It uses two detectors instead of one and a considerably higher neutron output source. The sonde is pressed against the borehole wall in order to minimize the borehole effects. The counting rates of the two detectors are computed as the ratio of the two signals (near to far); after corrections, they are converted to porosity according to certain calibrations. Figure 34 shows an example of this log. It can be simultaneously recorded with other logs such as density, as shown in the figure.

Density log

The subsurface device for measuring density is a sonde with a pad which is pressed against the borehole wall by a springarm. A gamma-ray source and two gamma-ray detectors are mounted in the pad. Gamma rays emitted by the source collide with the electrons of the atoms of the formation material which they traverse. In this atomic reaction some gamma rays lose energy before reaching the detectors, while others are absorbed through the different atomic absorption processes. Consequently, a smaller number of gamma rays arrive at the detectors than the number originally leaving the source. Heavy materials have more electrons in their atomic structure than light materials. Thus the heavier the material, the fewer gamma rays that reach the detectors. Therefore, the counting rates of detectors should be proportional to formation density.

The near detector is sensitive to the density of material immediately adjacent to the face of the pad, such as mud cake and other irregularities. The far detector counting is principally affected by formation density. The signals of both detectors are combined automatically to give a density correction, which is applied to the uncompensated density information from the detector (Wahl et al, 1964). Then both the compensated density and the applied compensation are recorded. Figure 35 is an example of a density log. A natural gamma-ray log and a borehole caliper curve can also be recorded simultaneously.

Formation porosity values can be derived from the recorded formation density by using the following equation:

$$\phi = \frac{\rho_{ma} - \rho_b}{\rho_{ma} - \rho_f}, \tag{17}$$

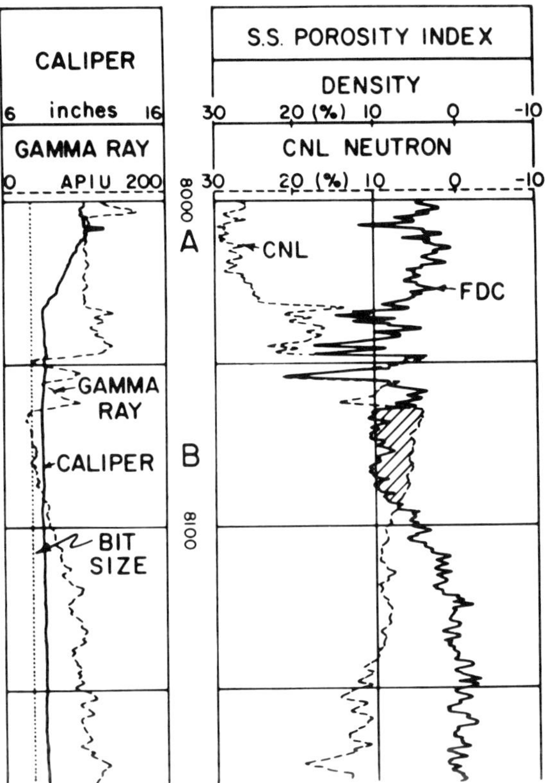

FIG. 34. Example of a neutron compensated log simultaneously recorded with a density log.

where

ρ_b = bulk density, as recorded by the log,

ρ_{ma} = matrix grain density,

and

ρ_f = density of fluid saturating the rock.

Acoustic log

Acoustic logs are based on the transmission of sound waves through the formation; they record the differential time Δt which an acoustic compressional wave takes to travel a given distance in the formation, measured in μsec/ft. Most modern devices use two transmitters and two pairs of receivers. The transmitters are pulsed alternately. When one of the transmitters is pulsed, part of the acoustic energy travels through the mud, strikes the borehole wall at the critical angle of incidence and is refracted, travels through the formation along the borehole wall, enters the mud again, and finally arrives at the receivers. Time Δt elapsed between the first arrival of the sound waves at the two receivers is measured. Δt values are read on both pairs of receivers and averaged.

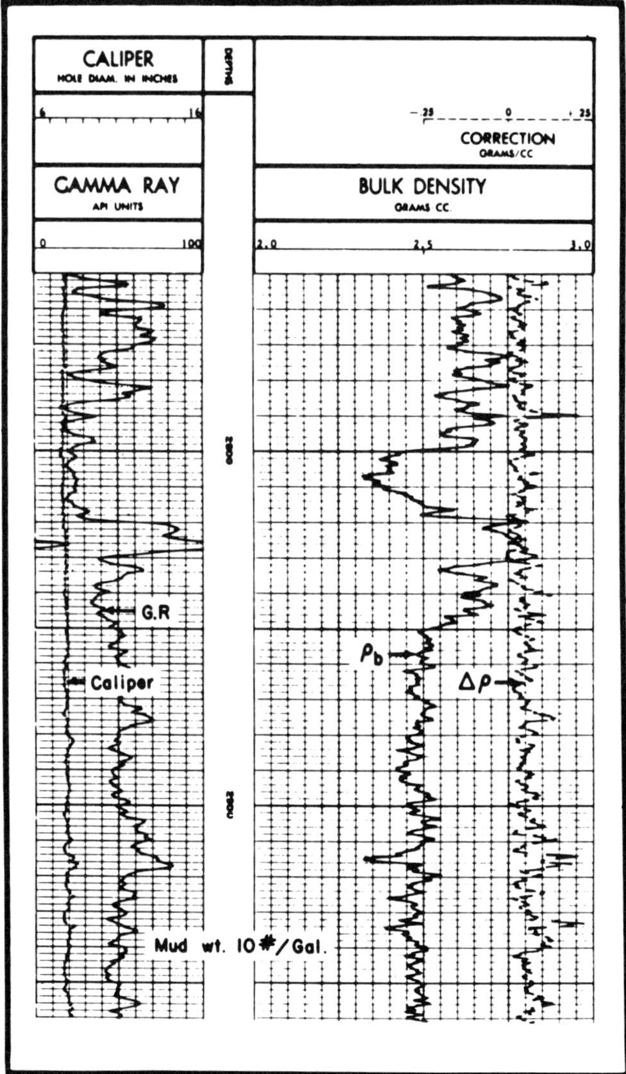

FIG. 35. A typical bulk density log.

One application of acoustic logs is to determine formation porosity. The Wyllie (1956) equation may be used for this purpose:

$$\phi = \frac{\Delta t - \Delta t_{ma}}{\Delta t_f - \Delta t_{ma}}, \tag{18}$$

where

Δt = compressional wave traveltime recorded,

Δ_{ma} =compressional wave traveltime of the matrix rock,

and

Δt_f = compressional wave traveltime of the fluid filling the pores of the rock.

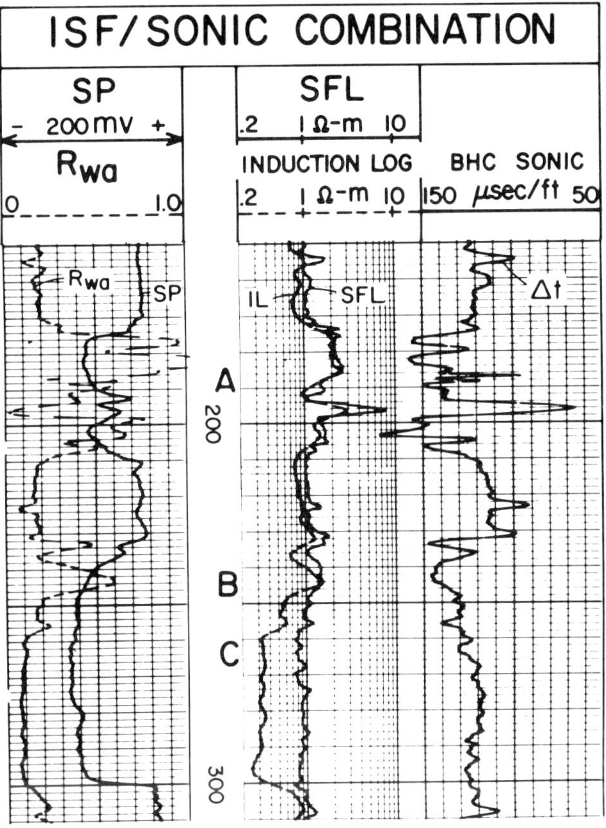

FIG. 36. A typical induction—spherically focused/sonic curve combination along with automatically computed R_{wa} curve.

Figure 36 is an example of the log where (along with the acoustic log) induction, spherically focused resistivity curves, and the SP curve were also recorded. An analog computer circuit in the panel was used to compute the apparent formation water resistivity (R_{wa}), where

$$R_{wa} = \phi^2 R. \tag{19}$$

Porosity values ϕ were computed from the acoustic curve [equation (18)], and the induction curve was used for resistivity R. The R_{wa} curve is commonly used for determining resistivity R_w of the formation water as well as fluid saturation. Water saturation is given by

$$S_w = \left[\frac{R_w}{R_{wa}}\right]^{1/n}, \tag{20}$$

where n is the saturation exponent, as defined in equation (12).

Summary

Well log technology is continually growing. Many logs are available today for exploration work. The scope of this book has justified a brief discussion of only the most commonly used ones. A list of the other logs of interest should include the electromagnetic, natural gamma-ray spectral, epithermal neutron, pulsed neutron capture, dipmeter, acoustic wave amplitude, ultralarge space electrical, and neutron activated logs. The reader is referred to the literature for further insight into these and other developments.

The importance of using well logs in exploration cannot be overemphasized. The cost of drilling dry holes is far too high. A good logging program for each new well and careful assimilation of the logged data into exploration studies can minimize the cost of drilling dry holes.

REFERENCES

Aguilera, R., 1976, Analysis of naturally fractured reservoirs from conventional well logs: J. Petr. Tech., v. 28, July, p. 764–772.

Allaud, L., and Martin, M., 1976, Schlumberger. Histoire d'une technique: *Berger-Levrault,* Paris.

Archie, G. E., 1942, The electrical resistivity log as an aid in determining some reservoir characteristics: Petr. Trans. AIME, v. 146, p. 54–62.

Arps, J. J., 1953, The effect of temperature on the density and electrical resistivity of sodium chloride solutions: Petr. Trans. AIME, v. 198, p. 327–330.

Belknap, W. B., Dewan, J. T., Kirkpatrick, C. V., Motts, W. E., Pearson, A. J., and Rabson, A. J., 1959, API calibration facility for nuclear logs: API Drilling and Production Practice, p. 289–316.

Clark, N. J., 1960, Elements of petroleum reservoirs: SPE of AIME.

Gomez-Rivero, O., 1975, Registros de pozos, parte 1. Teoria e interpretacion: Mexico.

——— 1976, A practical method for determining cementation exponents and some other parameters as an aid in well log analysis: The Log Analyst, v. 17, Sept.–Oct., p. 8–24.

Monti, R. L., 1969, Usan efectivos perfilajes: Petr. Interam. Nov., p. 60–64.

Von Gonten, W. D., and Osoba, J. S., 1969, A mcthod of predicting saturation exponents in logging: 44th Annual Fall Meeting, SPE paper 2530.

Wahl, J. S., Tittman, J., and Johnstone, C. W., 1964, The dual spacing formation density log: AIME 39th Annual Fall Meeting, SPE paper 989, Houston.

Winsauer, W. O., et al., 1952, Resistivity of brine-saturated sands in relation to pore geometry: AAPG Bull., Feb., v. 36, p. 253–277.

Wyllie, M. R. J. and Gregory, A. R., 1953, Formation factors of unconsolidated porous media: Influence of particle shape and effect of cementation: Petr. Trans. AIME, v. 198, p. 103–110.

Wyllie, M. R. J., Gregory, A. R., and Gardner, L. W., 1956, Elastic wave velocities in the heterogeneous and porous media: Geophysics, v. 21, p. 41–70.

REFERENCES FOR GENERAL READING

Alger, R. P., Locke, S., Nagel, W. A., and Sherman, H., 1971, The dual spacing neutron log—CNL: 46th SPE of AIME Fall Meeting, SPE paper 3565, New Orleans, Oct.

Atkins, E. R., Jr., 1955, The theory and instrumentation of radiation logging: AIME symp. formation evaluation, SPE paper 586G, Houston, Oct.

Calvert, T. J., Rau, R. N., and Wells, L. E., 1977, Electromagnetic propagation. A new dimension in logging: 47th Annual California Regional Meeting of SPE of AIME, SPE paper 6542, Bakersfield, April.

Clavier, C., Hoyle, W. R., and Meunier, D., 1969, Quantitative interpretation of TDT logs: SPE of AIME 44th Annual Fall Meeting, SPE paper 2658, Denver.

Dewan, J. T., Johnstone, C. W., Jacobson, L. A., Wall, W. B., and Alger, R. P., 1973, Thermal neutron decay time logging using dual detection: 14th SPWLA Annual Symp., paper P.

Doll, H. G., 1950, The microlog—A new electrical logging method for detailed determination of permeable beds: Petr. Trans. AIME, v. 189, p. 155–164.

Grosmangin, M., Kokesh, F. P., and Majani, P., 1961, A sonic method for analyzing the quality of cementation of borehole casings: Pet. Trans. AIME, v. 222, p. 165–171.

Hoyer, W. A., 1961, Induced nuclear reaction logging: J. Petr. Tech., August, v. 13, p. 797–802.

Kokesh, F. P., Schwartz, R. J., Wall, W. B., and Morris, R. L., 1965, A new approach to sonic logging and other acoustic measurements: J. Pet. Tech., March, v. 17, p. 282–286.

Lawson, B. L., and Cook, C. F., 1970, A theoretical and laboratory evaluation of carbon logging: Part II. Theoretical evaluation of oxygen interference: SPWLA Trans., paper B.

Lock, G. A., and Hoyer, W. A., 1971, Natural gamma-ray spectral logging: The Log Analyst, Sept.–Oct., v. 12, p. 3–9.

——— 1974, Carbon-oxygen (C/O) log: Use and interpretation: J. Petr. Tech., Sept., p. 1044–1054.

Meador, R. A., and Cox, P. T., 1975, Dielectric constant logging, A salinity independent estimation of formation water volume: 50th Annual Fall Meeting of SPE of AIME, SPE paper 5504, Dallas, Oct.

Mercier, V. J., Morris, R. L., Golwitzer, L. H., and Moran, J. H., Radioactive and electrical logging: reprinted from World Oil.

Morris, R. L., Grine, D. R., and Arkfeld, T. E., 1964, Using compressional and shear acoustic amplitudes for the location of fractures: J. Petr. Tech., June, v. 16, p. 623–632.

Pardue, G. H., Morris, R. L., Golwitzer, L. H., and Moran J. H., 1963, Cement bond log. A study of cement and casing variables: J. Petr. Tech., May, v. 15, p. 545–555.

Patten, E. P., Jr., and Bennet, G. D., 1963, Application of electrical and radioactive well logging to ground-water hydrology: U. S. Dept. of the Interior, U.S. Government Printing Office, Washington, D.C.

Poupon, A., and Gaymard, R., 1970, The evaluation of clay content from logs: SPWLA Symp., paper G.

Russell, W. L., 1941, Well logging by radioactivity: AAPG Bull., 25, Sept., p. 1768–1788.

Schlumberger Corp., 1969, Log interpretation principles.

——— Review of Schlumberger well logging and auxiliary services: document no. 2.

——— 1969, Log interpretation charts.

——— 1958 Introduction to Schlumberger well logging: document no. 8.

Schuster, N. A., Badon J. D., and Robbins, E. R., 1971, Application of the ISF/SONIC combination tool to Gulf Coast formations: Gulf Coast Asso. of Geol. Soc.

Scott, H. D., and Smith, M. P., 1973, The aluminim activation log: SPWLA Trans., paper F.

Seismograph Service Corp., Capabilities of logging services: Birdwell, A Division of Seismograph Service Corp.

Smith, R. D., Jr., and Schultz, W. E., 1974, Field experience in determining oil saturations from continuous C/O and Ca/Si logs independent of salinity and shaliness: SPWLA Trans., paper K.

Suau, J., Grimaldi, P., Poupon, A., and Souhaite, P., 1972, The dual laterolog R_{x0} tool: 47th Annual Fall Meeting of the SPE of AIME, SPE paper 4018.

Tittman, J., 1978, Radiation logging: Physical principles, *in* SPWLA reprint volume. Gamma ray neutron and density logging: Petr. Eng. Conf., Univ. of Kansas, 1956.

Tittman, J., and Wahl, J. S., 1965, The physical foundations of formation density logging: Geophysics, v. 30, April, p. 284–294.

Tittman, J., et al., 1966, The sidewall epithermal neutron porosity log: Trans. AIME, p. 1351–1362.

Tixier, M. P., Alger, R. P., and Tanguy, D. R., 1960, New developments in induction and sonic logging: J. Pet. Tech., May, v. 12 (5), p. 79–87.

Wahl, J. S., Nelligan, W. B., Frentrop, A. H., Johnstone, C. W., and Schwartz, R. J., 1970, The thermal neutron decay time log: SPE J., Dec., v. 10, p. 365–379.

Winn, R. H., 1955, The fundamentals of quantitative analysis of electric logs: Symp. on formation evaluation AIME, SPE paper 584–G, Houston, Oct.

Chapter 7

WELL LOGS IN EXPLORATION

Introduction

Seismic and well log data are widely used in petroleum exploration to map the subsurface. The two data sources are complementary: seismic profiles provide an almost continuous lateral view of the subsurface, whereas well logs yield fine vertical resolution of the geology at the borehole.

In contrast to seismic, other surface techniques, including gravity, electrical, magnetic, and magnetotelluric (MT) methods, have lower resolution and are mainly useful for reconnaissance mapping in relatively unexplored basins to define the gross tectonic configuration of the sedimentary rocks. Seismic profiles can resolve, with relatively high precision, the structural and stratigraphic changes from the arrival times and amplitudes of the reflection events. Significant advances in seismic data acquisition and processing methods, derived from modern systems and control theory as well as from the computerized solutions of the acoustic wave equation, have led to breakthroughs in mapping complex configurations, as described in the previous chapters.

The bandwidth of seismic data constrains the vertical resolution of the subsurface; high-frequency data are essential for delineating subtle traps. Also, the seismic expressions of anomalies cannot be interpreted uniquely in terms of the geologic variables. Well logs can be helpful in the interpretation of seismic profiles in both respects, viz., they can unambiguously provide, at the borehole, a high-resolution estimate of many essential geologic variables.

The use of well logs ranges from small-scale field studies, where dense well control yields information of short wavelength, to large-scale regional studies, where sparse control is utilized for long-wavelength changes. Explorationists construct structural and stratigraphic cross-sections and correlate geologic formations with seismic reflection events from logs. The sonic and density logs are used for interpretation of seismic amplitudes in terms of the subsurface stratigraphy, for wavelet deconvolution, for identifying rock and fluid changes on seismic data, and for modeling of changes in rock type, porosity, and fluids. Petrophysicists quantify logs for reservoir evaluation and for field development. They normally limit their investigation to a few discontinuous sections of the rock column which are suspected to contain hydrocarbons.

By means of digital computers, the thoroughness of petrophysical analysis can be extended to the entire penetrated section and to all the wells of interest in the study area. The computed variables can be used to study the reservoir characteristics, source and seal rocks, trap configurations, and depositional environment. They can also be displayed to highlight the areas of change, either vertically within a well or laterally from one well to the next. Such displays can help explorationists in analyzing a large bulk of processed data and in creatively translating the results into petroleum prospects.

The log data base is large; over 1.2 million wells were drilled in the United States alone during the period 1946–1976, and the rate of drilling wells is increasing. The type and number of logs vary a great deal from one well to another, due to need and cost considerations and also because of continuous improvements in logging techniques. Digital computers can handle the log data base adequately. It is important that computer usage also be directed to quantitative interpretation of the logs, because, no matter how small the area of exploration may be, such usage will minimize drilling of costly dry holes. In this chapter, some aspects of digital well log processing and their interpretational considerations in petroleum exploration are presented.

Digital Processing

Well logs provide measurements of physical, chemical, or physio-chemical properties of the media surrounding the borehole. For best results, these measurements should be converted into appropriate geologic/geophysical variables, such as rock type, porosity, fluid saturation, permeability, and synthetic seismograms.

The logging program for a well is determined by several factors: (1) the type of rocks encountered, i.e., clastic, carbonate, igneous, etc., (2) degree of rock consolidation and compaction, (3) drilling fluid used, i.e., fresh water, salty mud, oil based mud, air, etc., (4) type of well, i.e., rank wildcat or field development, and (5) cost of logging service. The logs of general interest for exploration work are resistivity, acoustic, density, and neutron; these logs also include the spontaneous potential (SP), gamma ray, and borehole caliper measurements. The physical basis of these logs and their relationship to rock properties were discussed in the previous chapter. Figure 1 shows a flow chart for digital processing; the various steps are discussed below.

(1) Digitization

It is possible to record logs digitally at the well site. However, most of the logs were, and many still are, recorded in a continuous analog form on paper or film. For quantitative studies, these continuous data have to be digitized. The quality of both the analog and digital data should be carefully monitored to ensure data accuracy. Some important considerations follow.

Quality check.—Continuous logs have many scales and photographic qualities. Early logging runs had inadequate quality control. The instruments were less sophisticated, and factors such as mud characteristics and borehole diameter were not as well understood. Even with modern advancements, errors due to miscalibrations, tool failures, galvanometer drifts, etc., are common. Therefore, logs should be carefully checked and adjusted before, during, and after digitization. Typical quality control problems are (see also Figures 2 and 3):

> **Grid distortion.**—This is a nonuniform stretch or shrinkage of the paper or film caused by film jams, depth shutter malfunctions, print skews, paper tear, and bad splicing.

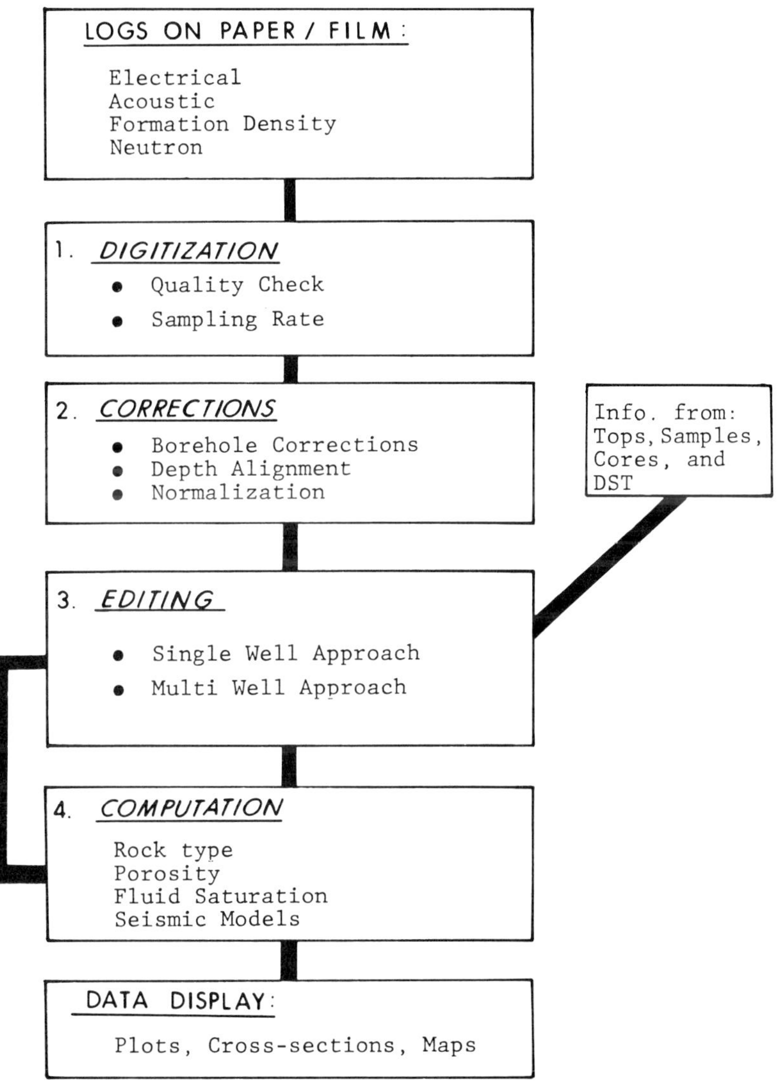

FIG. 1. Flow chart for digital well log processing.

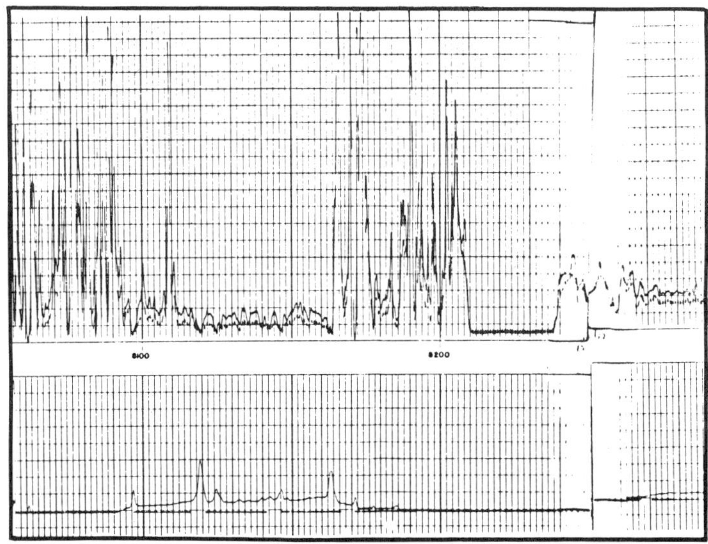

FIG. 2a. A microlaterolog showing grid distortion, poor optical quality, and overlapping curves.

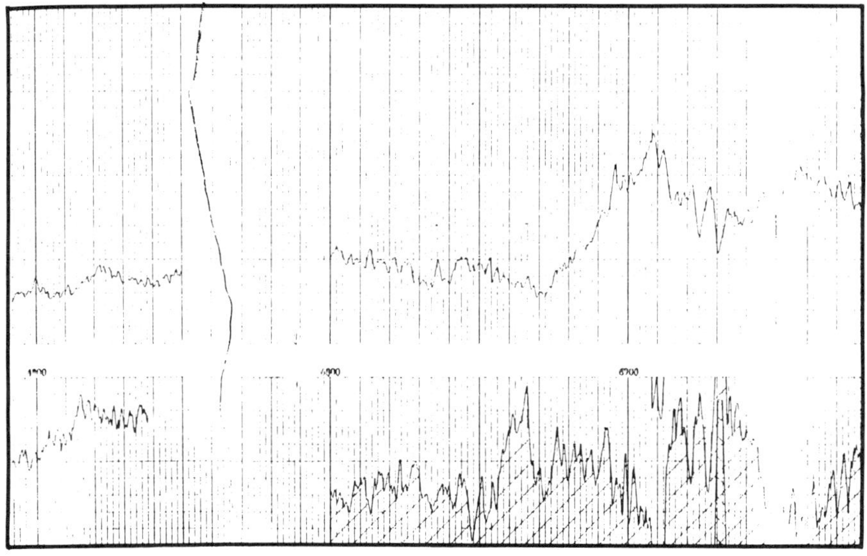

FIG. 2b. An example of tearing (dark, jagged vertical line), missing data, and double off-scale wraparound.

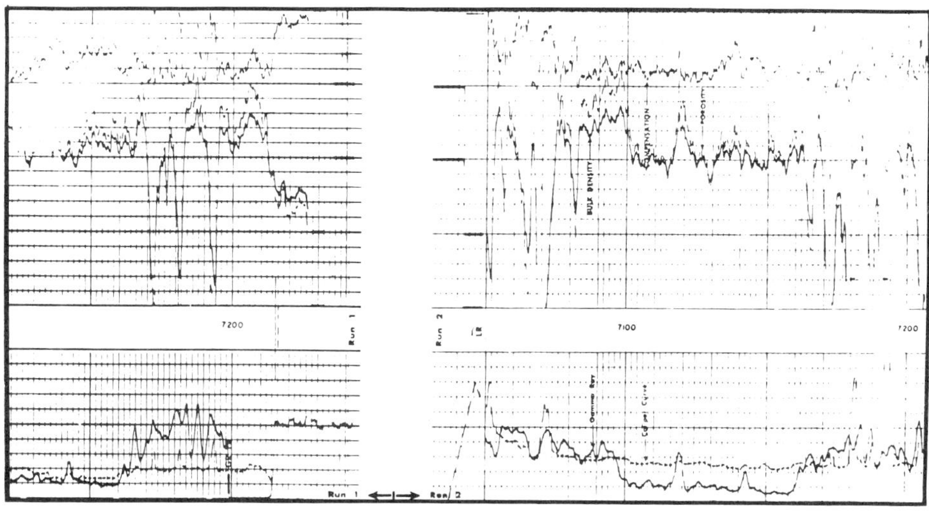

FIG. 3a. A portion of density log over two log runs. Note the change in curves between 7150–7200 ft interval.

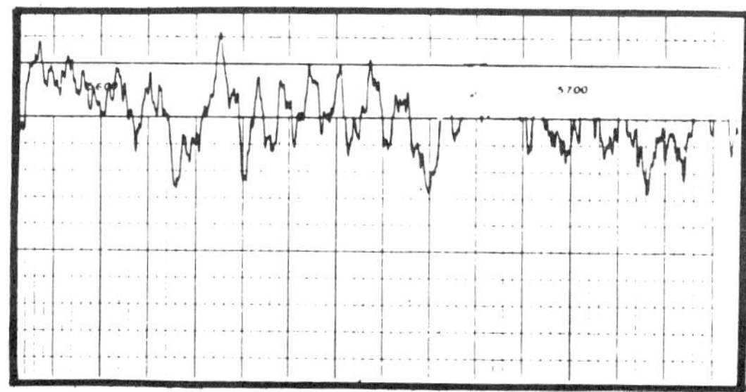

FIG. 3b. A segment of gamma-ray curve going off the track.

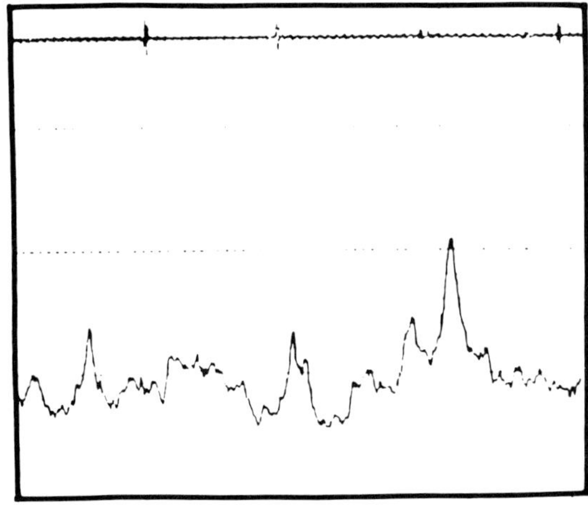

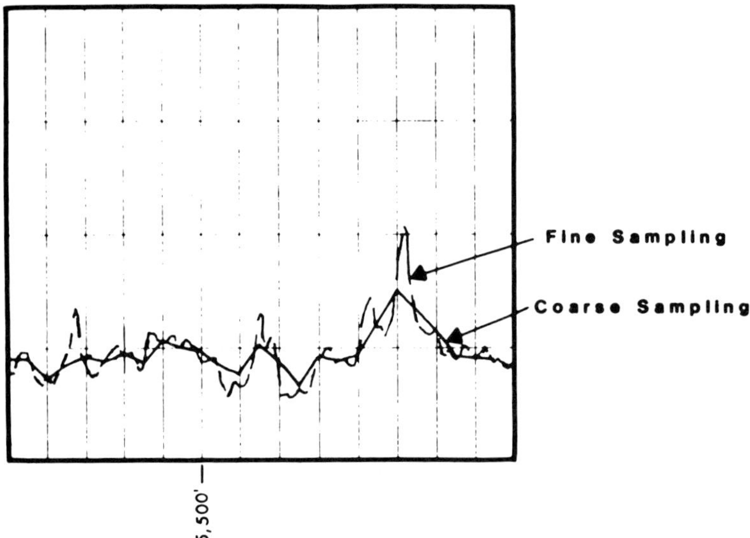

FIG. 4. The "aliasing effect" is illustrated by the dashed (based on 0.5 ft sampling rate) and solid (based on 5.0 ft sampling rate) tracings in the lower plot. The actual curve, shown in the upper plot, has been significantly altered by the coarser sampling.

Poor optical quality.—Portions of the curve may be indistinguishable because of poor printing or recording. This situation is generally caused by improper light exposures aggravated by rapid amplitude changes.

Missing or bad data.—Frequently, logs have data gaps or the photographic quality is too poor for digitization. Such log segments should be flagged for possible recovery in the editing step. Common causes are tool sticking, off-scale galvanometer failure, or failure to provide sufficient depth overlap between logging runs.

Overlapping curves.—When two or more curves are recorded on the same track, they can be intermingled and become indistinguishable.

Scales and wraparounds.—The scale of the logged variables, known as the engineering scale, can vary from log to log and is sometimes changed within the same logging run. Also, large-amplitude excursions outside the normal plotting scale range, displayed as "wraparounds" on the track, may be hard to decipher. Older electric logs tend to have several consecutive off-scale excursions before returning to the main scale.

Sampling rate.—Digitization is a form of data entry with a goal of faithfully transcribing the original data. Only an infinite sampling rate can do this perfectly, so the question becomes one of how much data distortion is acceptable. Too coarse a rate, even though the points sampled are accurate, results in an "aliasing effect" which causes intermediate log values to be interpolated and erroneously interpreted (see Figure 4). Thus, important log information may be distorted for any future evaluation.

The sampling rate is determined from an estimate of the minimum wavelength of significant energy in the continuous curve. Studies using power spectrum analysis indicate that a sampling rate of two points per foot is adequate for most well logs used in exploration. If a coarser sample rate is desired for analysis, then appropriate band-pass filtering after proper digitization should precede the desired coarser resampling.

Figure 5 shows an SP curve plotted from digital data. The curve in Figure 5a has significant energy at the wavelengths of 1 ft per cycle and higher; hence it was digitized at the necessary minimum of 2 samples per foot. For subsequent computations at coarser sampling rates, these data were filtered by appropriate low-pass filters and then resampled. Figures 5b, 5c, and 5d are examples of coarser resampling; the filter and the coarser resampling parameters are shown in the table beside the figure.

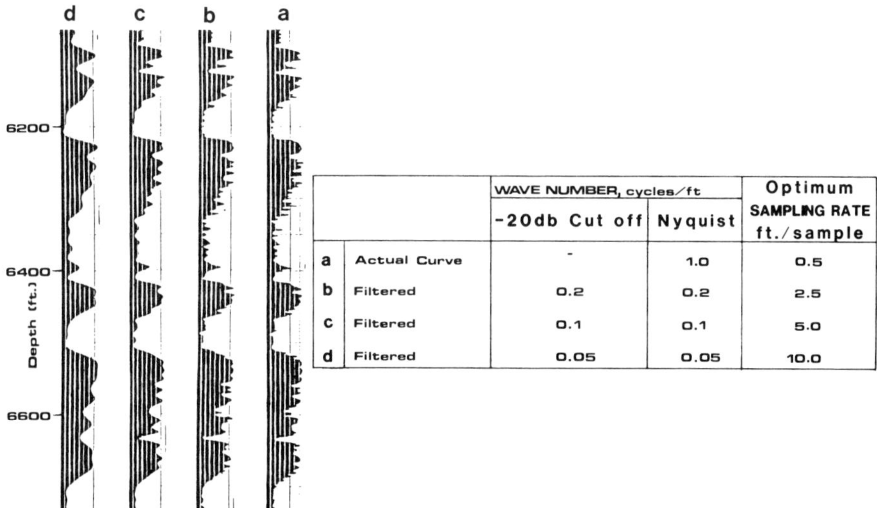

		WAVE NUMBER, cycles/ft -20db Cut off	Nyquist	Optimum SAMPLING RATE ft./sample
a	Actual Curve	-	1.0	0.5
b	Filtered	0.2	0.2	2.5
c	Filtered	0.1	0.1	5.0
d	Filtered	0.05	0.05	10.0

FIG. 5. In this example of SP curve, optimum sampling rates are indicated for different filter parameters.

(2) Corrections

Based on petrophysical analysis of logs from nearly 2000 wells containing over 34 million curve-feet of data, Neinast and Knox (1973) concluded that more than 50 percent of all well logs were erroneous. Their work points out the importance of log editing, although the problem may not always be so severe.

Logs can have a variety of errors. Some errors are caused by hole washout, electronic failure or drift, tool malfunction, incorrect tool design, and human error. There are such problems as depth alignment, discrepancies between different logging runs in the same borehole, and improper scaling of the log as a result of either error in calibrating the instrument drift or intermittent poor readings resulting from system malfunctions. Some of the procedures for correcting commonly occurring log errors are presented below.

Borehole corrections.—This correction is for tool sensitivity with respect to temperature, mud characteristics, and hole diameter, which are generally measured at the time of logging. Charts published by the logging companies can be used for this type of correction. An example of hole diameter correction to a gamma-ray curve is shown in Figure 6.

Depth alignment.—Depth discrepancies between two curves can lead to erroneous interpretations when the analysis involves simultaneous use of two or more curves. A simple method of alignment is to overlay two curve segments from different logging runs (for example, gamma-ray curves from sonic and neutron logs) over the same depth interval on a light table and shift one with respect to the other until the two curves are aligned. This is a widely used technique despite being time-consuming and subjective.

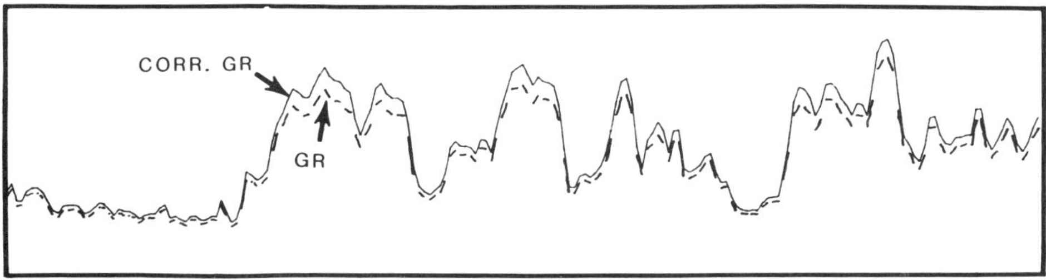

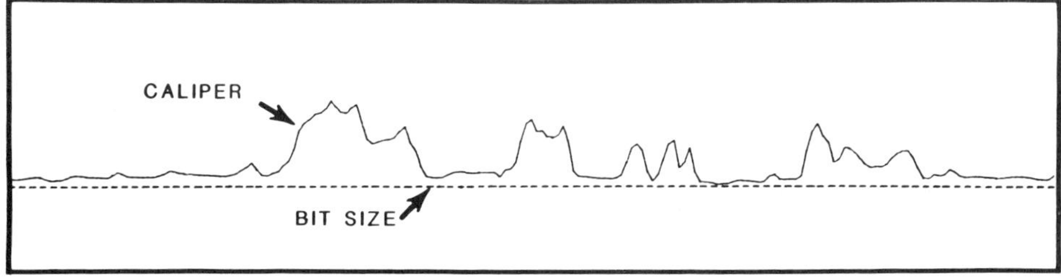

FIG. 6. A gamma-ray curve before and after hole diameter (caliper) correction.

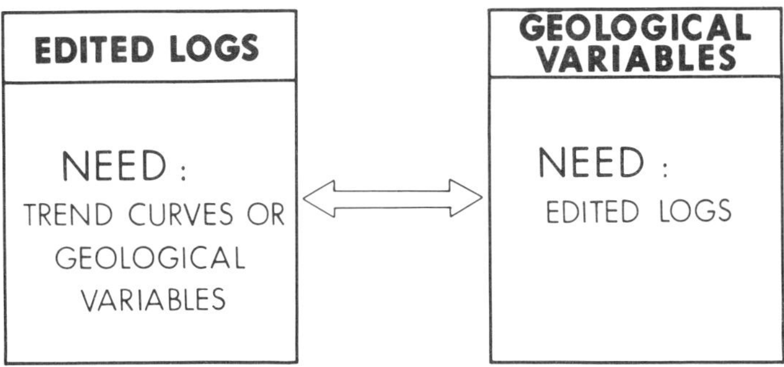

FIG. 7. The closed-loop problem of log editing and determining geologic variables.

Statistical techniques using crosscorrelation of two curve segments can be used to automate the estimates of relative depth changes. The data must be digitized prior to the depth alignment step in this case. Kwon (1977) suggested a statistical approach using spectral analysis of the data.

The relative depth changes should be determined for contiguous segments covering the interval of interest. Next, all the curves of one of the two logs should be appropriately stretched or squeezed, segment by segment, to depth match the second log (reference log).

Normalization.—This is a scale correction for such errors as miscalibration or instrument drift. Accuracy of scales is quite important for quantitative analysis. For example, errors in the neutron porosity scale can yield an erroneous interpretation of the reservoir characteristics of the rocks.

A collection of the same type of curves from many wells representing measurements across a common stratigraphic interval is necessary for this correction. Neinast and Knox (1973) described a variety of approaches for normalization by using the parameters derived from histograms, cross plots, or overlays. Generally, the histogram approach is adequate, where the mean and standard deviations of the measured values are estimated for each curve segment over a stratigraphically correlated interval. Zero shift and amplification factors are computed by comparing an individual curve histogram with a composite histogram of many curves for the stratigraphic interval.

This procedure should be applied to several contiguous stratigraphic intervals prior to choosing the normalizing factors for each interval. The factors should exhibit consistency across contiguous intervals because the conditions causing such errors are not likely to change abruptly from one layer to the next.

(3) Editing

This is a difficult correction. Log analysts tend to "redline" across poor data zones on the basis of their experience in the geologic province and their knowledge of rock properties. Such a procedure is subjective and often erroneous.

Some a priori knowledge of rock types, porosity, and fluids is necessary for editing logs. As illustrated in Figure 7, the problem is circular in the sense that good log readings are needed for computing these variables. Therefore, it is better to combine the two problems, viz., editing and estimating the exploration variables, solving them simultaneously and iteratively improving the solutions as the variable estimates are upgraded. Two separate approaches based on multivariate analyses are discussed next.

In both cases, the curves used are caliper, gamma-ray, resistivity, and available porosity curves (acoustic, density, and neutron). For simplicity, the methods discussed are for poor data caused by hole diameter problems; the concepts can be extended to other situations. It is assumed that the porosity curves (considered as output variables in the regression equations) need editing, and that the gamma-ray and resistivity curves (considered as input variables) have been corrected previously for caliper changes by using the logging company charts.

Single well approach.—The logged interval is divided into a suitable number of *zones* to keep the variations in depositional environment and rock types to a minimum. The zones should have enough data samples for regression statistics, but keeping the number of unknown parameters (rock types, fluids) to a minimum may necessitate shortening the zones.

The rock types are quantified in the form of a *lith curve,* which can be generated initially from sample logs, cores, SP and gamma-ray curves, and other available data sources. The lith curve is subsequently upgraded as the log analysis progresses. Computerized cross-plotting techniques should be used to define the boundaries for rock type discrimination.

Next, a *flag curve* is generated from the caliper curve to tag the good and bad data samples. Again, cross plotting is helpful in defining the discrimination boundaries to identify the bad data which require editing.

Regression analysis can be initiated after the lith and flag curves are generated. For a given rock type (using the lith curve) in each zone, the regression coefficients are derived from the good data (using the flag curve) in the zone. The functional form of the regression equation

$$Y(z) = F\,[X_1(z),X_2(z),X_3(z)]$$

can be expressed in terms of the sum of polynomial, cross-product, and logarithmic terms depending upon the conceptual physical model. In the equation, Y is the output variable (one of the porosity curves), z is depth, and X_1, X_2, X_3 are the input variables (gamma ray, resistivity, and sometimes the depth itself). Once the regression coefficients have been determined for a given rock type, the bad data for the same rock type are edited by substituting the appropriate values of the input variables in the regression equation.

The above procedure, consisting of determining the regression coefficients and editing bad data, is repeated in a systematic manner until all rock types, zones, and porosity curves are edited.

Multiwell approach.—A significant limitation of the single well approach is the requirement that the zones be short in order to minimize the number of unknown parameters (rock and fluid types). This often results in poor statistics for fitting the regression coefficients. The problem can be resolved through a multiwell approach where the statistical base is provided by an ensemble (group) of curves from many wells. Now the zones can be quite short depending upon the number of wells. It is assumed that all the wells in the group have the same set of logs.

The computational procedure is essentially the same as in the single well approach. For lith and flag curve generation as well as for fitting the regression coefficients and editing, the data for a given curve type for all the wells in the group are combined and processed together within each stratigraphic zone. Conceivably, only three or four wells may be used for initial processing. But as the data from additional wells become available, the results can be continuously upgraded to improve the estimates of rock types, fluids, and porosity by iterating the whole computation procedure while using the most up to date knowledge of the parameters as the starting point.

The multiwell concept transcends the overall log analysis methodology. Generally, explorationists construct the geologic framework of the subsurface by analyzing data for the rocks of certain age groups. In this process they try to understand the relationships of various rock layers through cross-sections and maps. Thus, as long as zone definitions are consistent with the interpretational needs, the multiwell ensembles provide appropriate data both for processing and interpretation steps.

Depending upon the depositional environment, rocks change much slower laterally than vertically. This favors the multiwell approach in the ''stationarity'' sense because the statistical attributes are derived from an ensemble of laterally similar rocks. Thus the sands of the same depositional cycle are likely to be analyzed together in the multiwell case, whereas different sands from several depositional cycles might be used together in the single well case. Because of this important difference, it may be possible to determine the rock types with a higher resolution in the multiwell case.

The hole washout problem is likely to vary from one well to another depending upon the hole conditions. Thus, for a given stratigraphic interval, good measurements may be available in some of the wells which may be useful in fitting the regression coefficients.

The multiwell approach has some limitations as well. The log suites, logging tools, drilling conditions, etc., are likely to change from one well to another, which can have unknown effects on the measurements. Proper normalization of the logs should help minimize some of the effects. Computer handling of ensemble data requires good program and data format design to keep track of the data samples in relation to their source. Finally, the hole washout problem may be so widespread for some rock types in a given zone that enough data may not be available for a regression fit. In such a case, the regression coefficients for that rock type may be used from the surrounding zones following a careful trend analysis.

Review.—Digitizing, correction, and editing procedures have been suggested for a set of log measurements. The resulting data can be considered reasonably correct with regard to electrical, acoustic, nuclear, and dimensional characteristics of borehole rocks subject to review with new knowledge. In addition, a computed lith curve of rock types is created. This curve may be refined iteratively as new knowledge is gained through the computation process.

A brief comment is included here on the use of additional data from wells when a full suite of four logs (electrical, sonic, density, and neutron) is not available. Increasing the number of subsurface control points by adding wells with partial data sets to wells with full data sets can be very useful for identifying laterally smaller (shorter wavelength) geologic changes. The first step is to process the wells with four logs. From such a data set a functional relationship between the geologic variables and a subset of data from three logs can be derived. A natural progression is to use two of the three porosity curves in the subset. The error statistics associated with the use of partial measurements should also be established. Having established functional relations between geologic variables and a log data subset, the next step is to combine the wells with three logs with the data from wells with four logs. At this stage the three log data set can be edited utilizing all the knowledge from the previous analysis. The concept can be extended iteratively to wells with two logs and even to those with one log as long as the associated errors do not become prohibitive.

(4) Computation

After the logs are corrected for measurement and digitizing errors, the next step is to compute the variables, hereafter referred to as the exploration variables, which are useful in geologic and geophysical interpretations. Some of the key exploration variables are:

Rock types (lithology): sand, shale, limestone, dolomite, anhydrite, salt, etc.
Porosity: matrix, fracture, vugular
Fluid saturation: gas, oil, water
Permeability

Other commonly used variables are synthetic seismograms, impedance logs, and velocity functions. These variables can become a basis for estimating other variables including the rock properties useful in seismic stratigraphic studies.

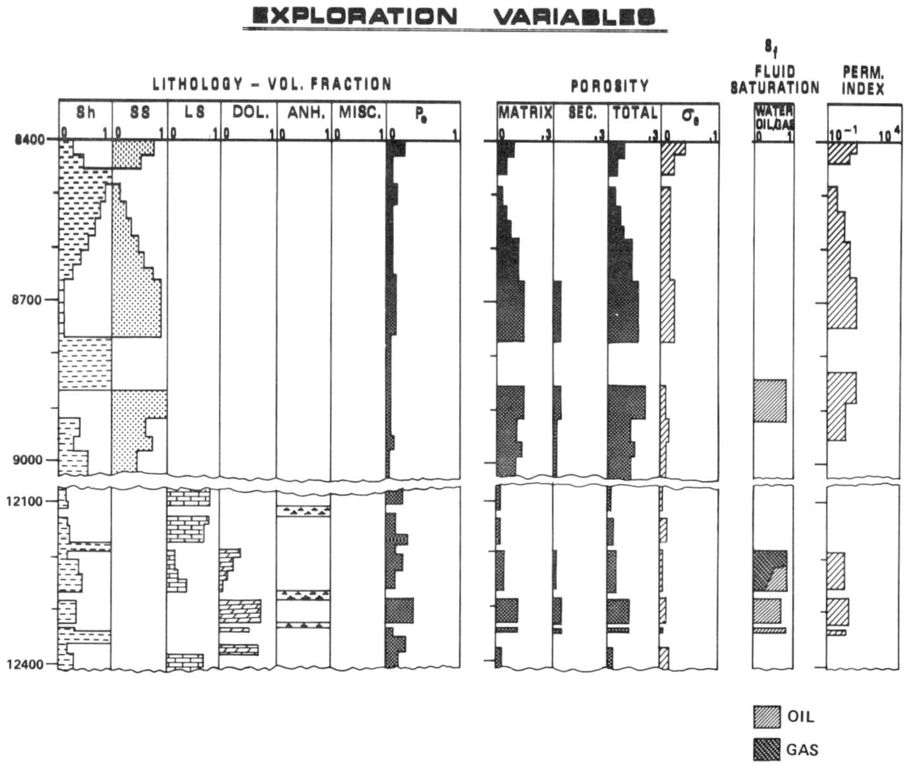

FIG. 8. An idealized diagram of some of the exploration variables which can be determined from well logs.

The key exploration variables are illustrated schematically in Figure 8. Shown in idealized block forms as functions of depth are volume fractions of the gross rock types, matrix and secondary porosities, fluid saturations, and the permeability index. The quantitative estimates of the rock types and porosities also include measurement and computational uncertainties in the form of standard deviation of error. Explorationists should transform the borehole measurements into this type of variable and attempt to upgrade their accuracy and level of resolution as the geologic knowledge of the area is improved.

Details of computational procedures are deferred to log analysis literature. The techniques are highly developed for the clastic and carbonate rocks, while the literature for the igneous rock environment is quite limited. Petrophysicists commonly use these techniques for field development or reservoir studies where they limit the computations to specific rock units. Explorationists need to extend such computations to whole logged intervals which include all the reservoir and nonreservoir rocks.

Some actual examples of the applications of well logs in exploration are given next. To maintain emphasis on the overall concepts, we have excluded uncertainty plots from these examples.

The first example of digitally processed logs is for the Cretaceous clastic rocks in the Denver-Julesberg basin. Shown side by side in Figure 9 are several measured curves and computed exploration variables. The measured curves shown are caliper, gamma ray (GR), spontaneous potential (SP), deep induction resistivity, acoustic, density, and neutron; all the curves have been depth aligned and edited for borehole effects. A similar plot of unedited measured curves should be routinely made and filed for reference purposes. Also, it is useful to flag the portions of the measured curves which are modified in the editing process.

The exploration variables shown in the above example are total porosity ϕ_T, apparent water resitivity R_{wa}, and depth of invasion. The total porosity was computed by cross plotting the neutron and density data assuming a dual-mineral matrix model. The empirical curves provided by the logging companies, such as the Schlumberger plots shown in Figure 10, were used for these computations. For interpretational convenience, the neutron (N) and density (D) curves in terms of limestone porosities are shown beside the total porosity curve T in Figure 9.

For the above set of data, true resistivity R_t (not shown) and depth of invasion were computed from the shallow, medium, and deep induction logs. R_{wa} was computed by using $m = 2$ as a first approximation for the cementation exponent in the Archie formula:

$$R_{wa} = \phi_T^m R_t .$$

R_{wa} is a good first indicator of hydrocarbons in the reservoir rocks. Using an R_{wa} plot, shown in Figure 9, the resistivity R_w of formation water can be inferred as a depth trend by joining the R_{wa} minima across the reservoir rocks. Large deviations in R_{wa} from this trend line are indicators of possible hydrocarbons. A commonly used formula for water saturation S_w is

$$S_w = \left[\frac{R_w}{R_{wa}}\right]^{1/n}$$

where fluid exponent $n \simeq 2$. Thus, $R_{wa} > 4R_w$ indicates hydrocarbons with less than

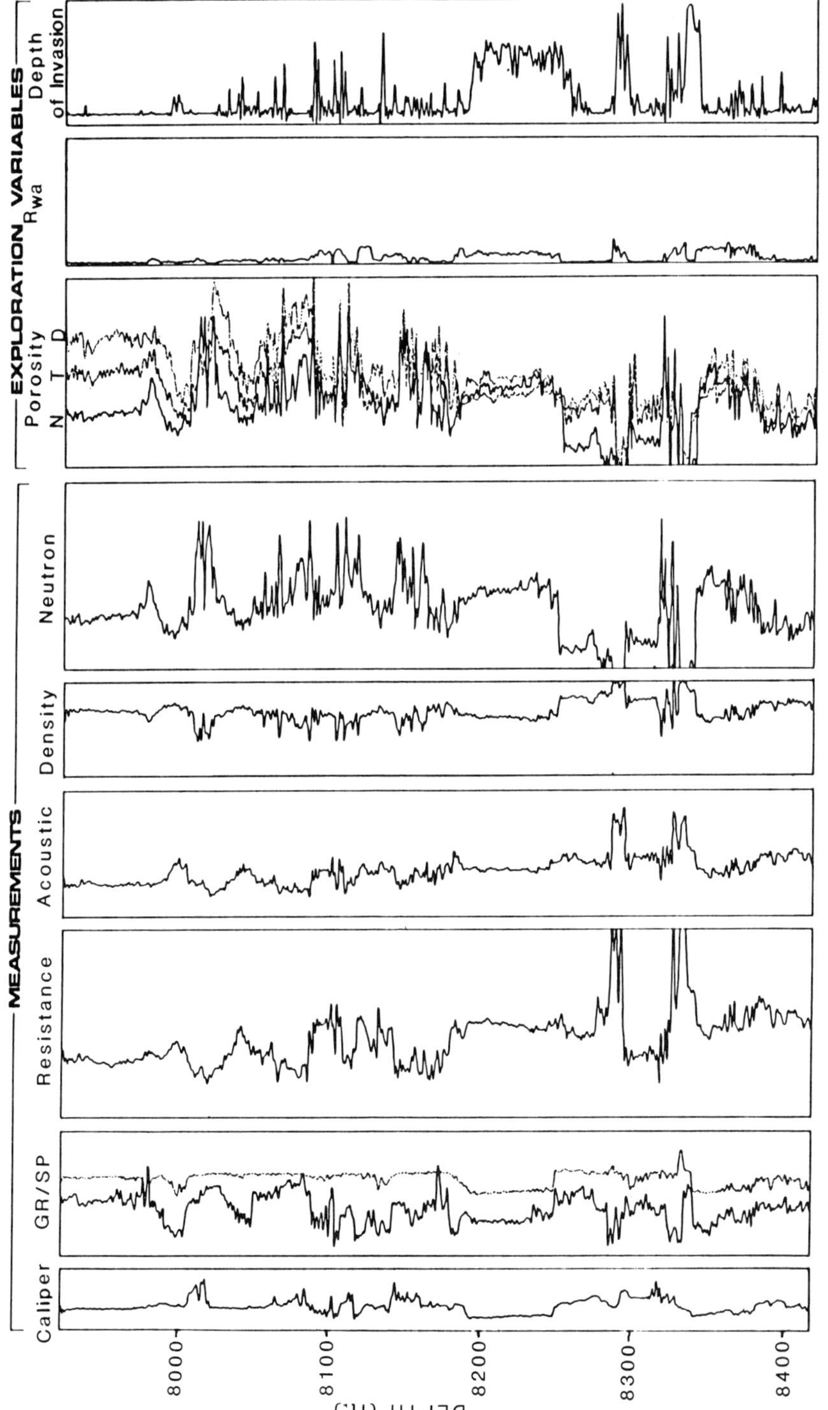

FIG. 9. An example of digitally processed logs for the clastic rocks of the Cretaceous age in the Denver-Julesberg basin.

POROSITY AND LITHOLOGY DETERMINATION

SALT WATER, LIQUID-FILLED HOLES

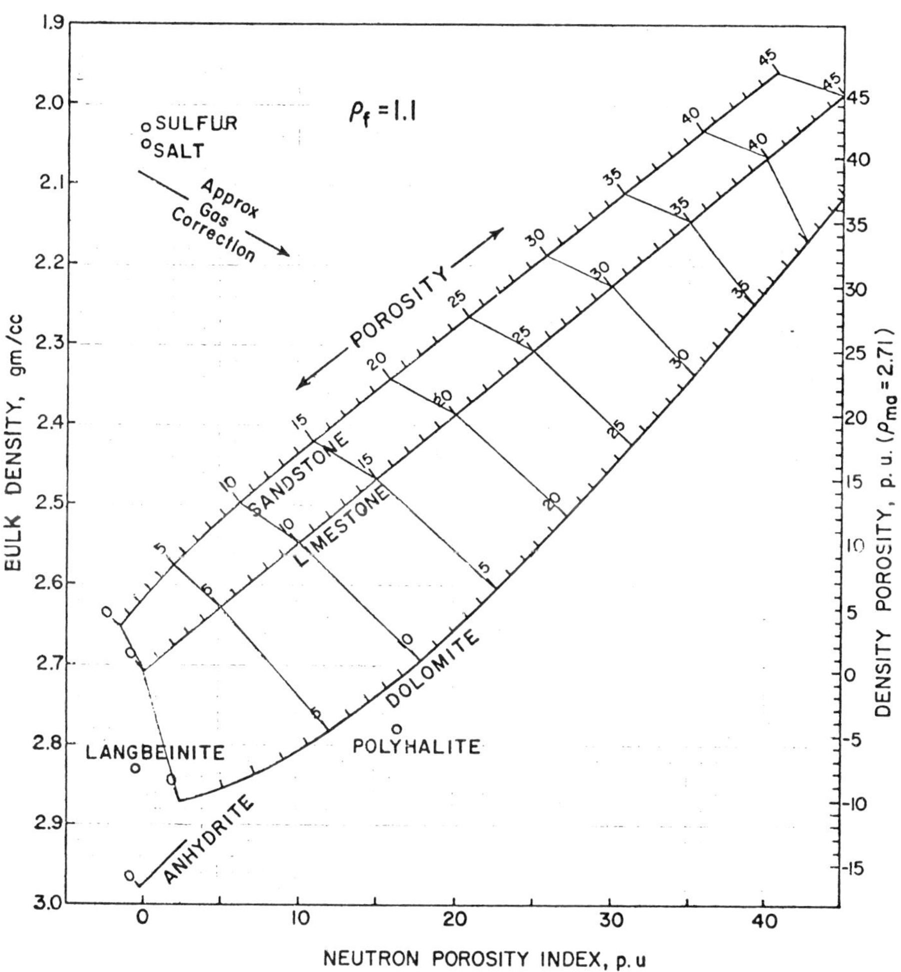

(Modified after Schlumberger Charts, 1978)

FIG. 10. A porosity-lithology cross plot for bulk density and neutron porosity measurements.

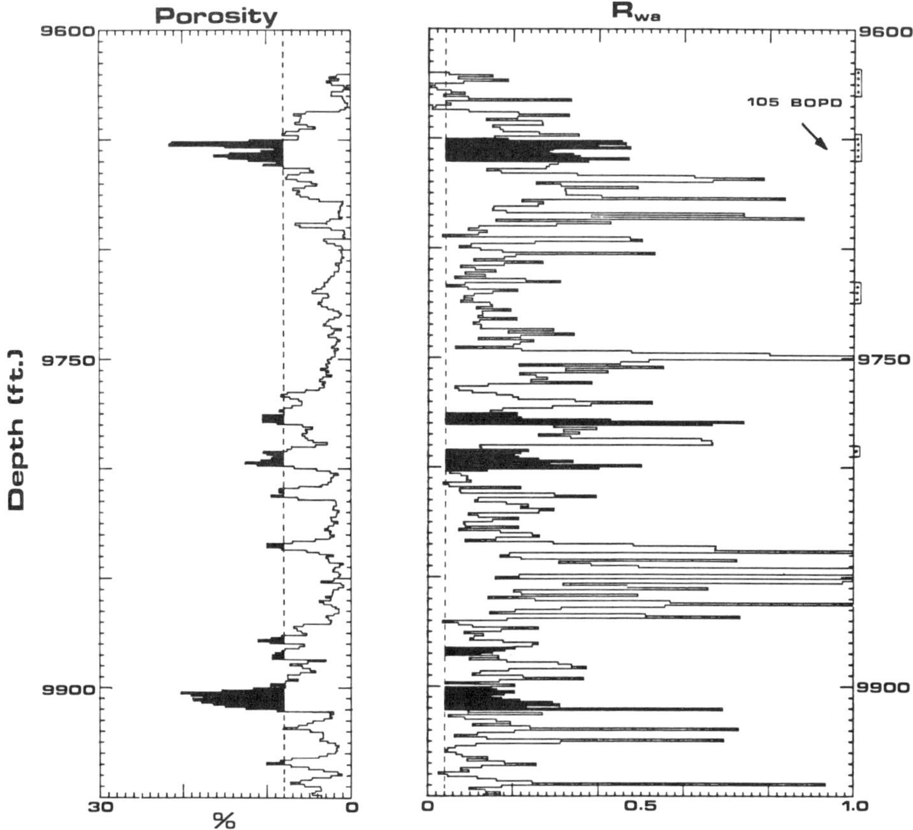

FIG. 11. The porosity-R_{wa} curves for the Mississippian carbonates in the Little Knife field in the Williston basin.

50 percent water saturation, which can be highlighted for a closer evaluation along with the other exploration variables.

The depth of invasion curve can be a reasonable "first look" indicator of the reservoir quality under proper logging conditions. However, more appropriate borehole data should be used for determining the permeability of reservoir rocks, especially when the porosity and fluid conditions are favorable.

The next example is for Mississippian age carbonate rocks from the Little Knife field in the Williston basin. Figure 11 shows the ϕ_T and R_{wa} curves, in which the zones with $\phi_T > 8$ percent and $R_{wa} > 4R_w$ are shaded to highlight favorable reservoir and fluid conditions. The uppermost shaded interval was tested and produced 105 BOPD. The potential of some of the other shaded zones should be investigated further.

Rock types can be identified in a variety of ways. Besides the common sand/shale dichotomy by the use of SP and/or GR curves, another simple method is to plot, as a function of depth, the apparent matrix density $(\rho_{ma})_a$ derived by cross plotting the density and neutron values on a chart, as shown in Figure 12. A plot of $(\rho_{ma})_a$ as a function of depth is shown in Figure 13 for the Williston basin data mentioned above; the matrix densities of sand, limestone, dolomite, and anhydrite

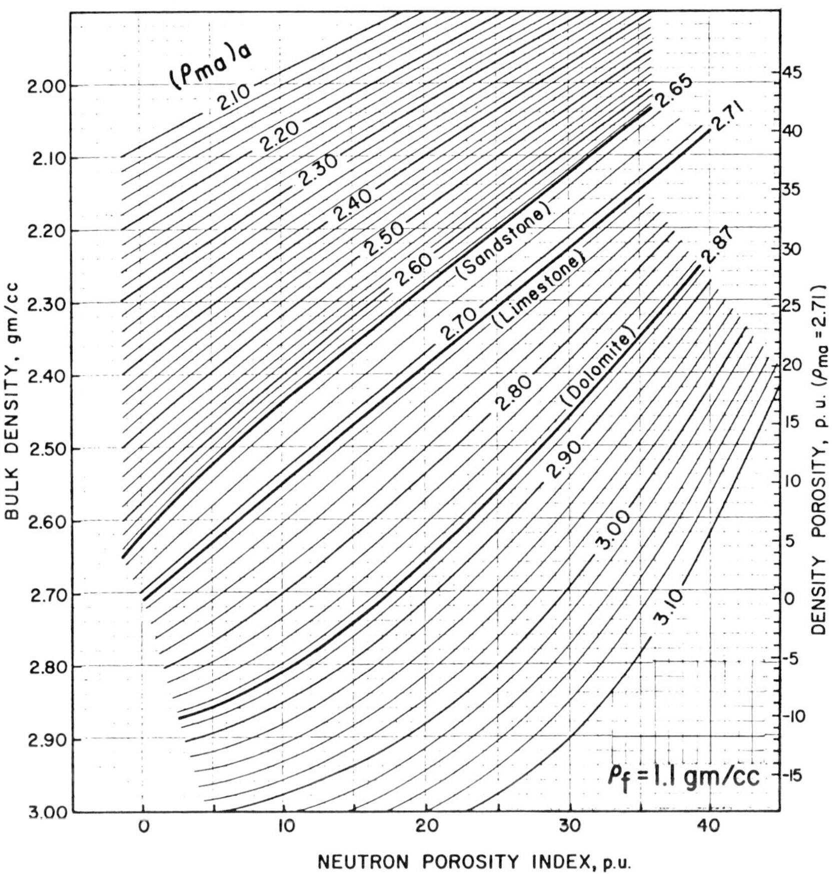

(MODIFIED AFTER SCHLUMBERGER CHARTS, 1978)

FIG. 12. Apparent matrix density from a cross plot of bulk density and neutron porosity measurements.

are also shown as dashed lines. This method works well for single mineral rocks. An improvement of this approach is the MID plot (Clavier and Rust, 1976) in which apparent matrix density values plotted against apparent matrix transit time add an extra dimension for rock type dichotomy.

Similar to the MID plot, another pattern recognition approach in which 3-D data are nonlinearly mapped onto a 2-D plot is the *M-N* plot (Burke et al, 1969). In this case

$$M = \frac{t_f - t}{\rho_b - \rho_f} \times 0.01,$$

and

$$N = \frac{(\phi_n)_f - \phi_n}{\rho_b - \rho_f},$$

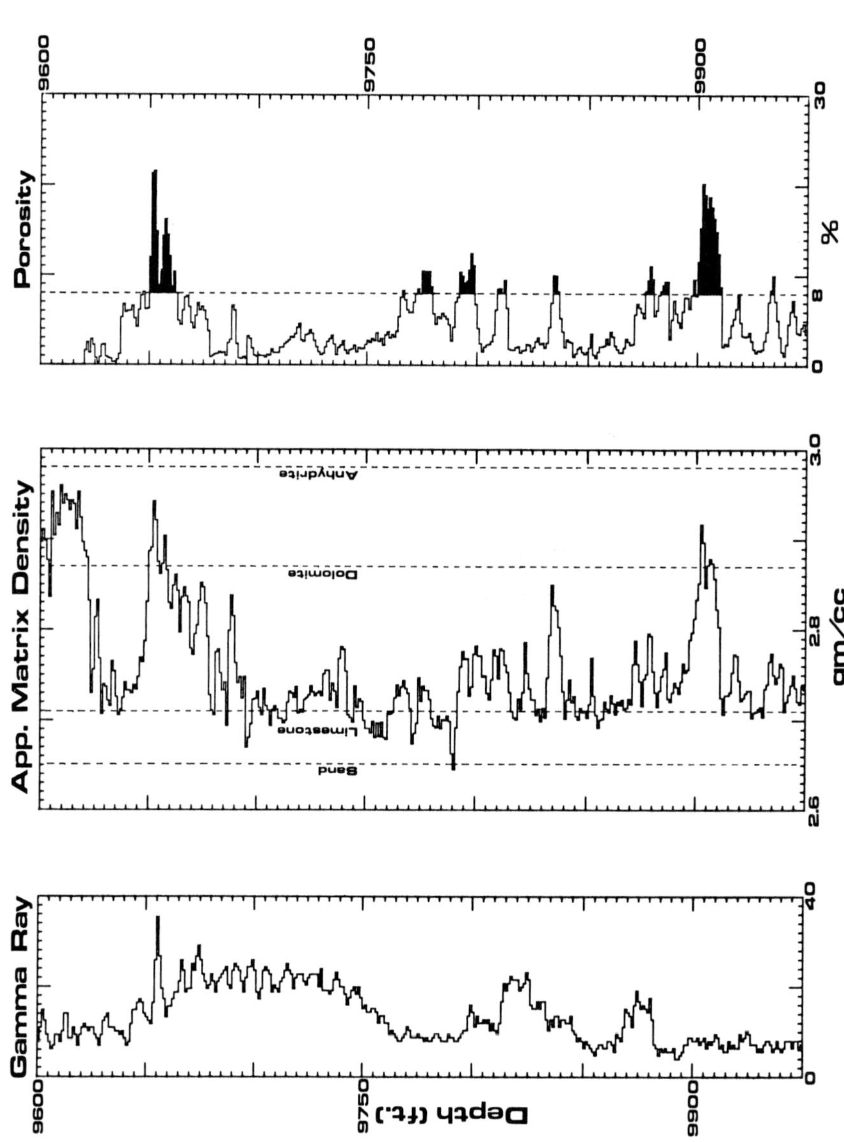

FIG. 13. Reservoir characteristics for a portion of the Mississippian carbonates in the Little Knife field.

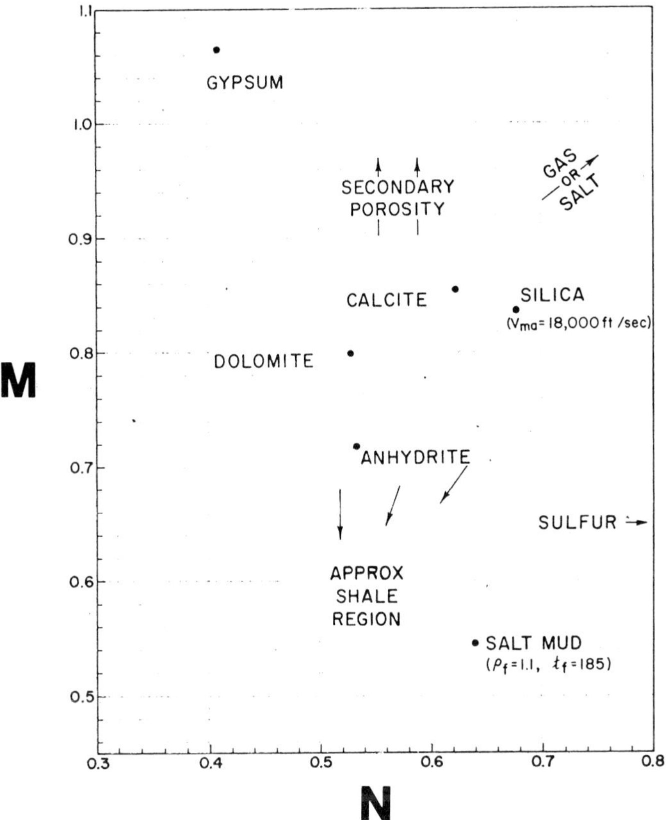

(MODIFIED AFTER SCHLUMBERGER CHARTS, 1978)

FIG. 14. The *M-N* plot for lithology determination from three porosity curves.

where t, ρ_b, and ϕ_n represent three porosity curves: acoustic transit time, bulk density, and neutron porosity. The subscript f denotes the fluid in the rock matrix.

For clean minerals like silica, calcite, dolomite, anhydrite, salt, and gypsum, the M and N values for a given fluid can be computed (Figure 14). Certain rock type patterns emerge when the *M-N* values from actual log measurements are plotted. For the Williston basin data shown in Figures 11 and 13, the *M-N* plot (Figure 15) is a cluster of points around anhydrite and limestone, indicating a predominance of these minerals. Another plot of these data (Figure 16) shows the *M-N* values as functions of depth along with the values of anhydrite, dolomite, and calcite dashed in as straight lines. The latter plot provides a better definition of rock types in comparison with the apparent matrix density plot shown in Figure 13. This is particularly true for dual-mineral rock types.

Data Display

An important consideration for digital well log processing is the ease with which large tasks of data manipulation can be handled for both computation and analysis. Imaginative displays can highlight the areas of change and provide a means for in-

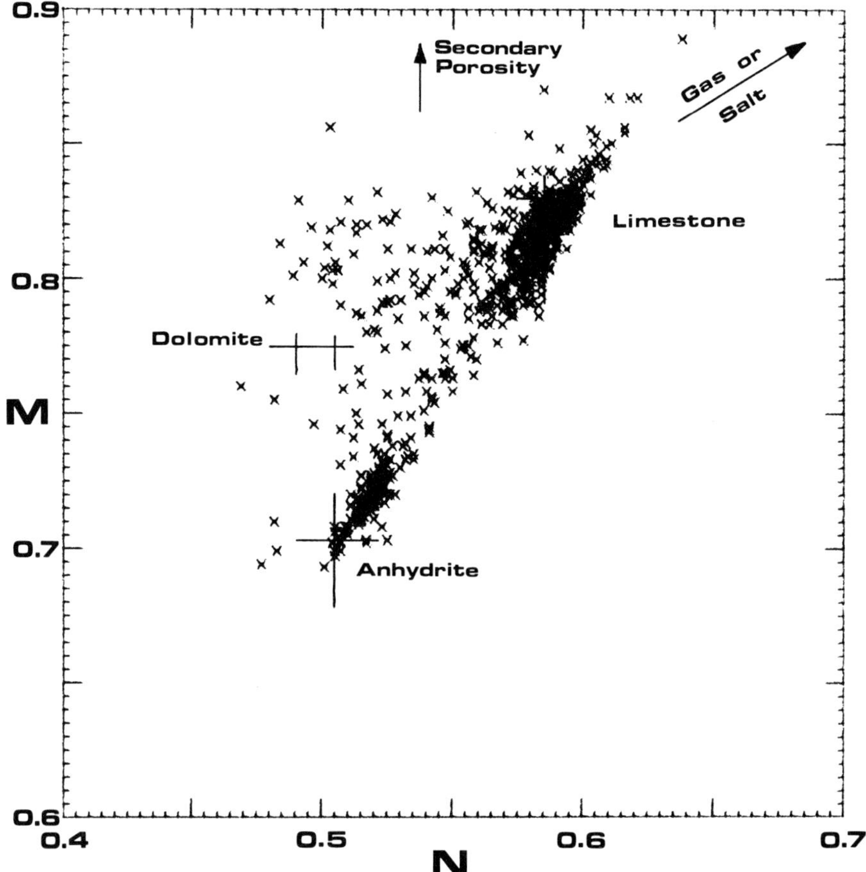

FIG. 15. The *M-N* cross plot of the data shown in Figure 13 indicates that this segment of rocks is mostly limestone and anhydrite.

terpreting a large bulk of data. Such displays are common for seismic data, where standardized sections are displayed in a variety of ways for a quick scan of anomalies.

Traditionally, as shown in Figure 17, geologic cross-sections are made from a series of well logs selected to portray a profile of the subsurface. The logs are photographically reproduced to desired scales and attached to a large sheet of paper to form a cross-section. These logs are stratigraphically correlated and interpreted for such things as structure, depositional environment, and lithology. In this interpretational effort, other subsurface data from core, drill cuttings, and drill stem tests are incorporated. Frequently, the quality of optically reproduced logs is quite poor, which makes the interpretation task much harder. Also, the logs in a cross-section may not be of the same type; the neutron porosity log may of necessity have to be correlated to sonic or electrical logs because of unavailability of the same types. To compound the problem, the scales may be different.

Such complications can be eliminated to a great extent when the displays are made from digitized well data. Geologic cross-sections are, in a way, highly subjective summaries of an explorationist's interpretation of the subsurface data, and it may

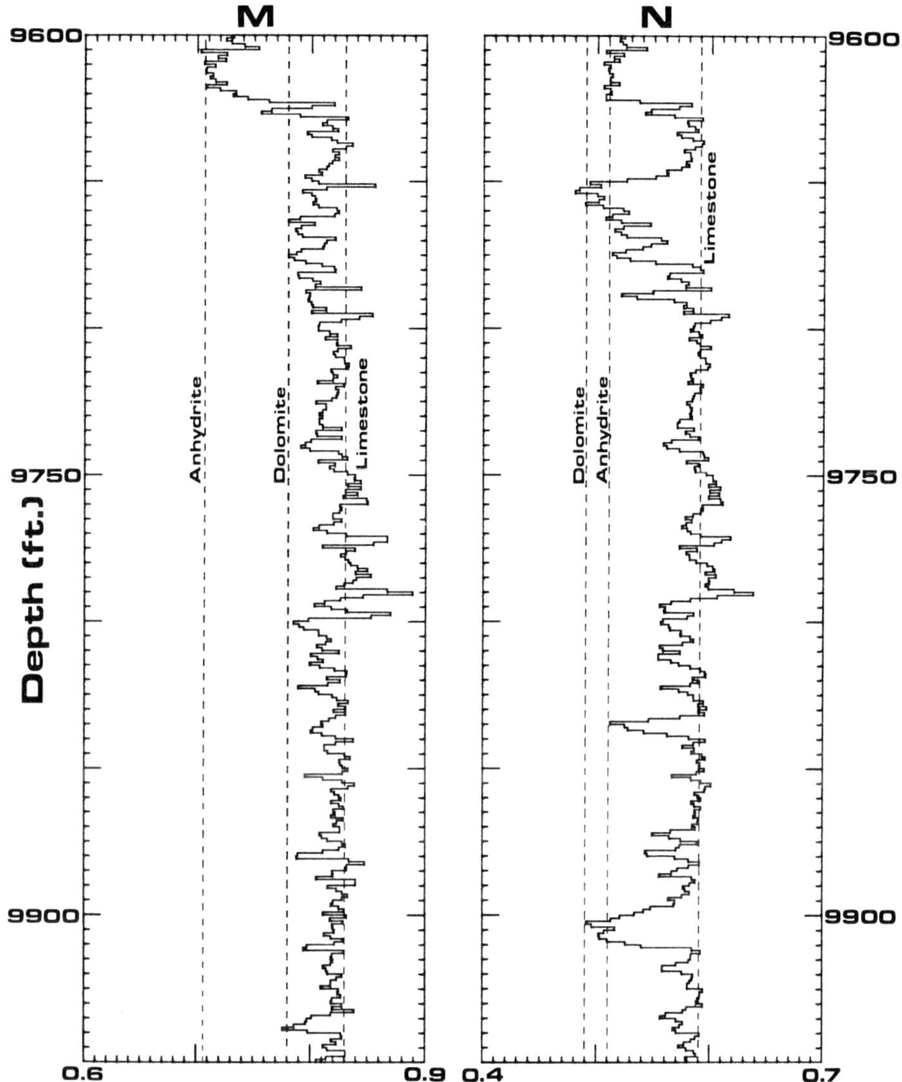

FIG. 16. The *M-N* parameters are plotted to show actual lithology as a function of depth.

not be desirable to change this to a totally automated and mechanized set of plots. Any change from a traditional cross-section to a computerized display should be accepted by the explorationist. The changes made in the general layout of the cross-section should be minimal at first. In time, the cross-sections can be upgraded step-by-step. Thus, starting from optically reproduced traditional plots, one can graduate through computers to the cross-sections made of the exploration variables, such as the one shown in the next example.

Figure 18 shows the same cross-sections of wells used in Figure 17, except that in this example the previously used compensated neutron formation density logs were

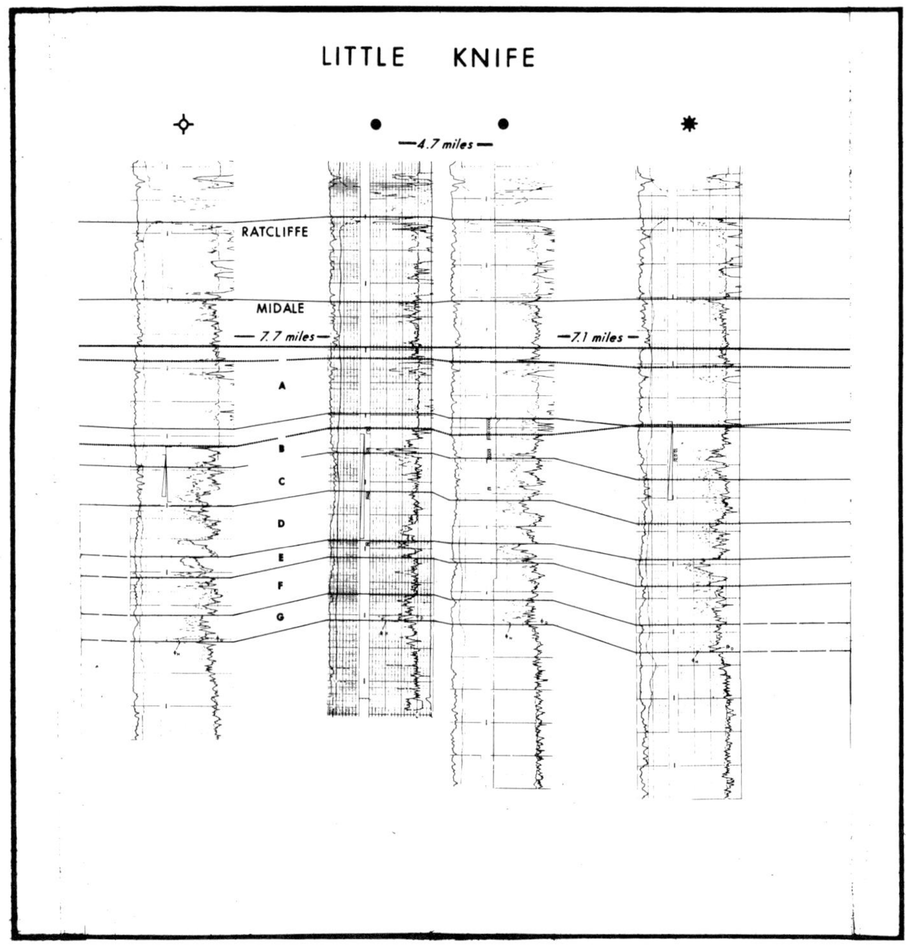

FIG. 17. A traditional regional cross-section showing correlation of the Mississippian formation in the Williston basin.

replaced by exploration variables computed from the measured data. The variables shown in Figure 18 are R_{wa} and total porosity, with the rock type annotated in the middle column. In this type of computerized display, not only is the optical quality superior (compare with Figure 17), but the geologic interpretation is much more direct since reservoir description, quality, and fluid saturation are directly described and need not be inferred. This can make the explorationist's task much less tedious. By appropriately shading the zones of good reservoirs (for example, $\phi_T > 8$ percent) and favorable fluid conditions ($R_{wa} > 4R_w$), one can quickly identify the zones which merit further review.

Figure 19 shows another cross-section where a number of wells from Williston basin are portrayed at a field scale detail. For each well column, the traditionally used gamma-ray and interval transit time curves are plotted along with the exploration variables ϕ_T and R_{wa}. The porosity curves are highlighted by yellow color for

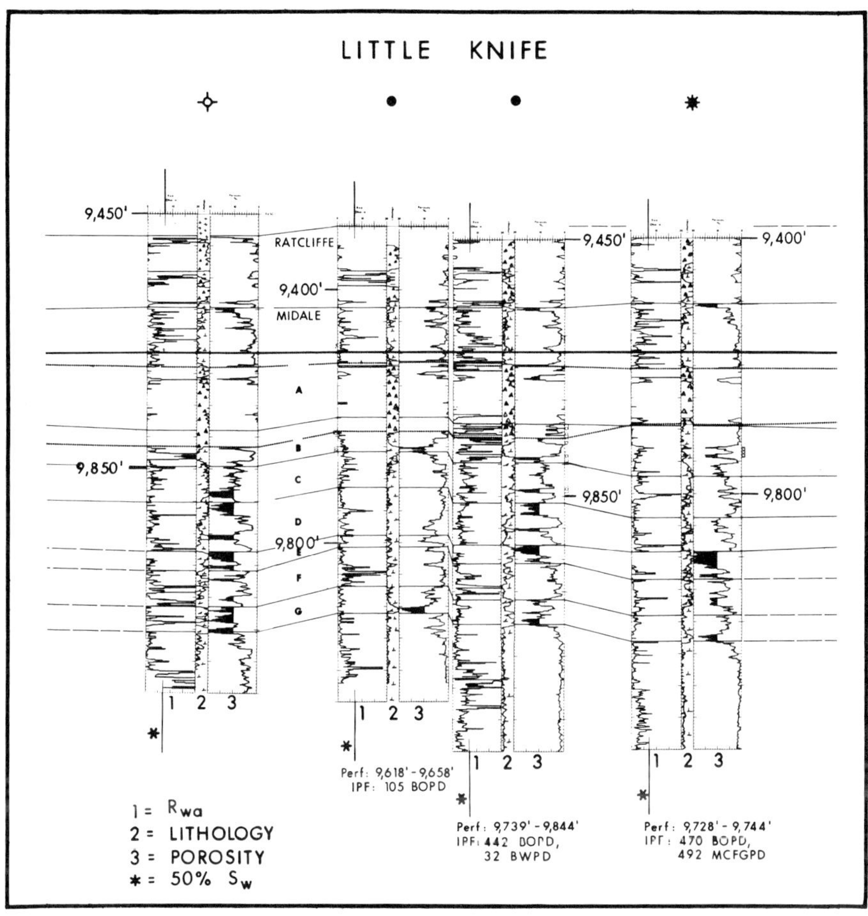

FIG. 18. A computer derived cross-section for comparison with Figure 17. This cross-section provides information about the rock characteristics and fluids directly, making the interpretational task of an explorationist much easier.

$\phi_T > 8$ percent, whereas the R_{wa} curve is highlighted in green only when both $\phi_T > 8$ percent and $R_{wa} > 4R_w$. Thus, the shaded R_{wa} curve points to potential pay zones. Some of the zones have already been confirmed by production as indicated by the well completion symbol to the left of each well column and by the production data shown below each well column. The untested zones should be evaluated further.

The two examples above are just a glimpse of the myriads of possible displays which are useful in petroleum exploration studies. Once the digitized data are in the computer, not only can the cross-sections of many different shapes and sizes be obtained quickly, but a variety of structure and isopach maps, trend curves, histograms, computation of net reservoir thicknesses, etc., can be prepared easily. This can enable explorationists to assimilate and analyze large amounts of data quite effectively.

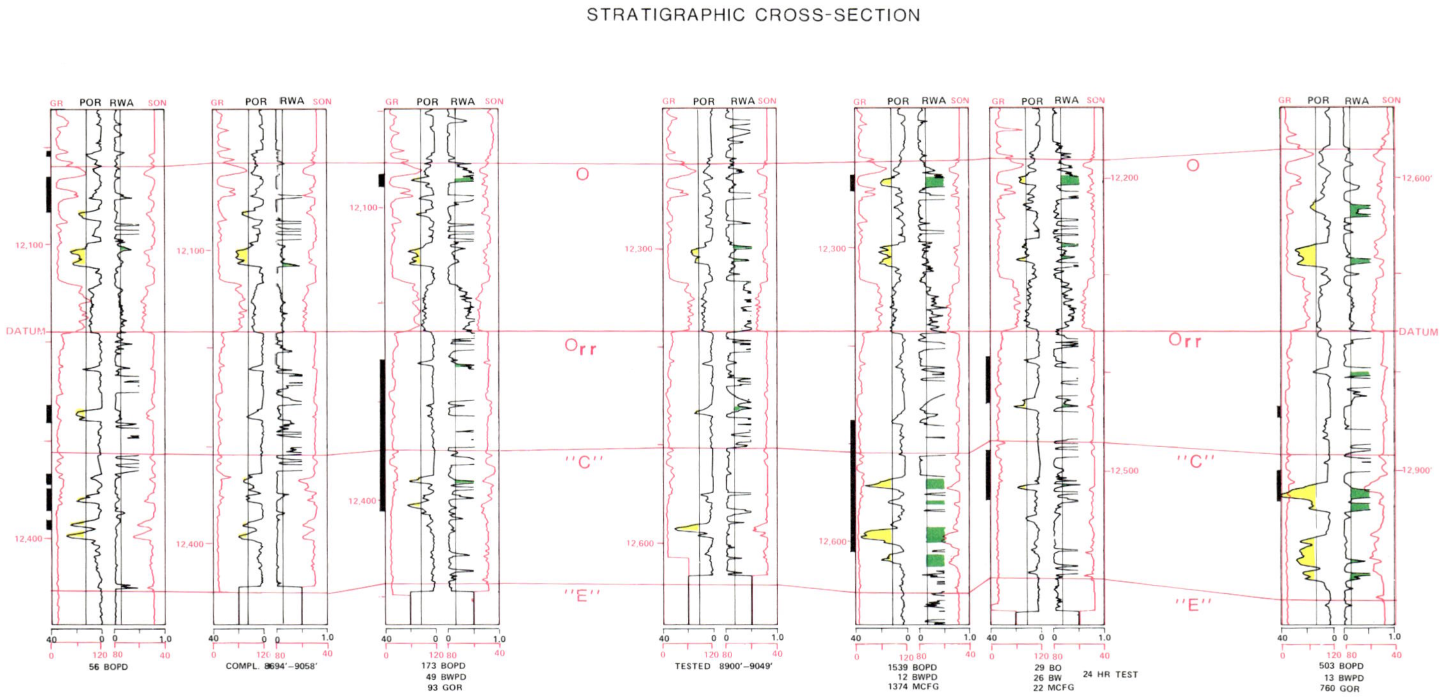

FIG. 19. A field scale stratigraphic cross-section of the Ordovician formation in the Williston basin. The curves (GR and sonic) used in traditional cross-sections as well as the computed curves are shown. The production data and perforated zones (shown as solid black columns to the left of each well) are also shown.

Summary

Petroleum exploration requires searching for geologic anomalies signifying changes in rocks, pore-fluids, and structural setting and relating the changes to the habitats of hydrocarbons. The search involves geologic reconnaissance using both surface and borehole measurements. Well logs are measurements of rock properties in the borehole which provide a high-resolution, unambiguous means for identifying changes in the subsurface.

The uses of well logs range from small-scale field studies to large-scale regional studies. Explorationists use well logs qualitatively as stratigraphic correlation devices and gross lithology indicators. In contrast, petrophysicists analyze a broad spectrum of logs quantitatively, but they normally limit their investigation to reservoir rocks. Using digital computers, the thoroughness of petrophysical analysis can be extended to the entire penetrated section and to all wells of interest in the study area. The computed variables can be displayed to highlight the areas of change either vertically or laterally. Displays of this type can aid explorationists in analyzing a large bulk of processed data and in creatively translating the results into petroleum prospects.

Physical and physico-chemical properties of rocks measured in the borehole by electrical, nuclear, and acoustic tools are indirect measures of the geologic variables. The necessary transformations tend to be empirical; they can be effectively upgraded by using modern signal processing and pattern recognition techniques developed for applications such as seismic, radar, speech recognition, and weather forecasting. Exploration for subtle hydrocarbon pools will require full use of the available information, including well logs. This would necessitate imaginative adaptation of known techniques as well as innovations specific to log analysis.

REFERENCES

Burke, J. A., Campbell, R. L., Jr., and Schmidt, A. W., 1969, The litho-porosity cross plot: SPWLA 10th Annual Logging Symp., May 25–28.

Clavier, C., and Rust, D. H., 1976, Mid-plot: A new lithology technique: The Log Analyst, November–December, p. 16–24.

Kwon, B. D., 1977, Spectral analysis of geophysical logs for correlation: Ph.D. thesis, Indiana Univ.

Neinast, G. S., and Knox, C. C., 1973, Normalization of well log data: SPWLA 14th Annual Logging Symp., Proc., May 6–9, paper I.

REFERENCES FOR GENERAL READING

Alger, R. P., 1957, Logging trends in carbonate rocks: The Mines Magazine, October, p. 97–100.

Alger, R. P., Raymer, L. L., Jr., Hoyle, W. R., and Tixier, M. P., 1963, Formation density log applications in liquid-filled holes: J. Petr. Tech., March, p. 321–332.

Allen, D. R., 1975, Identification of sediments—Their depositional environment and degree of compaction—from well logs, *in* Compaction of coarse-grained sediments, I: G. V. Chilingarian and K. H. Wolf, Eds., p. 349–401.

Archie, G. E., 1942, The electrical resistivity log as an aid in determining some reservoir characteristics: Petr. Trans., January, p. 54–62.

——— 1952, Classification of reservoir rocks and petrophysical considerations: AAPG Bull., v. 36, p. 278–298.

Arnoult, E. B., Jr., 1974, Logical considerations in well log digitizing: The Log Analyst, March–April, p. 3–17.

Banthia, B. S., 1977, Technical note: Some practical aspects of log interpretation of gas bearing formations: The Log Analyst, March–April, p. 39–41.

Bateman, R. M., 1977, The fluid identification plot: SPWLA 18th Annual Logging Symp., June 5–8.

Bertrand, M. A., 1967, Porosity and lithology from logs in carbonate reservoirs: SPE of AIME 42nd Annual Fall Meeting Preprint, October 1–4, no. SPE 1867.

Best, D. L., 1978, A computer-processed wellsite log computation: SPWLA 19th Annual Logging Symp., June 13–16.

Breitenback, E. A., 1969, An interpretative log analysis computer system: SPE of AIME 44th Annual Fall Meeting preprint, September 28–October 1, no. SPE 2574.

Campbell, R. L., Jr., 1972, Recent advances in log evaluation: World Petroleum, February, p. 22–30.

Clavier, C., 1977, The theoretical and experimental bases for the "dual water" model for the interpretation of shaly sands: SPE 52nd Annual Fall Tech. Conf. and Exhibition, October 9–12.

Doll, H. G., Dumanoir, J. L., and Martin, M., 1960, Suggestions for better electric log combinations and improved interpretations: Geophysics, v. 25, p. 854–882.

Fertl, W. J., 1978, R_{wa} method: Fast formation evaluation: Oil and Gas J., September 11, p. 73–76.

Fertl, W. J., and Hammack, G. W., 1971, A comparative look at water saturation computations in shaly pay sands: SPWLA 12th Annual Logging Symp., May 2–5.

Fons, L., 1969, Geological applications of well logs: SPWLA 10th Annual Logging Symp., May 25–28.

Goetz, J. F., Prins, W. J., and Logar, J. F., 1977, Reservoir delineation by wireline techniques: The Log Analyst, September–October, p. 12–40.

Hallenburg, J. K., 1973, Interpretation of gamma-ray logs: SPWLA 14th Annual Logging Symp., May 6–9.

Head, M. P., 1977, Wellsite computer analysis: A program for complex lithologies: SPE of AIME 52nd Annual Fall Tech. Conf. and Exhibition, October 9–12.

Kithas, B. A., 1976, Lithology, gas detection, and rock properties from acoustic logging systems: SPWLA 17th Annual Logging Symp., June 9–12.

Knoring, L. D., and Dech, V. N., 1971, The determination of porosity from geophysical exploration data: Doklady Akad. Nauk SSSR, v. 202, p. 819–822.

Kokesh, F. P., 1951, Gamma-ray logging: Oil and Gas J., July 26.

Krug, J. A., and Cox, D. O., 1976, Shaly sand cross-plot: A mathematical treatment: The Log Analyst, July–August, p. 11–15.

Labo, J., 1978, A practical introduction to borehole geophysics: Presented at the 48th Annual International SEG Meeting, November 1, in San Francisco.

Laptev, V., 1978, Evaluation of oil wells by logging methods in the U.S.S.R.: SPWLA 19th Annual Logging Symp., June 13–16.

Leeth, R., 1978, Log interpretations of shaly formations using the velocity ratio plot: SPWLA 19th Annual Logging Symp., June 13–16.

Lindley, R. H., 1961, The use of differential sonic-resistivity plots to find movable oil in Permian formations: J. Petr. Tech., August, p. 749–755.

Lindseth, R. O., 1969, A large volume computer system for digitized well log data: SPE of AIME 44th Annual Fall Meeting Preprint, September 28–October 1, paper no. SPE 2575.

Martin, M., and Dumanoir, J. L., 1956, Determining true resistivity: World Oil, July, p. 95–106.

Martin, M., Murray, G. H., and Gillingham, W. J., 1938, Determination of the potential productivity of oil-bearing formations by resistivity measurements: Geophysics, v. 3, July, p. 258–272.

Masters, J. A., 1978, Deep basin gas trap, Western Canada: AAPG Bull., v. 63, February, p. 152–181.

Mayer, C., and Sibbitt, A., 1980, Global, a new approach to computer-processed log interpretations: SPE of AIME 55th Annual Fall Tech. Conf. and Exhibition, September 21–24.

McCoy, R. L., and Smith, R. F., 1979, The use of interlog relationships for geological and geophysical evaluations: SPWLA 20th Annual Logging Symp., June 3–6.

Merkel, R. H., MacCary, L. M., and Chico, R. S., 1976, Computer techniques applied to formation evaluations: The Log Analyst, May–June, p. 3–10.

Morris, R. L., and Biggs, W. P., Using log derived values of water saturation and porosity: Schlumberger Corp., Houston.

Pickett, G. R., 1970, Applications for borehole geophysics in geophysical exploration: Geophysics, v. 35, p. 81–92.

Porter, C. R., and Carothers, J. E., 1971, Formation factor—Porosity relation derived from well log data: The Log Analyst, January–February, p. 16–26.

Poupon, A., Clavier, C., Dumanoir, J., Gaymard, R., and Misk, A., 1970, Log analysis of sand-shale sequences—A systematic approach: J. Petr. Tech., July, p. 867–881.

Poupon, A., Hoyle, W. R., and Schmidt, A. W., 1970, Log analysis in formations with complex lithologies: SPE of AIME 45th Annual Fall Meeting Preprint, October 4–7, paper no. SPE 2925.

Ransom, R. C., 1977, Methods based on density and neutron well-logging responses to distinguish characteristics of shaly sandstone reservoir rock: The Log Analyst, May–June, p. 47–61.

Revett, L. W., 1976, Comparison of permeabilities from logs and sidewall cores to predict production: SPWLA 17th Annual Logging Symp., June 9–12.

Riboud, S., 1971, Well logging techniques: Proc. of the 8th World Petr. Congress, review paper 7.

Rivero, O. G., 1976, A practical method for determining cementation exponents and some other parameters as an aid in well log analysis: The Log Analyst, September–October, p. 8–24.

Schlumberger, 1972, Log interpretation, Volume I—Principles: Houston, TX.

——— 1974, Log interpretation, Volume II—Applications: Houston, TX.

——— 1978, Log interpretation charts: Houston, TX.

Thomas, D. H., 1977, Seismic applications of sonic logs: The Log Analyst, February, p. 23–38.

Throop, W. H., Hammar, R. F., and Tinch, D. H., 1977, Localized processing of well-log data: SPWLA 18th Annual Logging Symp., June 5–8.

Tittman, J., and Wahl, J. S., 1965, The physical foundations of formation density logging (gamma-gamma): Geophysics, v. 30, p. 284–294.

Tixier, M. P., 1962, Modern log analysis: J. Petr. Tech., December, p. 1327–1336.

Tixier, M. P., Alger, R. P., and Tanguy, D. R., 1960, New developments in induction and sonic logging: Petr. Trans. of AIME, v. 219, p. 362–370.

Tixier, M. P., Alger, R. P., Biggs, W. P., and Carpenter, B. N., 1963, Dual induction-laterolog: A new tool for resistivity analysis: SPE of AIME 38th Annual Fall Meeting Preprint, October 6–9.

Truman, R. B., Alger, R. P., Connel, J. G., and Smith, R. L., 1972, Progress report on interpretation of the dual spacing neutron log (CNL) in the U.S.: SPWLA 13th Annual Logging Symp., May 7–10.

Vajnar, E. A., Kidwell, C. M., and Haley, R. A., 1978, Low resistivity masks high potential of Gulf Coast sands: World Oil, February 1, p. 49–54.

Wahl, J. S., Tittman, J., and Johnstone, C. W., 1964, The dual spacing formation density log: J. Petr. Tech., December, p. 1411–1416.

Waldschmidt, W. A., Fitzgerald, P. E., and Lunsford, C. L., 1956, Classification of porosity and fractures in reservoir rocks: AAPG Bull., v. 40, May, p. 953–974.

Winsauer, W. O., Shearin, H. M., Jr., Masson, P. H., and Williams, M., 1952, Resistivity of brine-saturated sands in relation to pore geometry: AAPG Bull., v. 36, February, p. 253–277.

COMPUTERS AND TECHNOLOGY

Chapter 8

GRAPHIC DATA BASES

The Information Problem

The information accumulated by the exploration industry is rapidly increasing. The problem of storage and handling of this voluminous data is rapidly growing as more information is gathered and processed for interpretation. During 1977, for example, the industry recorded an estimated 10^{14} to 10^{15} bits of seismic data in digital form. Packed at 1600 bpi in the most efficient form, these data occupy nearly three million magnetic tapes, each 2400 ft long.

The seismic data recorded on these tapes are only the starting point in a series of highly sophisticated processes that culminate in graphic displays such as maps, cross-sections, and charts. These graphic displays are visually interpreted to map the subsurface geologic conditions existing in the geographic area represented by the data.

In this chapter, we discuss a graphics approach to exploration data base management. Also, we present some aspects of exploration activity in the perspective of an electronic data processing (EDP) exercise.

The Seismic Data

Acquiring and processing seismic information is only one facet of exploration activity; however, this should serve as an example of the procedures through which all other data advance in the process of becoming a meaningful resource to the exploration industry. The procedures outlined below and shown in Figure 1 concern a marine seismic operation.

Acquisition (source data)

During an average day, a marine seismic survey vessel records on magnetic tapes approximately 300 megabytes of new data. These data reference 2000 discrete locations, with each location involving several data types and subreferencing 48 or more additional locations. For reasons associated with cost and confirmation, these raw data must be retained, stored, and referenced indefinitely. Therefore, organization of these data in a manner allowing additional input, retrieval, and storage is imperative.

Processing (intermediate information)

Many reduction and analytical processes are performed upon the raw data at a data processing center. These processes result in approximately six megabytes of "intermediate information" for every 300 megabytes of new data acquired by a seismic survey. This intermediate information is stored in the forms of digital data and photographed cross-sections, all of which are referenced to the source data and archived for storage and retrieval purposes.

Interpretation (stored data)

The processed data are next interpreted to correlate the observed continuities (seismic reflection horizons) with geologic structures. Additional signal processing, modeling, and/or reprocessing of source data may follow this initial interpretation.

The division between processing and interpretation is becoming increasingly hazy. However, in terms of data handling, the above procedures produce some 50 kilobytes of derived data existing as mapped information and statistical attributes.

The information at this point includes, at a minimum, the seven reference notations listed below.

(1) Coordinates of a minimum of 200 locations.
(2) Identifiers to the surface location and associated geologic beds.
(3) Recording instrument characteristics and data processing parameters required as confirmation to quality of derived data.
(4) Local water depth and relative gravity.
(5) Indicators of geologic fault location and reliability grading of the observed information.
(6) Numerical values of observed time and depth.
(7) Statistical estimates of local velocity, frequency, and amplitude variation.

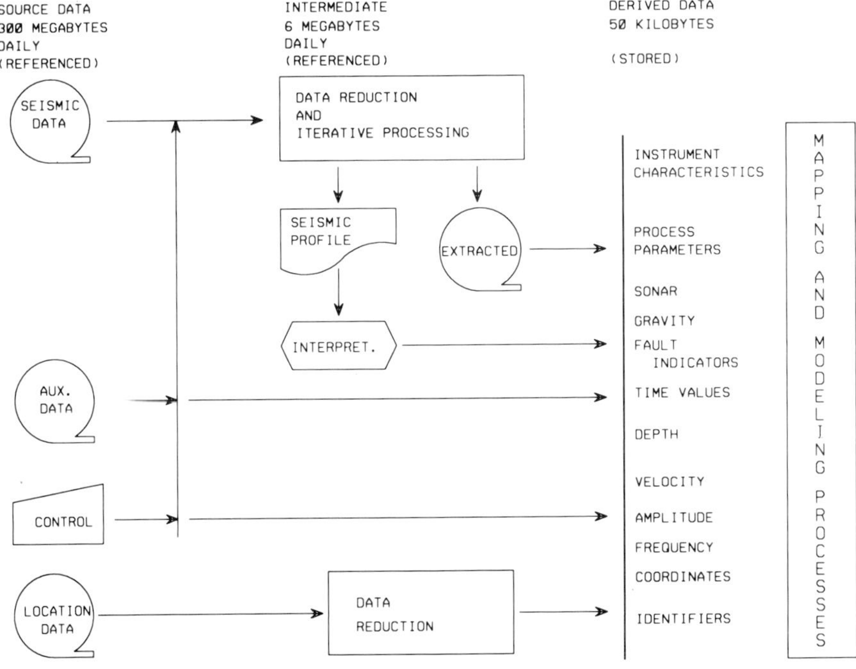

FIG. 1. Seismic data.

In EDP terminology, the processed and interpreted seismic data are considered to be "stored data." However, only initial interpretations have been made, and, as a data set, these stored data are upgraded as further information is acquired and processed. Ultimately, the seismic data are integrated with data from other sources, such as well logs, to produce a comprehensive interpretation.

Analysis of the seismic case

The sequence of events outlined above presents a view of typical activities within the exploration industry. In terms of data handling, this sequence is geared to the production level of an acquisition crew and the volume and types of information produced from the source data.

The EDP exercise illustrated above is based on 1977 production levels and processing techniques and does not truly represent the real problem of data management in the exploration industry today. Since 1977, in response to economic stimuli, the industry has changed gears. Small and subtle traps are sought, and more positive evidence of structure is required; hence, research into seismic hydrocarbon indicators has been accelerated.

Three developments have had the greatest influence on EDP activity within the exploration industry:

(1) New seismic instrument designs are allowing, with a relatively small increase in time and expenditure, acquisition of data at double or triple the volume mentioned above.

(2) Signal extraction and modeling techniques have been developed that effectively double the volume of relevant data per unit area.

(3) Three-dimensional acquisition and processing techniques now being adopted literally add another dimension to the reference, addressing, and display of information. Hence, the data base is considerably enlarged.

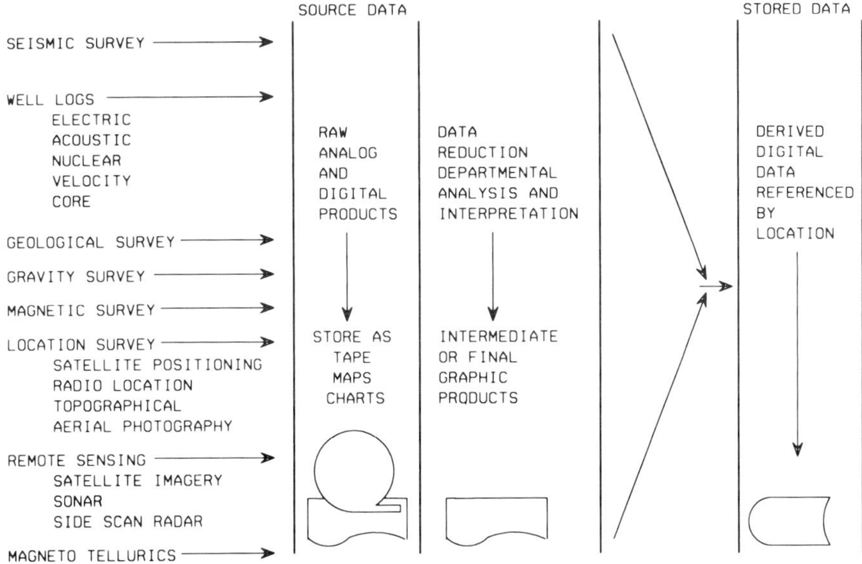

FIG. 2. Exploration data.

Other Exploration Data

The foregoing description of seismic activity illustrates the data flow of only a single facet of the exploration process. A similar set of procedures is undertaken by other earth science disciplines in various segments of the industry (Figure 2). Each acquires, reduces, processes, and interprets data based upon available evidence. Currently each discipline is reacting to the requirements for increased volume and precision.

Seismic departments generally encounter the most voluminous source data; however, the output data from other disciplines are equivalent in volume and importance to seismic data. For example, well logs produce the most positive evidence of subsurface conditions and serve as a base reference for most other types of data. Well log information is gathered from sensor or tool measurements of specific attributes recorded at a well. Sensor measurements produce analog charts or direct digital samplings of physical conditions as a function of depth. To a computer, these source data appear as some 25,000 samples (50 kilobytes) of data per log. Several logs for different specific attributes are recorded at each well, and the reduction and interaction of these measurements produce the final stored data. Remote sensing techniques are becoming increasingly common in the industry; they involve a volume of source data approaching that of seismic acquisition.

The abundance of information produced by the exploration industry is overwhelming both data management techniques and data management budgets. The various departments within the industry are attempting to handle the data management problem individually. However, all the data should ultimately come together for an evaluation process, which requires a different set of rules for data handling. While efficient storage and indexing remain imperative, the goal is to format the diverse sets of data in a manner allowing easy assimilation and interaction, so that the resulting displays can be interpreted effectively.

Integration of the various data sets is referred to as a "local data bank." Historically, the local data bank is a very crowded room containing many map racks, file cabinets, strip charts, section profiles, drafting tables, and overwhelmed explorationists who follow, with some anxiety, the progress of drilling operations resulting from their efforts. The sheer volume of information produced by today's technology prohibits consideration of all evidence without some means of mechanical assistance.

Data Base Management Systems

The problem is no longer simply data management but now includes data assimilation. This two-fold problem has caused petroleum organizations to turn increasingly to the EDP industry for assistance. Adaptation of data base management and interactive graphic systems for the exploration data problem offers a beginning to a management solution.

Data base management systems originally addressed business and commercial applications for data organization and presentation purposes. The data base approach proved successful as an aid to corporation control. In simple terms, a system comprises the five components listed below (Figure 3).

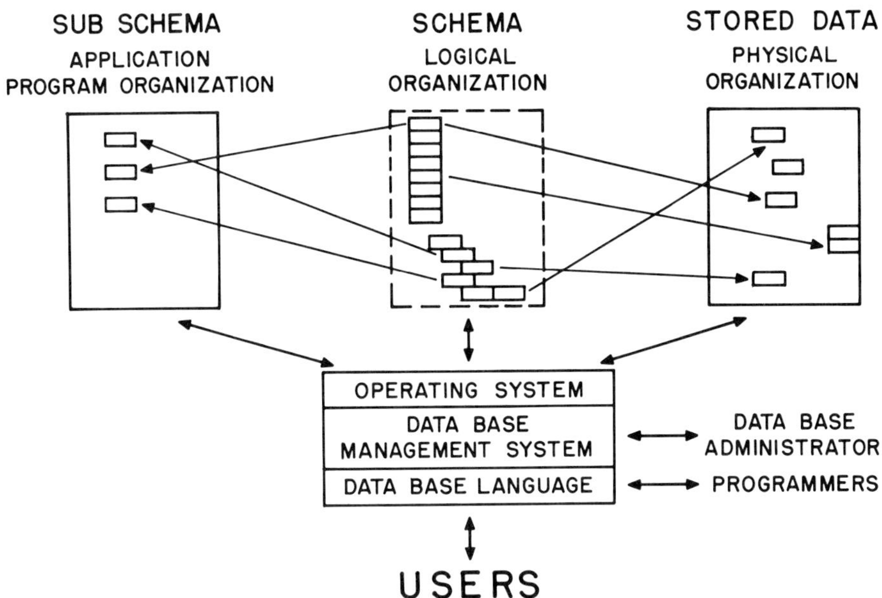

FIG. 3. Data base components.

Source data.—Data or information from which the stored data are derived. Source data may be in either digital or analog form and may be consigned ultimately to an area referenced by the computer, but they will not be entered into the data base.

Stored data.—Information in digital form that is stored, referenced, and indexed within the computer system; they represent a reasoned distillation of the source data. The stored data are often referred to as the data base or the data sets.

Data model (schema).—Organization of the stored data within the data base management system. The data model acts as a road map of the stored data organization.

Subschema.—Organization of the data as it appears to the application programmer or user. This organization resembles the normal filing system of the user.

Data management system.—It is a software system designed to accept and organize the stored data according to an overall plan.

The primary intent of a data base management system is to conceal from the user the mechanics and inherent difficulties of the computer. Hence, the organization of data appears to a specific user as that defined in the subschema, irrespective of the actual physical location of the data within the computer. The management system itself consists of a group of high-level languages developed for specific applications. Some of these languages are described in general terms below.

Data and schema description languages allow the data base administrator to describe the ''logical organization'' of the data; the languages translate this organization into ''physical storage,'' which is irrelevant to the programmer and user.

Application and manipulation languages allow the programmer access to the data for application programs while remaining ignorant of its physical location and/or condition.

Query and report generators retrieve information for routine or unique reports.

Interactive or on-line languages allow English-like conversations between the user and the computer.

Security languages maintain the integrity and safety of the data.

Teleprocessing and other communication languages allow users in distant locations access to the system.

Data dictionary languages are inventories of the data within the data base and are available to the users (people) and the system (software).

Several commercially packaged data base management systems are currently on the market, each with its own file structure, schema, and data definition language. Initial attempts to use these packages for exploration purposes were not successful without modification, primarily because of a failure to recognize the special requirements of exploration data.

In general, four properties characterize exploration data:

(1) Exploration data are gathered by apparently diverse disciplines that produce a set of measurements and observations having the single factor of geographical location in common; highly complex relationships exist within the abundance of information, both factual and deduced, produced by these measurements and observations.
(2) Most often output is required in large-scale graphic form such as maps and charts rather than in textual form such as reports and statistics.
(3) Users require lengthy interaction with the data for the purposes of editing, updating, and modification. This interaction involves fairly large subsets of data and demands regular visual confirmation of an action.
(4) Input devices tend to be highly specialized, involving complex procedures for conversion to digital form.

Another essential requirement, especially for data integrated from several disciplines, is the ability to establish the origin and confirm the veracity of the data presented. The abundance of data generated from the various disciplines may become confused with ''truth'' merely because of the amassed volume. Therefore, it is necessary that the ancestry, or source and development, of the data be apparent and traceable to the appropriate discipline to assure quality.

In designing systems to meet the requirements listed above, the EDP industry has applied major innovations to the file structure and equipment normally used by the classical data base systems. The majority of these innovations arise from adoption of visual addressing by means of graphic systems, which provides a sophisticated form of interaction between the user and the computer.

Graphic Systems

Graphic systems involve the manipulation and display of vector information on the face of a cathode ray tube (CRT). As such, graphic systems have found applications in the construction of visual materials for many years. Currently, graphic systems are beginning to be utilized for a variety of other applications.

A graphic system consists of one or more graphic screens or a single graphic screen supported by an alphanumeric screen. Interaction with the graphic screen is either by cursor manipulation of on-screen cross hairs or by the direct application of "light pens" to the face of the graphic screen. Control instructions are initiated from a menu, or datalog, of symbols and mnemonics placed on a free-moving tablet. The menu area corresponding to the appropriate instruction is activated by a cursor. Recent developments indicate the menu control method may be replaced in the near future by voice entry systems that respond to recorded utterances.

The majority of graphic systems in use are, for economic reasons, of the storage tube type. With storage tube equipment, lines are "painted" on the screen of the CRT by a series of commands indicating the coordinates of the start and stop points for a line. This method of graphic display is in vectors, the lines displayed between points. The benefit of the vector technique is that the coordinates are readily handled by a computer and may be easily translated into map form. This method is relatively slow in producing an image and limited by the number of coordinate pairs required to describe a complex drawing.

Another method recently introduced uses "raster" tubes that allow each point on the screen of the tube to be addressed separately by the computer. Images are formed by changing the degree of brightness at each point. This method allows higher resolution of complex shapes such as seismic traces and produces an image more rapidly than the vector method. However, the image is not based directly upon coordinate information, and hence a translator system is necessary to convert the original vector information into the dot raster display.

Ultimately, both of these types of display equipment may be replaced with the newly emerging equipment, such as gas plasma screens or holographic devices.

Exploration Schema Graphic Files

All information in an exploration data base concerns events that have happened on the earth's surface. These events vary from historical data and property documentation to technical data derived from earth science disciplines. As noted earlier, the only characteristic these events have in common is the geographic location at which they occurred. This information exists either as graphic data in vector form (recognizable by the human eye as maps which allow ready assimilation of large data sets), or as digital data (acceptable to a computer for arithmetic operations). The optimum configuration is a dual file system, with the graphic data in one file for reference purposes and the nongraphic text data necessary to application programs in a separate file. A link must be maintained between the two files to assure cross reference for identification and application purposes.

As shown in Figure 4, the graphic files may be considered a series of overlay maps, each representing a specific exploration department, discipline, or data class. The map file itself addresses a "twin file" containing the text or nongraphic information. Any location within the series of overlays finds a linked address in the text file. A large number of such overlays are normally available and visible to the user.

One or several of these overlays may be used as a general purpose graphic work area for the construction of visual material either by user interaction or generation of location data from a separate program. For construction purposes, data can be obtained from the formal data bank or data base; however, strict control must be

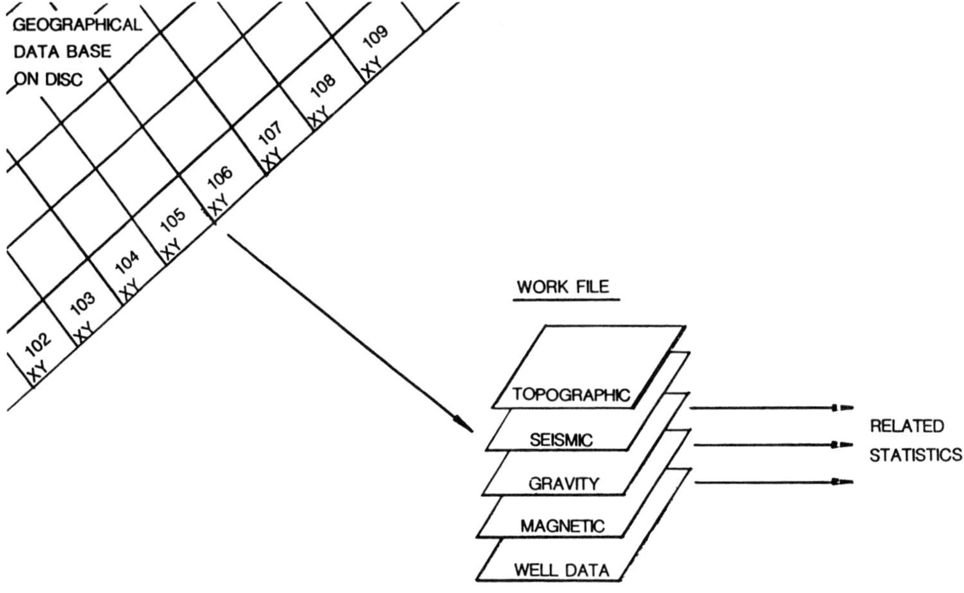

FIG. 4. Basic plat.

exercised over the return of information to the data base. Any data contained within the graphic files may be viewed on the screen as a single map or as a composite drawing consisting of several of the overlay files.

Ultimately some or all of the graphic files may be available to a body of users from various departments or disciplines. Each user may require annotation on the map of specific information in a specific manner to fit department standards. Annotation data are, therefore, maintained within the text file and later converted to graphic displays as directed by the user.

Exploration Schema Text Files

The normal function of text files is to store large quantities of statistics or text so as to allow rapid formatting for generation of printed reports. In the exploration data base, a second, equally important function is communication with and support of the graphic files.

Several techniques are available for file description and handling, normally involving a balance between speed and volume of file accessibility, with selection being a function of the end-use requirement. The exploration industry is a multidiscipline environment, and the most appropriate file description is what is known as a "table driven" system. Under a table driven system, the content of the text file is defined externally. The data are defined by means of a list of items (or a table) described by the user and relating to the discipline. Hence, the statistics or text stored for a location appear as a block of coded numbers which are meaningful only when referenced by the table. This table is available to the user as a menu or maintained as a reference elsewhere in the system. The table driven concept may be extended to a logical conclusion whereby all functions of the data base are controlled by tables. Examples of the tables essential to a data base system are listed below.

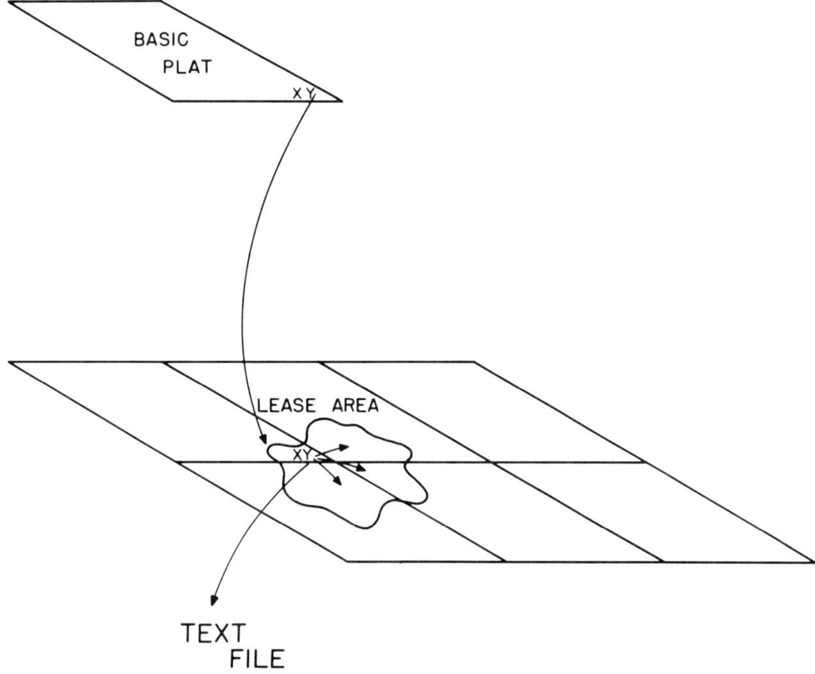

FIG. 5. Adjacent plats.

Symbols All symbols used in the graphic files
Lines—Line thickness and types
Font—Lettering of annotation
Text—Definition of content of specific text file
Annotation—Definition of items required for map posting
Criteria—Logical links between data sets
Form—Specific items required for report generation

Input, output, and reference in a table driven system may suffer slightly in access time. However, several benefits are apparent. The most important benefits are (1) the data base becomes independent of a specific user; (2) a user may tailor output to meet departmental requirements; (3) storage capacity is greatly enhanced by eliminating redundant graphic symbols from storage; (4) information is predefined in tables so control is exercised over entry to the data base, allowing standardization within each department; and (5) the collected tables may be regarded as a data dictionary that defines the content of the data base.

Geographical Index

In defining a logical data base, the map itself is coded to a geographical index. The basic unit of this index is a rectangle (or plat) describing a geodetic projection system and named by the projection coordinates or the reference latitude and longitude (Figure 5). The geographical location is the primary index. By referencing a coordinate and appropriate class, all information relevant to a location may be retrieved. The more familiar departmental indices are retained as secondary references. Hence, an ''area'' is defined as a boundary polygon with associated identifiers. Upon request the polygon is applied to the several plats involved, and the data within the polygon boundaries are retrieved and appear to be stored in that fashion to the user. This same technique is applicable to reservoir or seismic data.

In terms of a hierarchy, the levels of apparent data organization are as indicated in Figure 6.

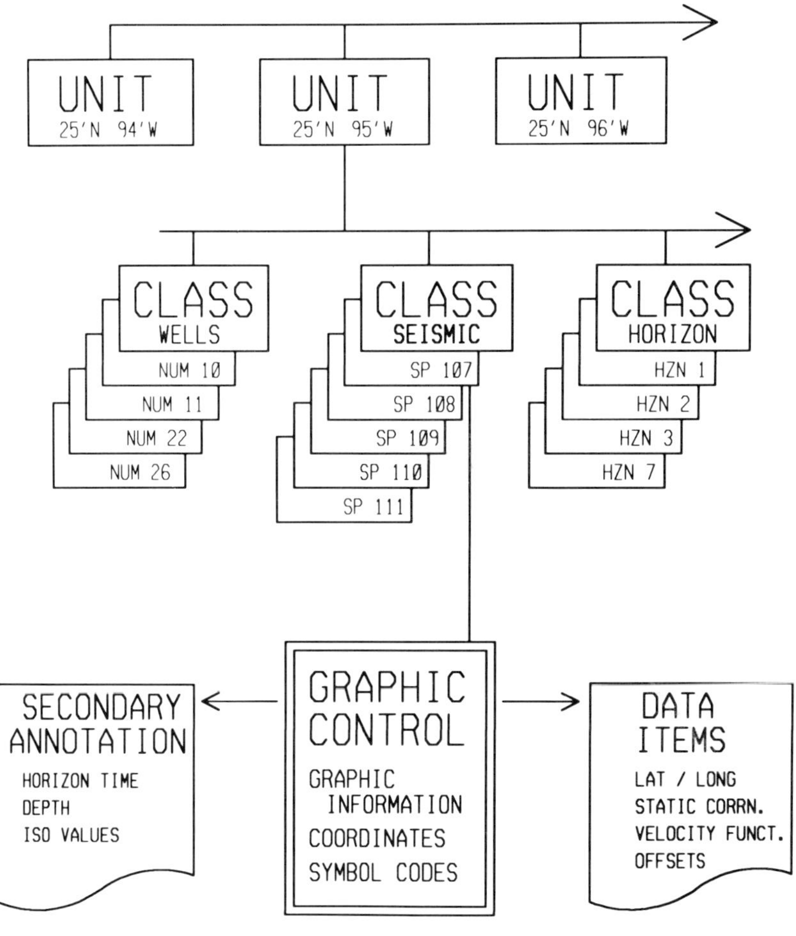

FIG. 6. Hierarchy.

Unit

The highest level of organization in an exploration data base is a file containing information on all locations within a predefined geographical rectangle, indexed by a reference coordinate (either projection or geographical). As an example, if a data base is established using a one degree unit, then unit 10W25N is the name of a file containing information on all locations between longitude 10 to 11 degrees west and latitude 25 to 26 degrees north.

Class

Each *unit* file consists of a series of subfiles representing a specific data classification or *class* of information. Each entry within a class file is a graphic control record containing location coordinates, a symbol code, and a pointer to the appropriate text file. At display time, the symbol code may be referenced to a table of symbols, and the requested symbol may be inserted at the correct location. Note that by use of alternate symbol tables the map may be tailored to the user's specific application or discipline.

Data items

The text file consists of a series of *data items* or statistics associated with the location. This information may be introduced directly into arithmetic operations, annotated upon the appropriate map, or accessed for report generation.

A data base so defined may, in concept, cover the entire world. In a practical sense, it may cover a large operating area and may contain a large amount of data requiring continual maintenance and may be subject to heavy usage.

Control within this type of a data base is exercised through three primary functions.

(1) Data base administrator.—The administrator is the person holding complete responsibility for definition, organization, security, and standards of the system.

(2) Department entry.—The content of the data base must be dictated by department or discipline involved. Hence, the classes of data are assigned to specific departments which hold responsibility for content, entry, and modification of data within the assigned classifications and under the formats defined by the data base administrator.

(3) Work file.—As shown in Figure 7, this is a detached file, the structure of which is identical to the main data base. However, the work file may cover any geographical shape or area specified by the user and contains all classes of information requested by the user.

The file may be an interpreter's work area allowing interaction between the various data sets, application programs, and direct plotting facilities. The file is also the editing area for the input and manipulation of preliminary data and the output area for compilation of graphic data prior to plotting. A sort program acting upon the location coordinates distributes the data into the geographically ordered file system and retrieves data sets as required.

Hardware

The facilities demanded by a data base are normally a function of the area to be covered and/or the application programs to be applied. Large-scale modeling of exploration information places a heavy demand upon the largest available computers.

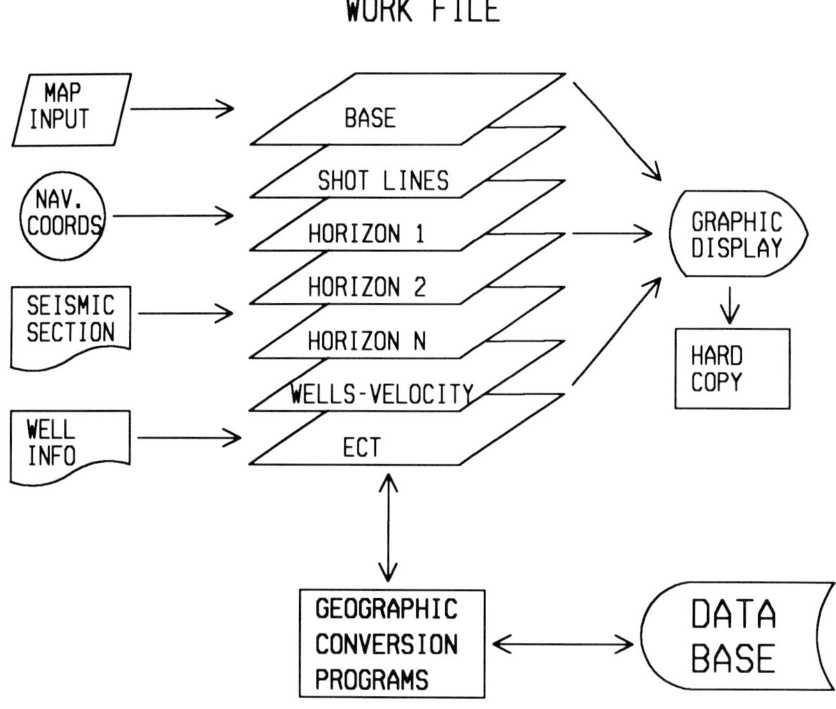

FIG. 7. Work file.

However, the specialized input/output requirements and lengthy interaction associated with exploration work are somewhat prohibitive in cost when carried out on a conventional computer. It is preferable to move these tasks off-line from the main system to an application computer, one of several minicomputer-based systems now on the market that can communicate with a main computer for the transfer of data files. The industry has gained much experience in the use of these systems for the purpose of data preparation and on-site processing. They are relatively inexpensive, can easily accept special peripherals, and can provide fairly powerful computing facilities in themselves. The resources of this computer are dictated by the variety and scope of the disciplines served. Figure 8 illustrates a typical system.

A central processing unit with 256 kilobytes of memory is considered minimum if local arithmetic processing is needed. The local or working data base organization should be resident within this system. Therefore, some large storage medium is required. A 300 megabyte disc drive is optimum. Some form of hard copy device is essential for the generation of work maps. This is probably one of several electrostatic plotters currently in use within the industry. Dot-matrix printers may be used in conjunction with a device for the conversion of line graphics to dot pattern. The resolution and accuracy offered is considered adequate for all preliminary work, and the operational costs involved are minimal compared to the high-resolution plotters used for final maps. Other peripherals including tape drives, card readers, and system

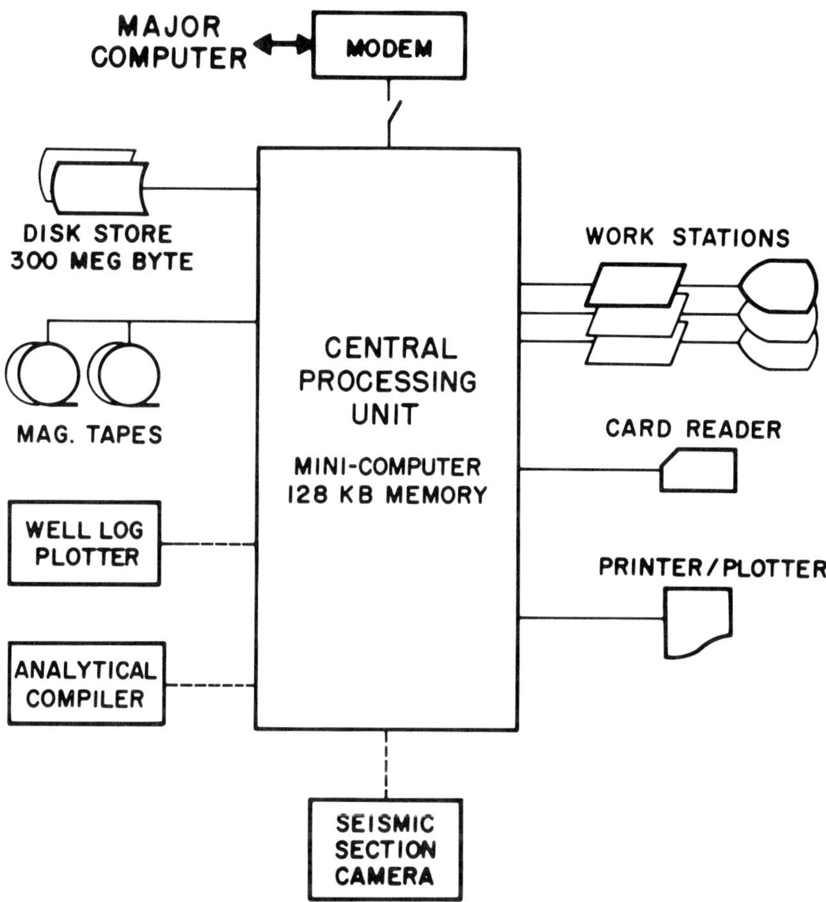

FIG. 8. Application computer.

consoles may be dictated by the scope of the operation. Additionally, the system accepts interfaces with peripheral devices specific to a given discipline, e.g., plotters, drum cameras, and analytical compilers.

At the input speeds associated with human interaction, such computers can, and normally do, support several users who may not be in the immediate area. Such users are not necessarily computer oriented, but they require some control over the system while maintaining a degree of privacy conducive to concentration. From this requirement has grown the concept of the "work station."

The work station is a configuration of hardware and software devoted to the requirements of the end user and is normally tailored to the discipline of that user. A typical station as illustrated in Figure 9 consists of the following units.

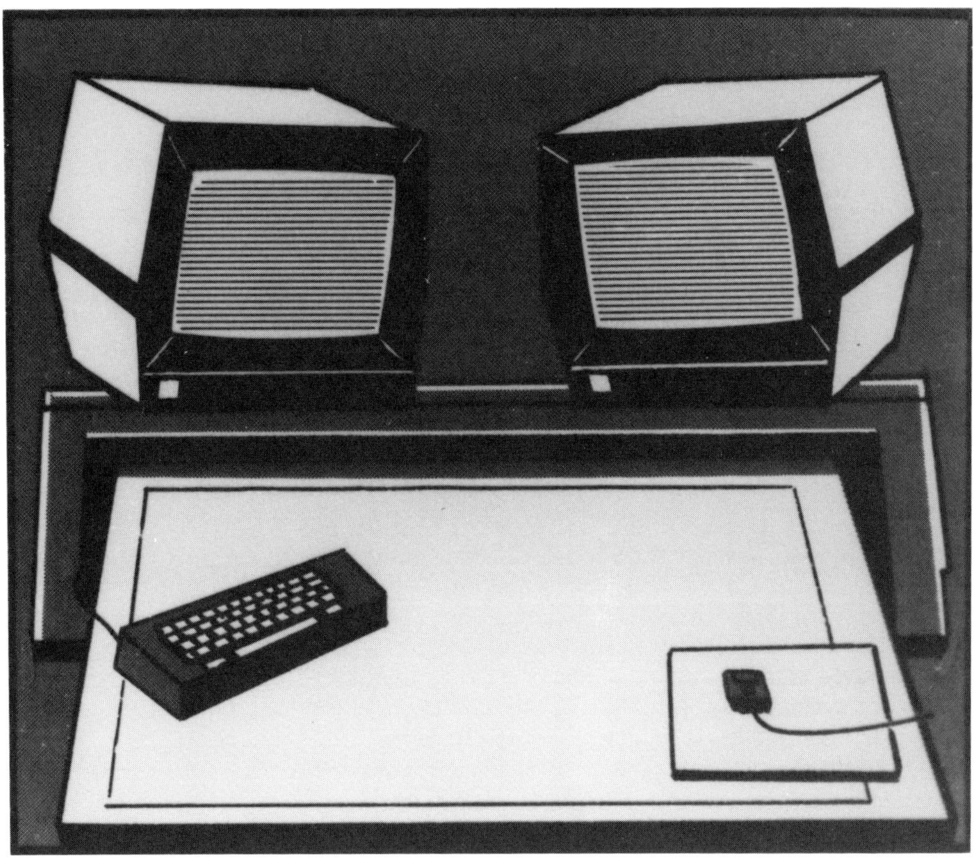

FIG. 9. Work station.

(1) General purpose, high-resolution digitizing table.
(2) Menu digitizer containing the symbols and instructions associated with the discipline.
(3) Keyboard for communicating alphanumerics in conjunction with the digitizing operation.
(4) Primary CRT for visual interaction with the graphic data.
(5) Secondary CRT for computer communication with the user and for the display of text information associated with the graphic data.

In conjunction with this equipment, a software set should be dedicated to the individual user. We can call this an input/output module or a discipline language. Its function is to allow simple dialogue in terms familiar to the user.

In summary, an ultimate organization of hardware is presented in Figure 10. The main-frame computer maintains the common data base, provides computer power when required, and handles communications with application computers that are dedicated to specific departments or disciplines. Interactive work stations are distributed on the basis of department data volume. An additional requirement is direct transaction terminals, for use where information is required but no further processing

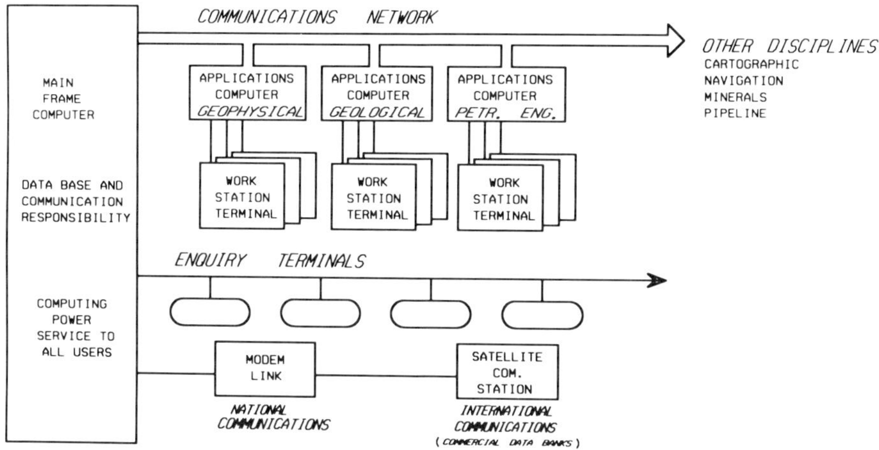

FIG. 10. Ultimate configuration.

is involved (simple inquiry functions with localized display formatting facilities). Finally, communication links are made through modems to external services such as public telephones and satellite communication or other data banks.

Within such an organization, the data base management system is maintained on each application computer. Hence, each department may exist as an independent entity in the creation of a local data base, files being transferred to and from the common bank as required. Input, output, processing, and interpretation remain under departmental or discipline control.

Application

The tangible measurement of success in a data base system is the ability to meet the requirements of the maximum number of end users. The fundamental criteria in the design are, therefore, the application for which it is intended and the volumes anticipated.

Applications may be divided into three broad groups: production mapping, interpretation, and analysis. There is considerable overlapping; however, each group is a proponent of some operational criteria that must be recognized in the system.

Production mapping

This area covers the generation and maintenance of routine maps for control and measurement purposes. In terms of single-edition maps, the exploration industry has probably the highest throughput of any similar group. Use of mapped information is currently limited by the speed at which maps are produced manually. Map production has been the prime mover toward data management systems and will place the heaviest demand upon such systems.

The major requirement is speed of input/output in conjunction with high map accuracy and the ability to select a specific data set from a very large range of possible combinations. Production flow usually follows the steps below.

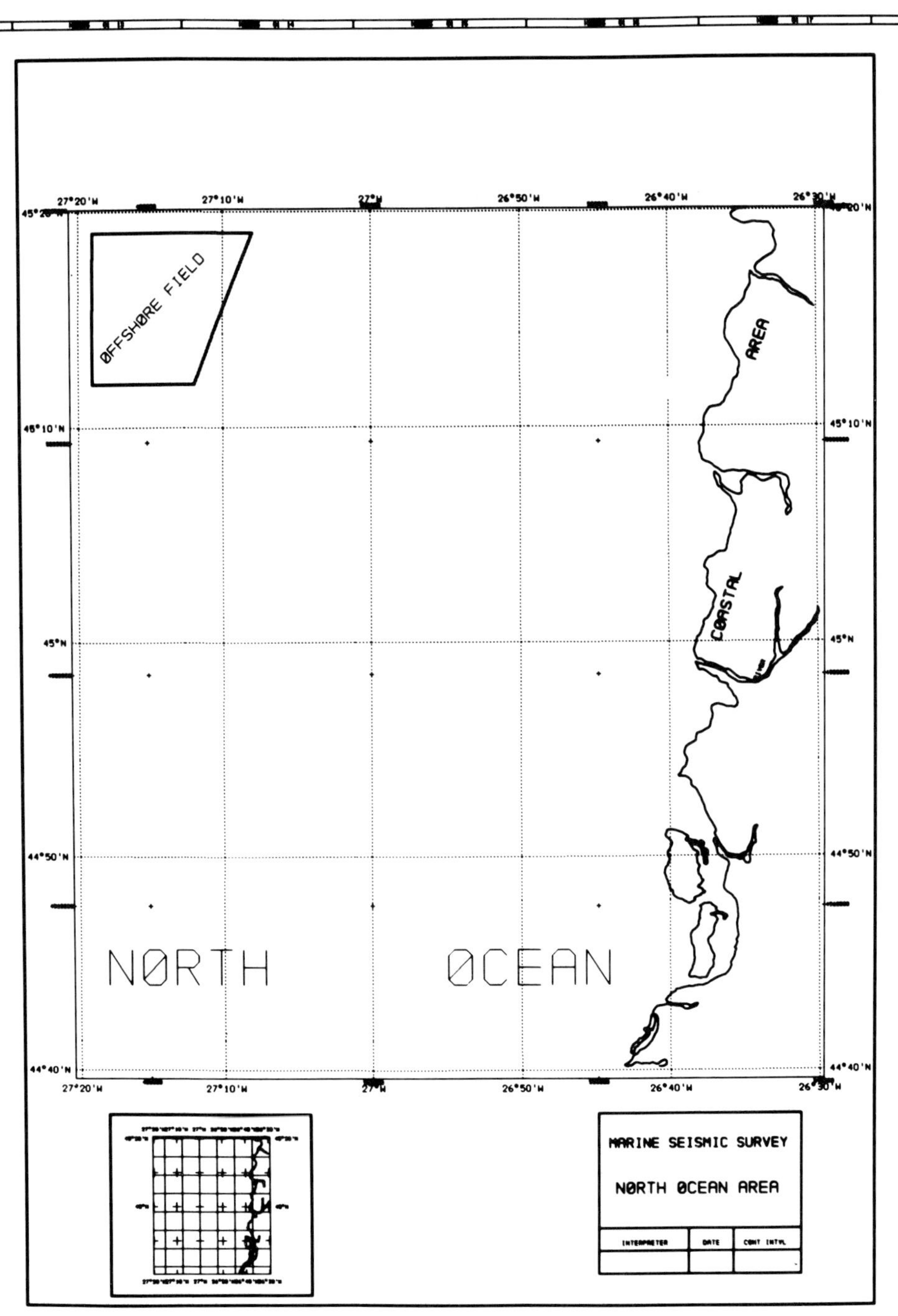

FIG. 11. Base map.

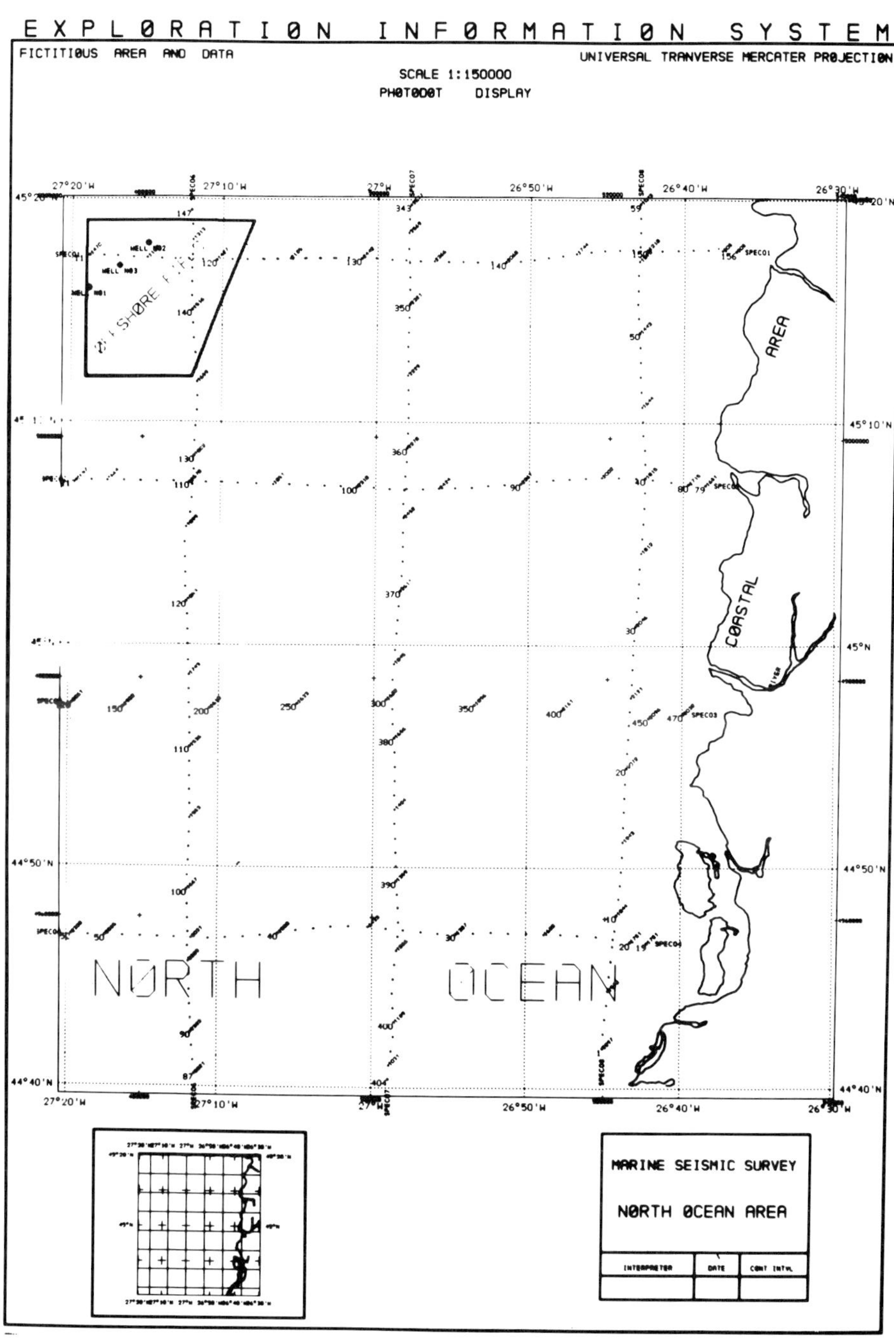

FIG. 12. Shotpoint map.

(1) Geodetic information is entered by the appropriate department and exists as a class of data viewable in mapped form (Figure 11).

(2) From tape or digitizing table, location data on the seismic survey are read in as a second data class viewable as an overlay and annotated by shotpoint identifiers (Figure 12).

(3) Seismic horizon data are digitized and entered as time values versus identified surface locations. These are stored as another data class consisting of statistical data associated with the location data by the common identifier name. If recalled from storage, seismic horizon data may be reassembled as a graphic representation of the original cross-section (Figure 13). However, within the system they are stored and referenced by the *x-y* location indicated on the map.

(4) Production of several versions of the line map is now available, where each is annotated by horizon time values. Additionally, the interval between any horizons may be mapped by simple arithmetic procedures. The seismic process may produce statistical estimates of rock-velocity distribution in the area. These "velocity functions" should form a separate class available for independent manipulation and interaction against time grids to produce depth maps.

(5) Any of the stored data sets may be submitted to a variety of arithmetic processes. The data may be randomly fitted to a regular grid and displayed as machine contours (Figure 14).

(6) From the data now stored, as many maps can be produced as required.

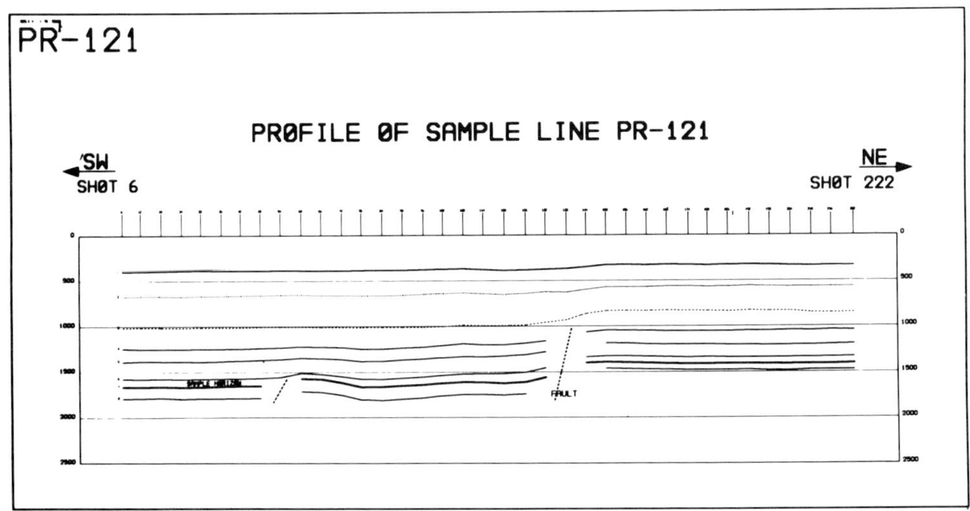

FIG. 13. Profile.

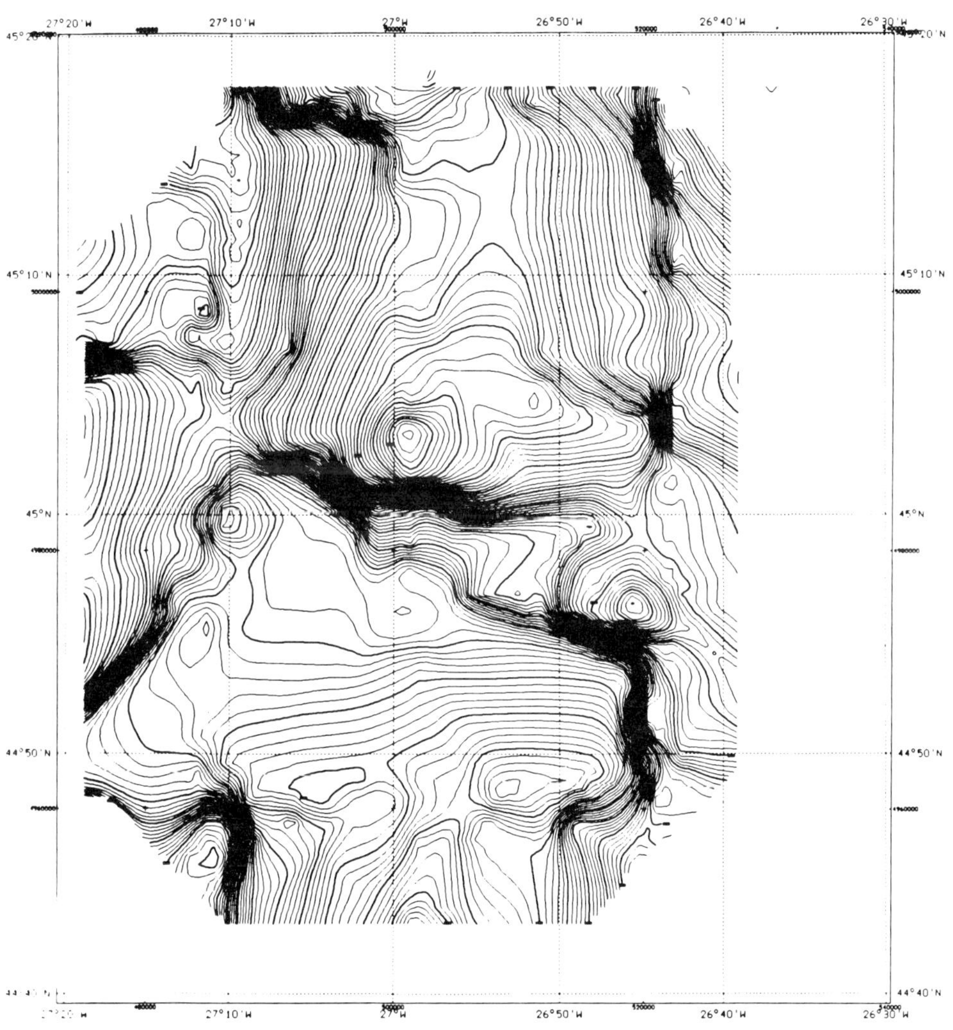

FIG. 14. Contour.

Interpretation

These activities cover the interaction and manipulation of data sets for the purpose of producing a geologic interpretation of a prospect. This is probably the fastest growing area of applications and normally involves the integration of data from several disciplines.

Machine-produced maps are essentially objective descriptions of the data. The process of interpretation introduces subjective opinion, and for this reason, the primary requirements of the user are ease of interaction with the data sets and availability of application programs that act upon the data. An example is the handling of geologic fault information.

(1) In Figure 15, three classes of data are compiled and displayed: seismic location, well site, and fault crossing. Geologic fault data are initially derived and input from the seismic profile. In the illustration, potential fault crossings are registered on lines 127, 119, 121, 120, and 122. Line 121 is illustrated in profile form in Figure 13. The fault map represents a discrete class of stored information that may be reinforced by data from surface geology or well survey classes but may ultimately reflect the opinion of the interpreter in correlating the line crossings.

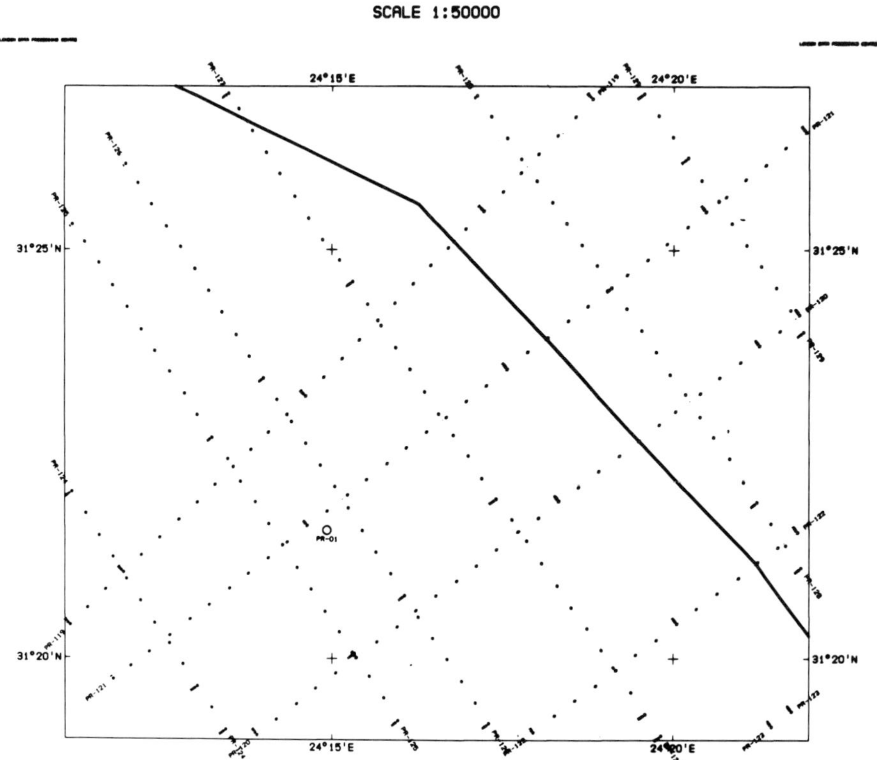

FIG. 15. Line map with fault.

(2) In Figure 16, the data class of horizon 3 has been subtracted from horizon 7, grid fitted, contoured, and displayed as an isochron of an interval below the nonfaulted horizon.
(3) The interpreted fault map is now used as a limiter in the grid fit operation to observe the effect of the fault upon the contoured data set (Figure 17).
(4) The derived velocity information is validated by information from the well class. It may now be smoothed or filtered (Figure 18) before applying against the time grids to obtain depth maps.
(5) From the stored information isometric views of a structure may be displayed (Figure 19).

At times, such displays are regarded as merely interesting visual effects. Significantly, the data from which they are created provide the model for volumetric calculations which establish reservoir dimensions and economic risk factors.

Analysis

This activity is concerned with the construction of small-scale earth models to illustrate or test the effect of parameter changes. In this category, projection coordinate computations are irrelevant; however, high precision is required in arithmetic computations, together with flexibility of the graphic system in presenting isometrics.

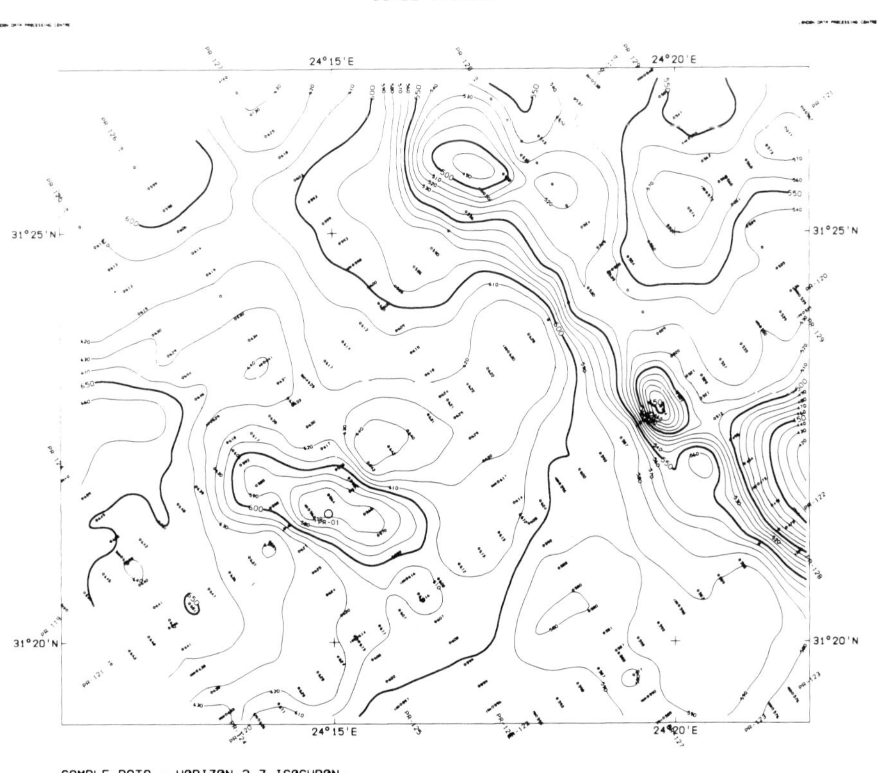

FIG. 16. Contour isochron 3 to 7.

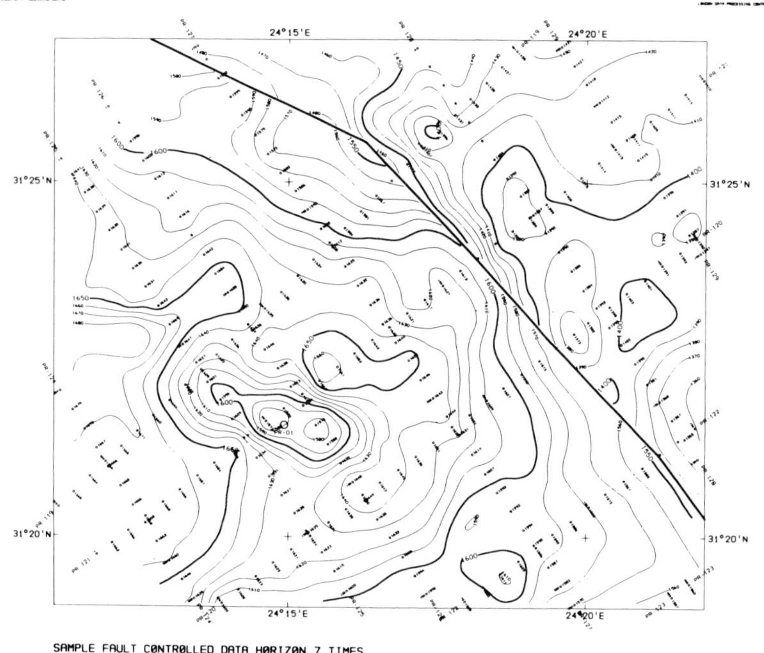

FIG. 17. Contour horizon 7 with fault.

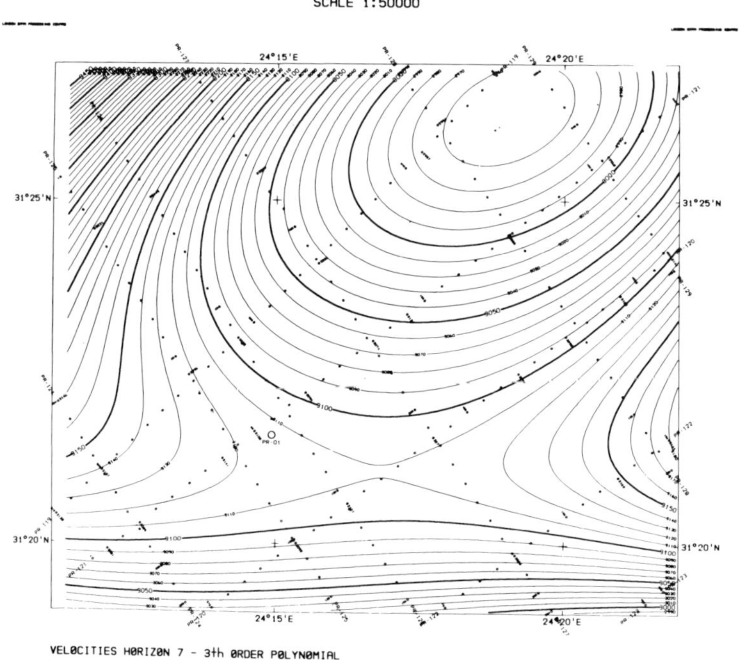

FIG. 18. Velocity contour.

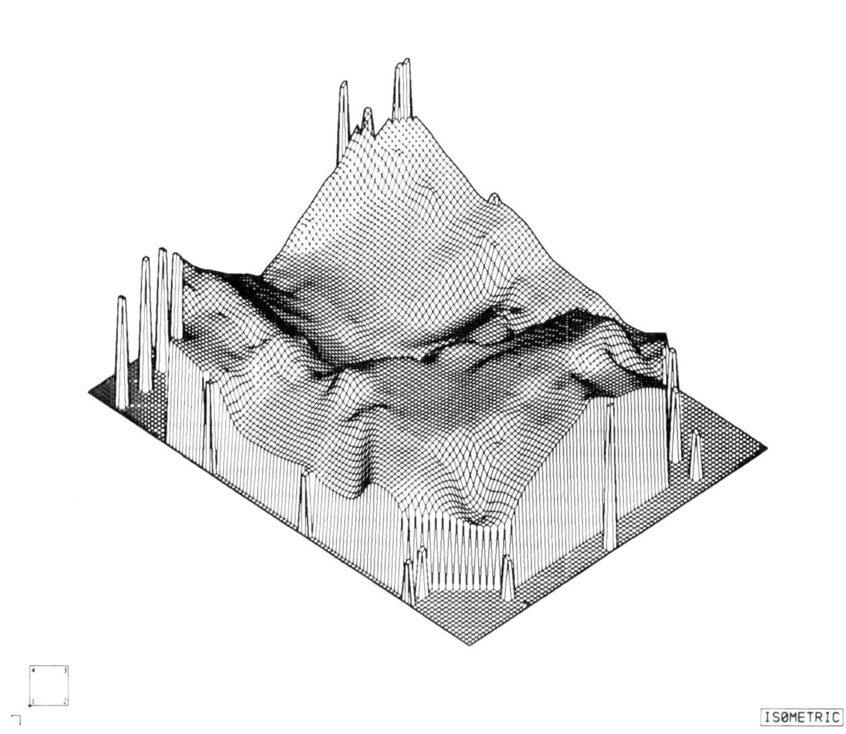

FIG. 19. Isometric.

This area of application is often considered as a tool for research. Development of 3-D seismic methods has placed heavy demand upon data base systems. For example, a 3-D seismic model should be constructed prior to the acquisition and processing of data. Such constructions are useful to test the feasibility of the seismic program. The volume and complexity of parameters involved in 3-D seismic data require that a model be built to describe the geometry of any survey to control and order the acquisition and processing of data.

Summary

The EDP groups within the exploration industry are evolving data base management systems that address their specific requirements, which include (1) integration of various types of data generated from many disciplines, (2) production of large-scale graphic aids, (3) user interaction with data files, and (4) conversion of specialized input into digital form for computer retention and manipulation. These systems are beginning to solve the unique problems that the commercial systems could not address. In due course, data base management should become an essential tool for success in exploration ventures.

REFERENCES

Association for Computing Machinery, 1973a, ''DBTG data description language:'' Report of the CODASYL Data Base Task Group, New York, May.

——— 1973b, ''DBRG data manipulation language:'' Report of the CODASYL Data Base Task Group, New York, May.

Canning, R. C., 1973, The cautious path to a data base: EDP Analyzer, v. 2, June.

GUIDE, 1972, ''The data base administrator:'' Data Base Admin. Proj., rep. of the Information Management Group.

Martin, J., 1975, Computer data-base organization: Englewood Cliffs, NJ, Prentice-Hall, Inc., chapters 18 and 36.

Chapter 9

TECHNOLOGICAL IMPACT OF EXPLORATION

Introduction

In November of 1979 the Society of Exploration Geophysicists held its 49th Annual International Meeting in New Orleans. The technical sessions which ran in parallel embodied a curious mix of technical applications and theory. Among the sessions were offerings on general and nuclear waste disposal, deconvolution and filtering, modeling, migration, 3-D seismic, potential field methods, and a symposium on pattern recognition. While some of these titles are self-descriptive, it is particularly revealing to note these session contents and observe their relationships to other and allied disciplines.

Techniques for waste disposal represent a general awareness of the environmental repercussions of virtually all human activities. Knowledge of the subsurface gained by geophysical means can be translated into sites which would be particularly tolerant for the safe repose of harmful waste products. Deconvolution operations and filtering are standard signal processing procedures which permeate systems and control theory, speech processing, radar, sonar and communication.

Modeling in the exploration context encompasses one of the more sophisticated computer simulation tasks which is currently addressed the description of wave propagation in complex multimedia environments representing the earth's near subsurface. Migration and 3-D seismic purely in technical terms can be described as imaging methods again addressed to a most difficult problem. Of course, the principles and objectives are quite analogous to those of computer tomography (CT scanning) the role of which in medical imaging was recently honored with a Nobel Prize (Cormack and Hounsfield, 1979).

Potential field methods draw on the most elegant formalisms of classical physics and inversion theory to develop procedures for defining the earth's structure. Pattern recognition and artificial intelligence are areas which are beginning to be considered by the explorationist for use in geophysical data interpretation. Thus the impact of exploration geophysics on these areas is yet to be known. Other papers offered at the meeting suggested the possibilities of sensing hydrocarbons directly by electrical transient soundings at great depth (Azad, 1979; Keller, 1979), and of using a proposed mechanism of dolphin echolocation to develop methods for directly measuring signal dispersion (Neidell, 1979).

All of this sweep and span of technology were presented against the seemingly

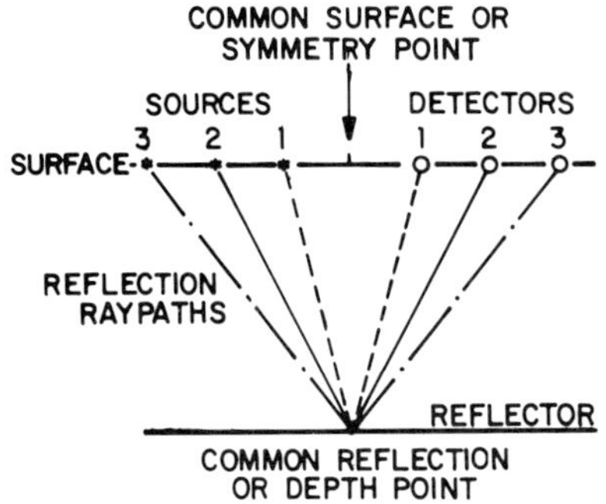

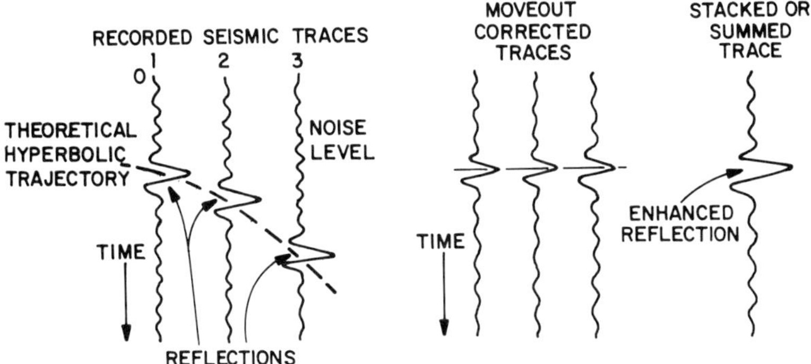

FIG. 1. Principle of the CRP method (after Mayne, 1962).

mundane background of developing surer and sharper tools for finding oil and gas. We may then ask, how long has exploration geophysics probed so widely and so deeply into the frontiers of science and technology, and what kind of people practice exploration geophysics? While students of the history of science should best treat these questions, the views which follow should provide some insights.

Limiting the scope of this chapter to exploration seismology, we first examine its roots and accomplishments. We then look more broadly at the impact of exploration geophysics on a broad spectrum of technical developments.

Exploration Seismology—The Technology

The common reflection point method (also known as CRP or CDP method) of imaging, which is the most widely used exploration seismic technique, is shown in Figure 1. The basic method was patented by Mayne (1962). We see from the figure that a local region in the subsurface is ''viewed'' from different angles, and the task of reconstituting these views to form an enhanced image of the point of interest is in fact analogous to the operation of the side-looking airborne radar (SLAR) and synthetic aperture methods in general (Harger, 1970).

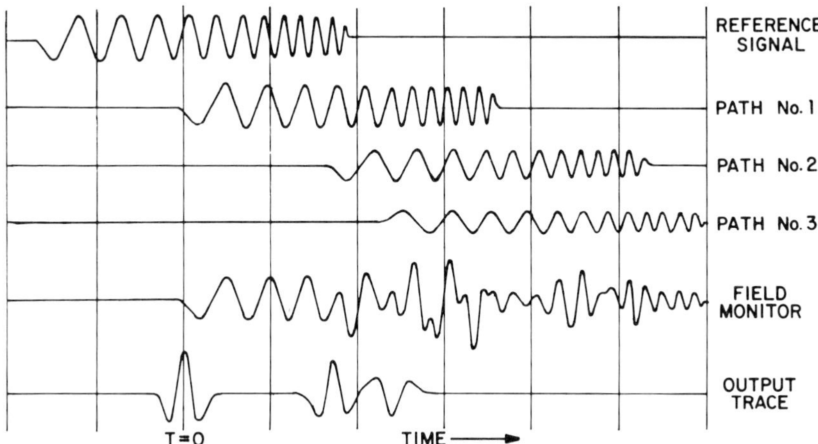

FIG. 2. Composition of Vibroseis record (after R. O. Lindseth, 1974, Review of recent advances in digital processing of geophysical data: SEG Cont. Ed. program).

In Figure 2 we note the principles of the Vibroseis® method (Crawford et al, 1960). This seismic energy source has achieved increasing acceptance by using low-amplitude signals over long periods of time to put energy into the ground with minimum disruption and environmental impact. The companion correlation process compresses the signals to give sharp, impulse-like signals of more readily recognized form. We should recognize in this sequence the principles of the chirp radars and sonars and other applications which employ swept frequency signals (Klauder et al, 1960; Altes et al, 1977).

Turning from the present to the past, we note the pioneering efforts of men like R. Fessenden and J. C. Karcher who, with others, made exploration via the reflection seismic method a viable tool some 60 years ago (Sweet, 1966). Inherent in their work were technological requirements whose recognition shaped much of our current technology. As an example, we cite the weak electrical signals whose precious information content had to be preserved and enhanced so that the subsurface might be defined. Also consider the problems of transmitting, sorting, enhancing, and displaying these signals so that they are accessible and recognizable to human intelligence.

Most companies engaged in reflection seismic operations soon developed a high degree of expertise in the design and fabrication of signal amplifiers. The Geophysical Society of Houston Museum, located in the lobby of Geosource, Inc., proudly numbers in its collection several early and ingenious signal amplifiers. It was against such a background that the physicist John Bardeen spent the early part of his career at the Gulf Oil Company Laboratories. He went on to a career at the Bell Laboratories where his work with weak signals and their amplification led him to share a Nobel Prize for fundamental works relating to transistors and to become

®Trademark of Conoco.

the only physicist to win the Nobel Prize twice. Bardeen wrote one of the finest treatises describing the theory and use of aeromagnetic survey data (Nettleton, 1976).

Of course one should not mention transistors and their technology without acknowledging the pioneering and even dominating role of Texas Instruments, Inc. It is easy to forget the origins of this giant in the field of electronics—a seismograph contracting company which of necessity built its own instruments. The parent company, Geophysical Services, Inc., is one of the leading exploration seismic service companies.

The problem of recording the very weak signals and their subsequent display also gave rise to some interesting directions of special significance to computing. Figure 3 shows in diagram form the decay over recording time of a seismic signal and the role of several mechanisms which contribute to this decay. To capture and work with such information, the seismic exploration industry progressed from primitive analog graphic displays with programmed gain expanders to recordings on tape. Such rec-

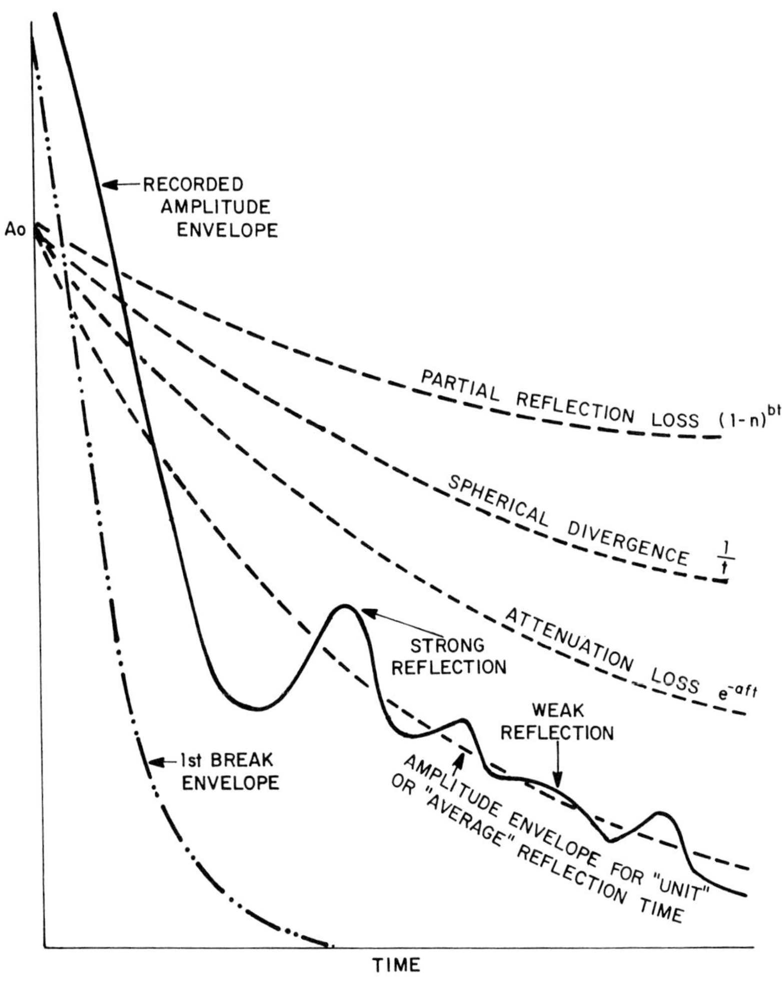

FIG. 3. Mechanisms of amplitude decay of the seismic trace (after Lindseth, 1974).

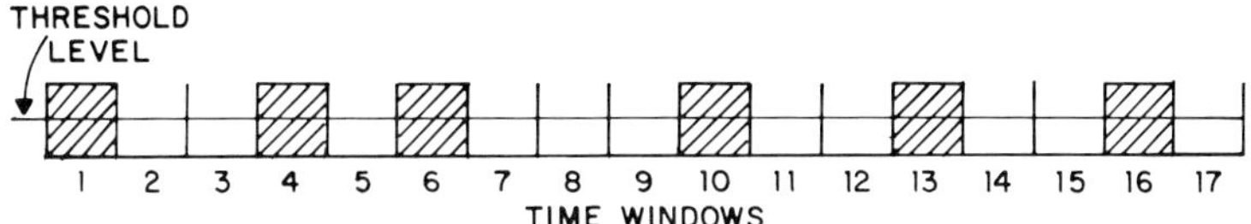

FIG. 4. Analog signal for recording on analog tape bearing binary encoding of digital samples.

ording started with amplitude modulated formats but rapidly changed over to frequency-modulated and finally digital approaches. Digital recording of appropriate samplings of the data could most conveniently accommodate the full potential dynamic range of seismic data which may be expressed by the ratio of 1:1,000,000 (approximately 120 dB).

Digitally recorded seismic data soon turned into an avalanche of data bits. More channels of recording and more frequent or more closely spaced recording meant greatly improved signal-to-noise ratios, but they also presented challenging problems. How were data samples acquired at a 0.25 msec rate on 1024 channels simultaneously to be transmitted? or recorded? or processed? or stored? Of course, this example taxes the current state of capabilities, but it can, in fact, be accommodated.

The matter of data transmission at high rates was treated by telemetry (Convert et al, 1976) using both wired and wireless systems. One can imagine that without telemetry a system recording from over 1000 data channels would require a cable having more than 1000 pairs of wires. The expense, physical dimension, and associated operational difficulties virtually rule out this alternative for day-to-day rugged use in the field. The impact of these considerations on electronic data transmission technology is signaled by the development of various sophisticated coaxial cable and wireless telemetry systems (see, e.g, O'Brien et al, 1979) currently offered off the shelf by a number of companies servicing the seismic exploration activity.

Storing the torrent of data arriving in real time on tape proved still another challenge particularly under conditions which characterize practical acquisition. Television tape recorders had the capability of handling the data volume since they were designed for the demands of image representation, but these were systems designed for use under almost sterile laboratory conditions. Globe Universal Sciences, Inc. adapted the principle of the television tape recorder and developed robust systems for field use. Direct interface to standard computing systems with recording densities of 8000 bpi per channel was possible with as many as 16 channels on a one-inch tape. Systems like these served on the North Slope, the North Sea, and the Sahara Desert years before the major computer vendors announced high density tape formats as "new" products.

The principle of high-density tape recording is suggested by Figure 4. Digital formats are encoded as analog "square" wave signals, and several channels of analog signals may be recorded in parallel on a single tape. This approach has obvious tolerance to high noise levels and is also quite resistant to problems of skew in recording. The skew problem, of course, commonly arises in adverse field conditions and is characterized by irregularity and/or misalignment of the rows of recorded bits on the tape, making the reading process, which expects regularity and alignment, quite difficult if not impossible.

The flow of seismic data nevertheless is quite alarming. "A Quadrillion Geophysical Data Bits Per Year" and "Geophysical D P Requirements Could Exceed the World's GP Capacity by 1985" warns Savit (1978, 1979). Figure 5 (taken from Savit, 1979) graphically shows the explosive growth of seismic computing requirements against the background of growth in general computing capabilities.

It is interesting to point out, as Savit does, that earth satellites as data transmission links currently just have a maximum link capacity of 250,000 bits per second, while new 500 or 600 channel seismic data acquisition systems generate about 10,000,000 bits per second. Even the slowest of the new models requires transmission rates of 1 to 2,000,000 bits per second.

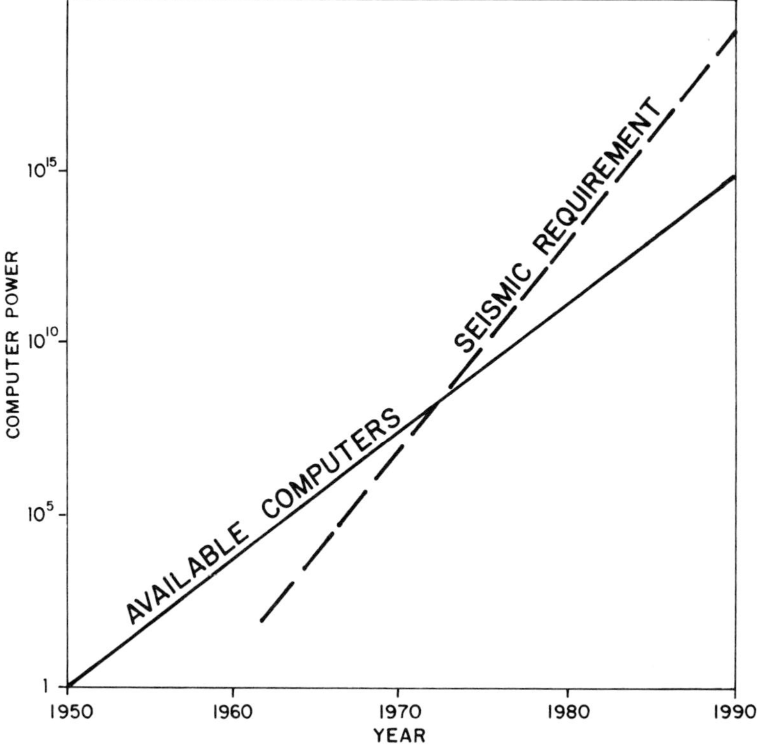

FIG. 5. The power of typical commercially available computers has been growing about one order of magnitude every 2.7 years and should continue to do so through the 1980s. Requirements of the seismic exploration industry have been growing and should continue to grow at a substantially faster rate (after Savit, 1979).

We should not be surprised then to see the largest and fastest of the super computers routinely employed in great numbers in the processing of seismic data. Seismic data processing clearly numbers as one of the largest scale scientific computing applications.

The computing community has recently recognized the potentially important role of the array processor (Robinson, 1979; Winningstad, 1978), but it has not properly recognized its origins. Array processors are peripheral arithmetic units designed to enhance computing effectiveness in dealing with a small subset of calculations. Since they achieve much of their efficiency by performing arithmetic operations in ''pipeline'' or overlapping manner, they appear to have some resemblance to parallel computer processing in general.

Perhaps the greatest distinction between general parallel processing and array processors is the matter of practicality. Array processors address a modest suite of operations and produce extreme efficiency when such operations are performed in conjunction with a general purpose computer. Fully parallel computing systems, on the other hand, such as the Illiac IV, Staran, and the Texas Instrument's Advance Scientific Computer (ASC) have yet to be used in fully efficient modes.

One of the earliest of array processors was built by IBM in cooperation with Western Geophysical Company of America specifically for the processing of seismic data. Since mid-1965, seismic data processing systems have as a matter of routine been equipped with some sort of array processors. The seismic computing industry remained the only substantial market for array processors until quite recently. Now, sonar and radar processing, tomography, and even general purpose scientific computing also constitute respectable markets. We note that Floating Point Systems, Inc., who first served the seismic data processing area exclusively with their product line of array processors, is now a major entity in making the CT scanner a viable medical imaging tool.

Advances relating to computing and data transmission continue at a rapid pace, again as evidenced by presentations at the 1979 New Orleans meeting of the SEG. One author reported on the effectiveness of a seismic processing system organized around an intelligent data bus (Vallhonrat, 1979), while another described the use of solid state memory modules to synthesize recording systems of essentially unlimited capacity (Shave, 1979). Inherent in these presentations are exciting new concepts, for example, the elevation in stature of the data bus or transfer unit from a passive unit to one exercising controls not only on its own functioning but reaching beyond to other parts of the computing system.

Of course seismic acquisition is costly and the resulting data are therefore of considerable value. Storing and accessing the tapes which contain these valuable data present problems of some magnitude in themselves. The major oil companies and larger independents have tape libraries consisting collectively of millions of reels of tapes. We see here additional impetus for the use of high-density formats both as an economy measure and a means of improving data accessibility.

In recognition of these problems, companies like Geodata Service, Inc. and the Record Service Center have developed storage and access systems which make it feasible to address files of hundreds of thousands of tapes in a matter of minutes. Mundane as such a contribution might seem initially, the benefits have accrued to other industries such as banking and insurance where tape files of information have begun to amass in great numbers.

Figure 6 shows a processed seismic section. For all the sophistication and technology involved, the final product of seismic acquisition and processing is a graphic display. It follows then that computer displays and digital graphics would become another province for technological innovation.

It is not unreasonable for a single seismic section to have a dimension of 3 × 5 ft and to consist of over 32 million bits of information. Seismic plotting systems can develop images 5 × 80 ft having 200 resolution elements/inch. Such volumes of data limit the usefulness of many conventional digital graphics systems. Interactive graphics using conventional displays have for this same reason also found limited use. Unconventional systems for dealing with seismic data employing digitizers, video displays, electrostatic plotters, film plotters, and similar hardware have been developed by companies like Geophysical Services, Inc. (Texas Instruments) and GeoSource, Inc., not to mention a variety of prototype units developed by various oil companies.

Innovative displays and the use of color have become quite important for assisting in the assimilation of these masses of data by the human seismic interpreters. Figure 7 shows the seismic section of Figure 6 where the amplitude levels of the data in absolute terms are superimposed in color over the original display. Other attributes

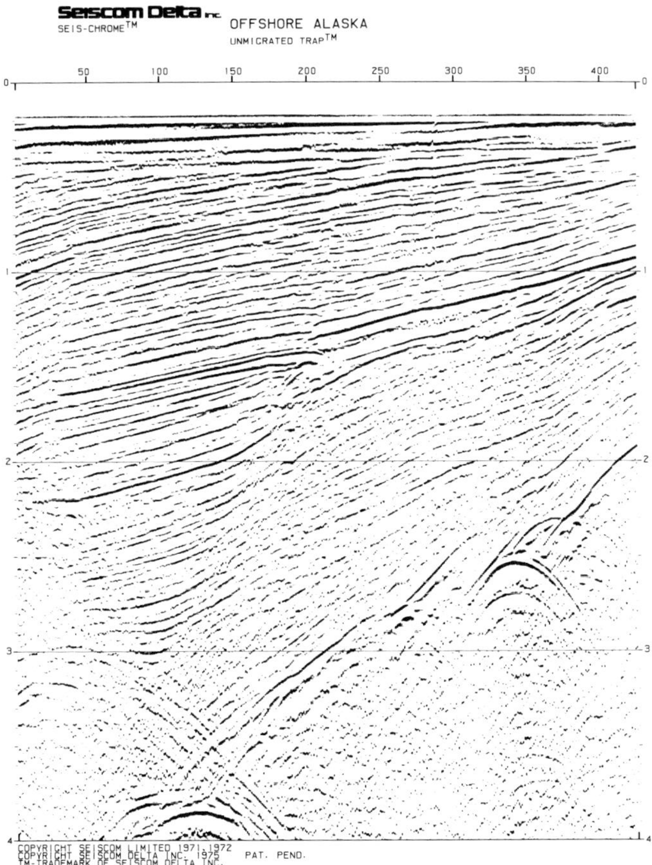

FIG. 6. Conventional display of processed seismic data.

can be similarly treated. The display is produced by Seiscom-Delta, Inc. using a laser film plotter. Figure 8 shows estimates of lithologic velocities derived from seismic data displayed in color by Western Geophysical of America using an Applicon ink jet plotter.

Exploration seismology has been an early and fertile recipient of virtually all new developments in computer graphics of every type and a source of innovations in this area as well. It is likely that it will remain so in the years to come.

We ought not to leave the field of exploration seismology without emphasizing the significant influence seismic exploration has had on developments in signal processing. Progress in this field accelerated greatly by the efforts of the MIT's Geophysical Analysis Group (GAG group) (Flinn, 1967) which was supported by industry sponsors. Deconvolution processes (Webster, 1979) and a variety of imaging

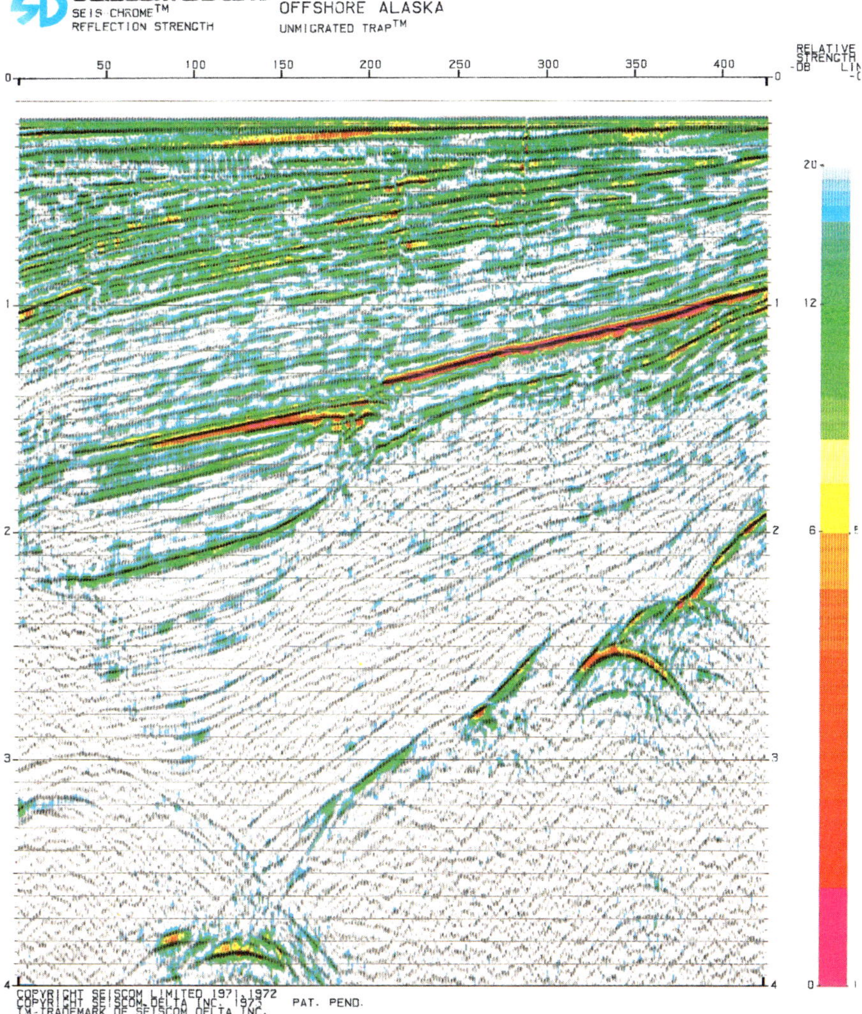

FIG. 7. Color display of seismic reflection strength (amplitude) superimposed on original display (offshore Alaska). (Copyright Seiscom Limited 1971, 1972; copyright Seiscom Delta Inc., 1975).

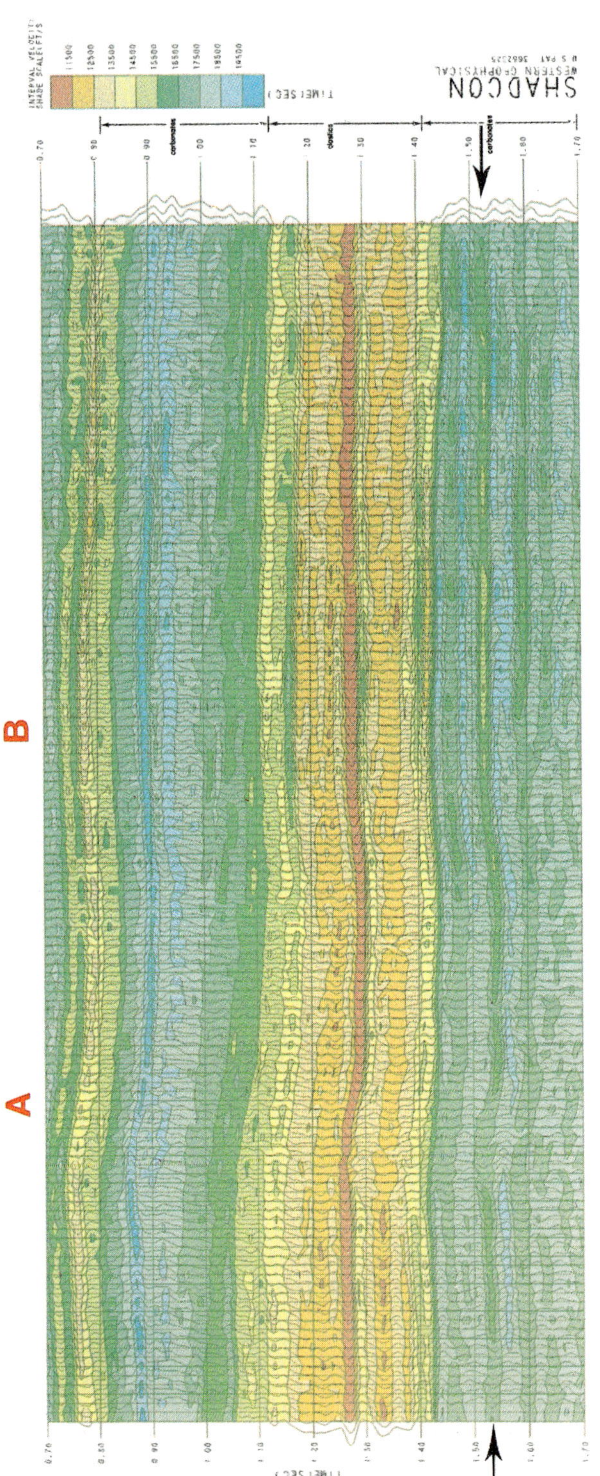

Color SHADCON Section

West Texas speculative data of Western Geophysical Company of America and Geo-Search Corporation.

FIG. 8. Ink jet color display of seismic derived velocity information. SHADCON section (Western Geophysical, U.S. pat. 3663235). West Texas speculative data of Western Geophysical Company of America and Geo-Search Corporation.

techniques (French, 1975) are the most widely recognized accomplishments of exploration seismic signal processing which have found important applications elsewhere.

Because of its very nature, seismic data processing is likely to serve as a strong catalyst for new developments in the currently active area of fast algorithms for one- and multidimensional digital signal processing. Turning now from such sidelights, let us take a brief look at exploration geophysics in a somewhat broader context and note again a history of breadth of technology and significant achievement.

Both exploration seismology and the study of earthquakes make use of the wave equation and its various solutions. A most distinguished group of investigators, including Rayleigh, Love, Stoneley, etc., have developed solutions of special nature which bear their respective names. These solutions are characterized by having applicability under various propagation models meant to represent the earth. The Journal of The Acoustical Society, Bulletin of the Seismological Society, Journal of Geophysical Research, and the Geophysical Journal of the Royal Astronomical Society are among the publications which document much of the recent progress which pursues this theme.

Describing wave propagation through the solid earth itself is somewhat complicated mathematically by at least one presumed density discontinuity in the internal structure. Figure 9 (after Harris, 1972) shows a schematic diagram of the internal structure of the earth. This problem background led the mathematician Stieljes to develop a theory of integration for handling both discrete and uniform distributions within a single formalism. His works were well documented in the mathematical literature in the late 19th century.

The COCORP group at Cornell University, under the direction of Kaufman and Olivcr (Cook et al, 1980), has tried to blend the exploration approaches to seismology with the approaches derived by academicians from earthquakes for crustal studies. They have provided important data for studies of the wave equation.

Recent works with the wave equation have also concentrated on numerical solutions. Studies such as those by Kelly et al (1976) have provided new approaches to treating reflections from the very boundaries which delimit the problem. In addition, the works of Claerbout (1971, 1972), which drew initially on the operational methods and approximations commonly employed in quantum mechanics, led to significant new directions and insights including a family of modeling and migration methods as well as improved approximations for differential operators. An industry sponsored group at Stanford University has resulted from these efforts.

Inverse problems and their formulation are a natural consequence of many geophysical investigations where a system structure must be inferred from remote active or passive measurements. Many of these problems have no unique solution for reasons such as, in many instances, the observations being fewer than the unknowns. Interpreting potential field data resulted in the work by Bjierhammer (1973) on the pseudoinverse method which sought to define solutions to ambiguity problems by least-squares statistical criteria. The work of Parker (1974) in handling such ambiguity and finding solutions represents a departure, where the plausibility of the family of models and the nature of their extremes narrows the ambiguity. Further work of Backus and Gilbert (1967, 1968) addresses the same sort of problem with the global objective of defining the internal structure of the earth.

Over the years the technology of geophysics has been directly called upon to serve

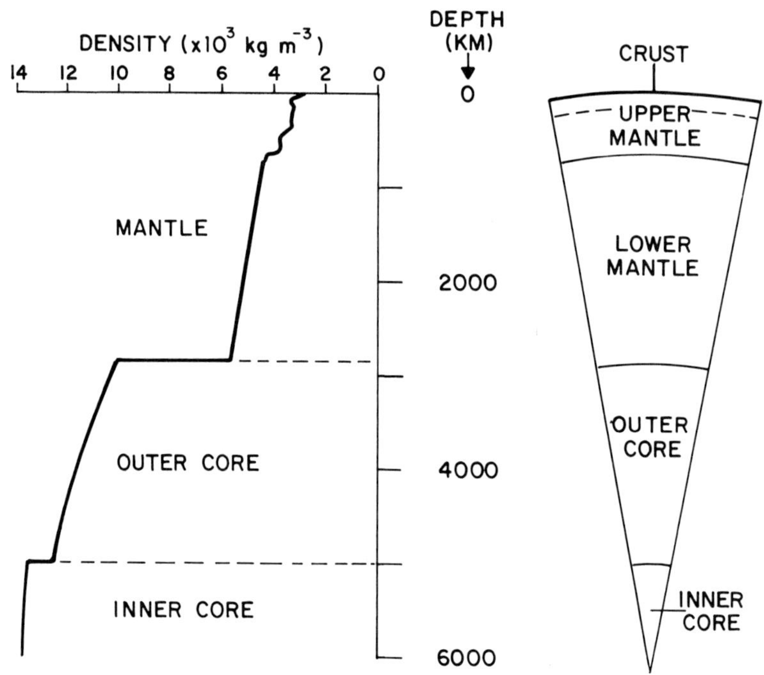

FIG. 9. The earth structure and densities (after Harris, 1972).

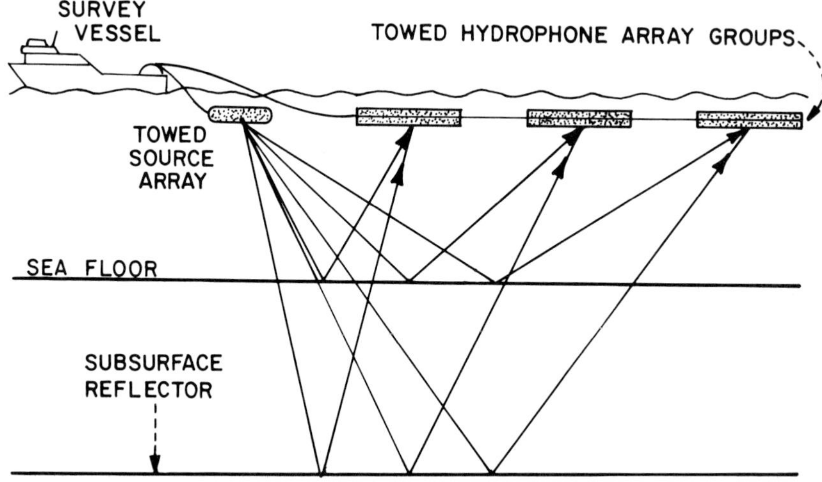

FIG. 10. A marine seismic survey utilizing a towed streamer cable having arrays of grouped hydrophones.

in seemingly unrelated applications. The military has been particularly adept at recognizing the immediate benefits of such technology transfer. The precedents for this interchange date back to the location of artillery emplacements by acoustic triangulation and submarine detection by flying an airborne magnetometer.

More recently, geophysical methods have proved important in underground tunnel detection, and the technology of the towed streamer used in marine exploration (Figure 10) has again aided in locating submarines.

A formal vehicle for effecting this cross-fertilization has emerged in the form of a biannual joint meeting sponsored by both the Society of Exploration Geophysicists and the United States Navy. The first such meeting, held in August, 1978, developed around the themes of "Acoustic imaging technology and on-board data recording and processing equipment." A second meeting, dealing with high-resolution seismic applications, was held in March, 1980.

Summary

The technological impact of exploration geophysics over the years has been quite significant and out of all proportion to the number of workers in this science. Making contributions which have significance to other fields has almost become a tradition.

In order to suggest the scope of contributions made by exploration geophysics, first, the field of exploration seismology was examined, since it includes perhaps the majority of resources and efforts allocated within exploration geophysics. Several other branches of exploration geophysics were then noted.

The sampling of achievements listed indicates that exploration geophysics remains at the forefront of both applied and theoretical scientific activity. Current progress in the field appears to be maintaining this tradition.

REFERENCES

Altes, R. A., and Skinner, D. P., 1977, Sonar velocity resolution with a linear-period-modulated pulse: J. Acoustical Soc. Am., v. 61, p. 1019–1030.

Azad, J., 1979, Seismic electrical transients combine to discover oil and gas fields: Presented at the 49th Annual International 49th SEG Meeting, November, in New Orleans.

Backus, G. E., and Gilbert, J. F., 1967, Numerical applications of a formalism for geophysical inverse problems: Geophys. J. Roy Astr. Soc., v. 13, p. 247–276.

——— 1968, The resolving power of gross earth data: Geophys. J. Roy. Astr. Soc., v. 16, p. 169–205.

Bjierhammer, A., 1973, Theory of errors and generalized inverses: Elsevier.

Claerbout, J. F., 1971, Toward a unified theory of reflector mapping: Geophysics, v. 36, p. 467–481.

——— 1972, Downward continuation of moveout-corrected seismograms: Geophysics, v. 37, p. 741–768.

Convert, J. P., Devault, J. L., and Pieuchot, M., 1976, Telemetry offers simplicity, accuracy: Oil and Gas J., October, p. 88–96.

Cook, F. A., Brown, L. D., and Oliver, J. E., 1980, The southern Appalachians and the growth of continents: Scientific American, v. 243, p. 156–168.

Cormack, A. M., and Hounsfield, G. N., 1979, Computed tomography: Nobel Prize address.

Crawford, J. M., Doty, W. E. N., and Lee, M. R., 1960, Continuous signal seismograph: Geophysics, v. 25, p. 95–105.

Flinn, E. A., 1967, Foreword to the M.I.T. Geophysical Analysis Group Report: Geophysics, v. 32, p. 411–413.

French, W. S., 1975, Computer migration of oblique seismic reflection profiles: Geophysics, v. 40, p. 961–992.

Harger, R. V., 1970, Synthetic aperture radar systems, theory and design: Academic Press, 239 p.

Harris, P., 1972, The composition of the Earth, *in* Understanding the Earth: I. G. Gass, P. J. Smith, and R. C. L. Wilson, Eds., MIT Press, 2nd ed., 383 p.

Keller, G. V., 1979, Electrical prospecting for oil: Presented at the 49th Annual International SEG Meeting, November, in New Orleans.

Kelly, K. R., Ward, R. W., Treitel, S., and Alford, R. M., 1976, Synthetic seismograms—A finite-difference approach: Geophysics, v. 41, p. 2–27.

Klauder, J. R., Price, A. C., Darlington, S., and Albersheim, W. J., 1960, The theory and design of chirp radars: Bell System Tech. J., v. 39, p. 745–808.

Mayne, W. H., 1962, Common reflection point horizontal data stacking techniques (common-depth-point): Geophysics, v. 27, p. 927–938.

Neidell, N. S., 1979, Observing seismic dispersion—Lessons from a theory of dolphin echolocation: Presented at the 49th Annual International SEG Meeting, November, in New Orleans.

Nettleton, L. L., 1976, Early magnetic interpretation: Geophysics, v. 41, p. 575–576.

O'Brien, J. T., Kamp, W. P., and Hoover, G. M., 1979, Theory of amplitude recovery from sign-bit recorded data with applications to seismic problems: 49th SEG Annual Meeting, New Orleans.

Parker, R. L., 1974, Best bounds on density and depth from gravity data: Geophysics, v. 39, p. 644–650.

Robinson, A. L., 1979, Array processors: Maxi numbers crunching for a mini price: Research News, Science, v. 203, p. 156–160.

Savit, C. H., 1978, Geophysical DP requirements could exceed the world's GP capacity by 1985: AFIPS Conf. Proc., v. 47, p. 63–66.

——— 1979, A quadrillion geophysical data bits per year: Preprint, AAAS Annual Meeting, Houston.

Shave, D., 1979, A trace-sequential seismic recording system with effectively unlimited capacity: Presented at the 49th Annual International SEG Meeting, November, in New Orleans.

Sweet, G. E., 1966, The history of geophysical prospecting: Science Press, v. 1 & 2, 326 p.

Vallhonrat, J. B., 1979, Data-bus centered seismic processing system: Presented at the 49th Annual International SEG Meeting, November, in New Orleans.

Webster, G. M., ed., 1979, Deconvolution: SEG Geophysics Reprint Series.

Winningstad, C. N., 1978, Scientific computing on a budget: Datamation, p. 159–173.

Chapter 10

PATTERN RECOGNITION APPROACH TO EXPLORATION

Introduction

The past three decades have seen the development of an approach, known as the "pattern recognition approach," to problems of retrieval of information from data. This approach has been successfully used in a number of applications (ICPR-V, 1980), such as in the classification of agricultural crops from remotely sensed satellite data; in the machine recognition of speech and handwritten text; in the computer-vision-based inspection of industrial objects; in radar, sonar, and other defense-oriented applications; in the classification of electrocardiograms and electroencephalograms; and in medical diagnosis. The purpose of this chapter is to introduce the main concepts underlying pattern recognition theory to those explorationists not familiar with the subject. For a survey of the basic techniques and their actual and potential applications to oil and gas exploration problems, the reader is referred to de Figueiredo (1982). [A selected bibliography follows this chapter.]

By a "pattern" we mean some form or structure present in a data set. In oil and gas exploration, examples of patterns of interest are the waveshape of a seismic trace in the neighborhood of a lithological interface; the pattern exhibited by an entire set of traces, in a variable area display of a common-depth stack, in the region of a fault; the configuration of equipotential contours on a map pertaining to an aeromagnetic survey of a basin; and the shapes of signatures of one or more well logs as a function of depth. These are only a few of the many types of patterns which occur in various exploration tasks.

A machine or man-machine system whose function is to recognize patterns appearing in a given application is called a "pattern recognition system." According to the specific objectives and methodology involved in recognition, most pattern recognition systems may be classified into *decision-theoretic* and *structural*. Some systems combine features of both these types.

Decision-theoretic pattern recognition systems are mainly concerned with pattern classification. They are based on the premise that what really interests the interpreter are the events which give rise to, or are associated with, the patterns being observed. For this reason, all patterns corresponding to any given event are lumped into a pattern class, and the various pattern classes Π_1, Π_2, ..., Π_M corresponding to different events E_1, E_2, ..., E_M which may occur in the context of a given application are constituted. Then a decision-theoretic pattern recognition system is designed to

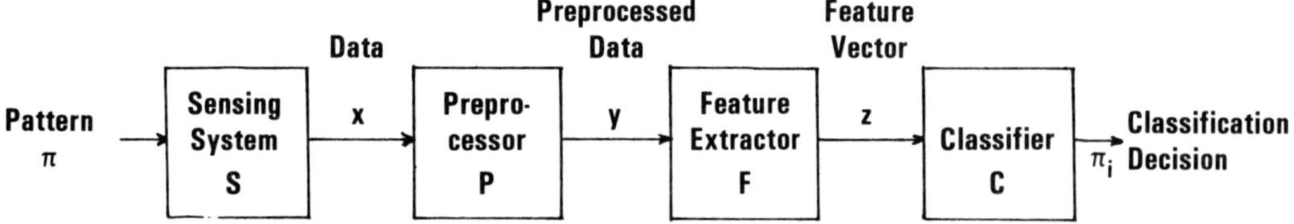

FIG. 1. Block diagram of a decision-theoretic pattern recognition system.

implement a strategy (called classification strategy) for assigning any observed pattern π to the class Π_i to which it is most likely to belong, and thus for detecting the event E_i which gave rise to it. The classification strategy is usually predicated on considerations from statistical decision theory or on geometrical considerations such as those in the partitioning of a data set in a vector space into subsets (classes) by means of separating hyperplanes as done by "perceptrons."

If the observed pattern has a complex structure, as in the case of a set of contours or surfaces delineating the boundaries of a reservoir in a two-dimensional (2-D) or three-dimensional (3-D) seismic display, then the scope of the pattern recognition problem goes beyond mere pattern classification to that of pattern analysis and interpretation. To attain the latter objective, the observed pattern is broken up into subpatterns, and each subpattern into sub-subpatterns. The procedure continues until a set of elementary patterns is reached. The goal of pattern analysis and interpretation is to develop an understanding of how the various component parts thus obtained are assembled into the observed complex pattern according to a set of meaningful rules. The *structural*[1] *pattern recognition systems* mentioned earlier are oriented toward this goal. The analysis and interpretation by these systems is performed based on formal language theory, relational structure theory, techniques from artificial intelligence, or on a mix of these approaches.

In the following, we briefly describe the fundamental concepts underlying the methodology characterizing the pattern recognition systems mentioned above and illustrate them by means of simple examples.

Decision-Theoretic Systems

The block diagram of Figure 1 shows the operations performed on the data by a typical decision-theoretic pattern recognition system. Let π denote a pattern belonging to one of M possible pattern classes Π_j, $j = 1, \ldots, M$, its class membership being unknown. A sensing system S maps the stimulus from π to a set of measurements, usually a vector $\mathbf{x}$ of high dimensionality. For example, $\mathbf{x}$ could be a waveform, continuous or sampled, or a set of waveforms, or a picture. The preprocessor P minimizes the distortion and noise present in $\mathbf{x}$ without necessarily analyzing the information that $\mathbf{x}$ contains, and yields (the vector) $\mathbf{y}$ at its output. The block F, called "feature extractor," retrieves most of the information from $\mathbf{y}$ relevant to classification. It consists of a set of preselected numerical (real-valued) functions f_i of $\mathbf{y}$, called "features," producing the components $z_i = f_i(\mathbf{y})$, $i = 1, \ldots, m$, of the

[1]The reader should be cautioned that the term "structural" used here is taken from the pattern recognition literature and is not used in the same sense as in structural geology.

"feature vector"[2] $\mathbf{z} = (z_1, \ldots, z_m)$ at the output of F. Finally, the classifier C maps $\mathbf{z}$ into the pattern class Π_i to which the observed pattern is most likely to belong, according to some appropriate decision criterion.

The above pattern recognition system may operate in a *training* or *operational* mode. In the training mode, the various functions making up the system are designed, and the parameters appearing in them are estimated, using a set of labeled sample patterns called a "training pattern set" or simply "training set." For example, the training set for a system intended to classify seismic traces might consist of seismic data from a region for which well data are available and hence whose classification is known. In the operational mode, the pattern recognition system performs automatic classification. In an *interactive* pattern recognition system, a human is in the loop, and both the above modes usually occur iteratively.

A few additional remarks on the individual blocks constituting the system of Figure 1 are in order.

The structures of the sensing system S and the preprocessor P are largely dictated by the physics of a particular application. On the other hand, the design and operation of the feature extractor F and the classifier C are amenable to analysis based (as stated earlier) on statistical or geometric considerations.

Among the general methods available for the design of the feature extractor F, we may cite those based on (1) the minimization of probability of misclassification; (2) maximization of probabilistic distances among classes; (3) factor and principal component analysis; (4) multiple discriminant analysis; (5) multidimensional scaling and nonlinear mapping procedures; and (6) cluster analysis—all of which have been amply discussed in the scientific literature.

The classifier design is often based on a strategy which minimizes the average risk of making a wrong classification decision, called the Bayes risk. In such a case, it is necessary to have enough training sample patterns available from which the statistics of the pattern structure may be gleaned. Other types of classification strategies include the so-called "nonparametric techniques," among which special mention should be made of the "nearest neighbor" and "nearest mean" classification strategies. The nearest neighbor strategy assigns to any given pattern being observed the class label of the training vector which is nearest to it, in the feature space, as illustrated in Figure 2. In the nearest mean strategy, for each class the mean of all training vectors belonging to that class is computed first. Then any new observed pattern is assigned the class label of its nearest mean training vector, as illustrated in Figure 3.

The following simple example, not related to exploration, illustrates the above concepts in some detail. Then two additional examples further illustrate the approach in the context of seismic trace event detection and well log data classification. A fourth example presents a scheme described by Hagen (1981) for porosity classification based on actual exploration data.

[2]By a vector space $\mathbf{Z}$ of dimension m, we mean the set of all m-tuples (called "vectors") $\mathbf{z} = (z_1, \ldots, z_m)$ of real numbers $z_1, \ldots, z_m$ for which the following three rules are defined. If $\mathbf{v} = (v_1, \ldots, v_m)$ and $\mathbf{w} = (w_1, \ldots, w_m)$ are any two vectors from $\mathbf{Z}$ and α is any real number, then: (1) $\mathbf{v} + \mathbf{w}$ is the vector in $\mathbf{Z}$ defined by $(v_1 + w_1, \ldots, v_m + w_m)$; (2) $\alpha\mathbf{v}$ is the vector $(\alpha v_1, \ldots, \alpha v_m)$; and (3) the (Euclidian) distance between $\mathbf{v}$ and $\mathbf{w}$ is the nonnegative number $\sqrt{(v_1 - w_1)^2 + \ldots + (v_m - w_m)^2}$. Vectors in a 2-D vector space may be represented by points on a plane, where for any vector $\mathbf{z} = (z_1, z_2)$, the numbers z_1 and z_2 represent the x- and y-coordinates of the point representing $\mathbf{z}$ on the plane.

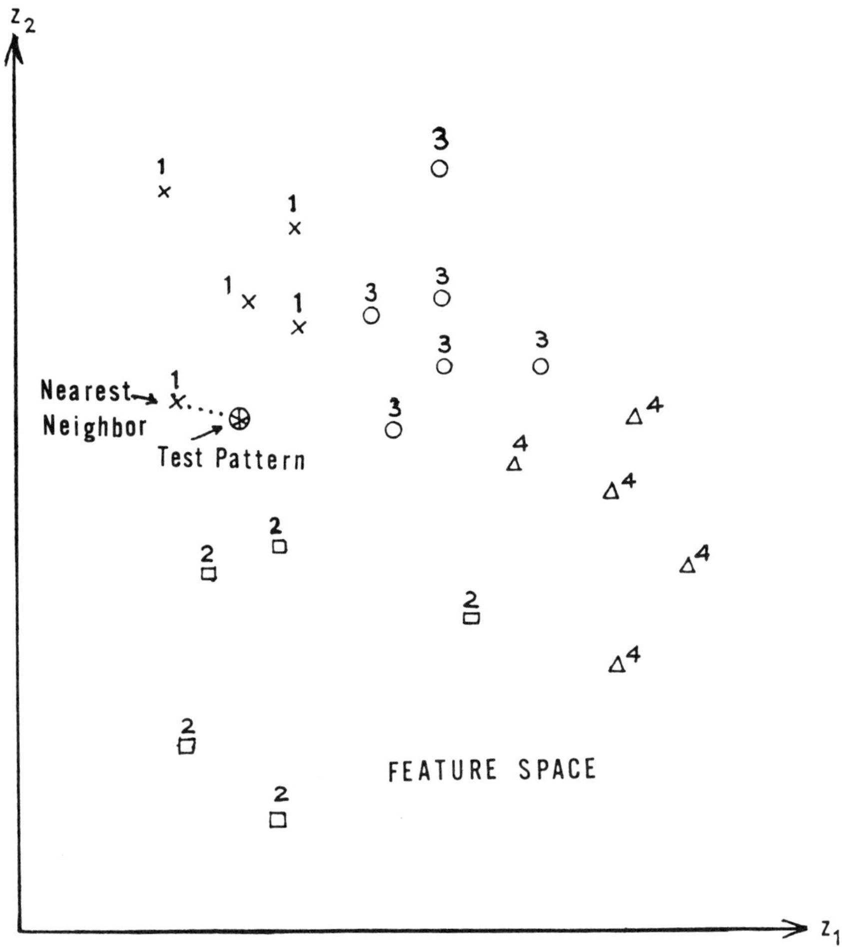

FIG. 2. Illustration of nearest neighbor classification rule. Note: Training vectors from classes Π_1, Π_2, Π_3, and Π_4 are labeled 1, 2, 3 and 4.

Example 1: Illustration of the concepts

Consider the hypothetical case of a teacher who, toward the end of the semester, wants to classify his students according to the grades which they are most likely to get in the final examination in the course that he is teaching. For this purpose, he wishes to design a machine which performs classification based on two features, namely, the student's intelligence quotient (I.Q.), denoted by z_1, and the number of hours per week that a student devotes to study the course material, denoted z_2. Thus, $\mathbf{z} = (z_1,z_2)$ is a 2-D feature vector which takes a specific vector value for each student.

In this problem, the students can be considered to be "patterns" and the events of interest $E_1, E_2, \ldots, E_5$ are the grades A (highest), B, C, D, and F (failing grade) which they may be getting in their final examination. Thus the students may be

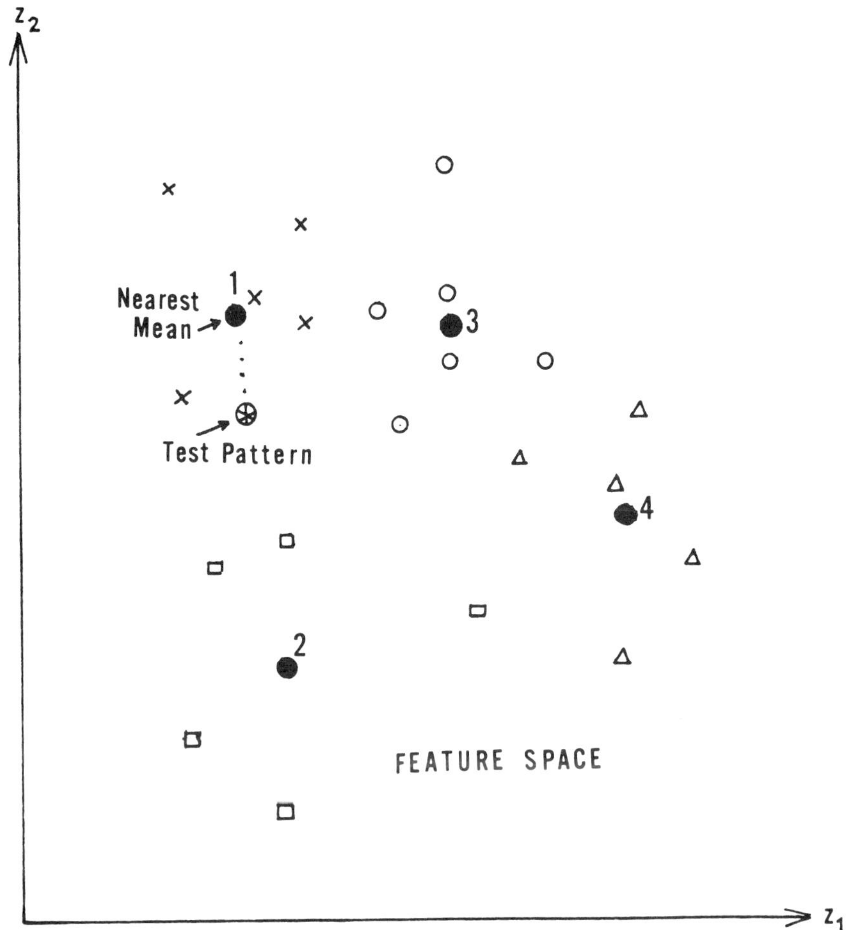

FIG. 3. Illustration of nearest mean classification rule. Note: Means of training vectors from pattern classes Π_1, Π_2, Π_3, Π_4 are labeled 1, 2, 3 and 4.

grouped into five classes Π_1, ..., Π_5 according to the event with which each will be associated. The objective of the classifier is to assign each student to the correct pattern class based on observations made on him. We assume these observations lead to the feature vector (z_1,z_2) mentioned above. The functions of the blocks *S*, *P*, and *F* of Figure 1 applied to the present problem may be identified as the various operations involved in obtaining the value of such a feature vector for each student.

In order to design the classifier, the teacher chooses a sample of 18 past students, e.g., from the class of the preceding year, and obtains from them their feature vector values as well as their final examination grades. These data are exhibited in Table 1 and also in Figure 4, where the feature vector values for the individual students are indicated by points labeled by their letter grades. The data are used as training data in the design of the classifier.

Table 1. Training data for the classifier of example 1.

Student no.	1	2	3	4	5	6	7	8	9
IQ	166	166	156	152	148	146	144	142	138
Study hours/week	2	3	2	3	5	1	4	9	1/2
Final exam grade	B	A	C	B	A	D	B	A	F
Student no.	10	11	12	13	14	15	16	17	18
IQ	136	134	130	124	122	118	116	110	104
Study hours/week	3	7	9	6	8	2	9	6	5
Final exam grade	C	B	B	C	C	D	C	D	F

The feature space of Figure 4 is partitioned into the decision regions $\Omega_1, \ldots, \Omega_5$ pertaining, respectively, to the pattern classes $\Pi_1, \ldots, \Pi_5$ according to some appropriate statistical or geometric criterion which takes into account the way the training vectors are distributed in that space. In Figure 4 we have drawn the decision region boundaries Γ_{ij}, $i = 1, \ldots, 5, j \neq i$, in such a way that complete separation of the training vectors according to their classes is achieved.

Now any student of the current semester may be classified by obtaining his or her feature vector and assigning to him or her the class label of the decision region within which the feature vector falls. For example, a student with an I.Q. of 140 who studies 5 hours per week would be assigned to the class Π_2 of students who will be getting the grade B because his feature vector (140, 5) is located in the region Ω_2.

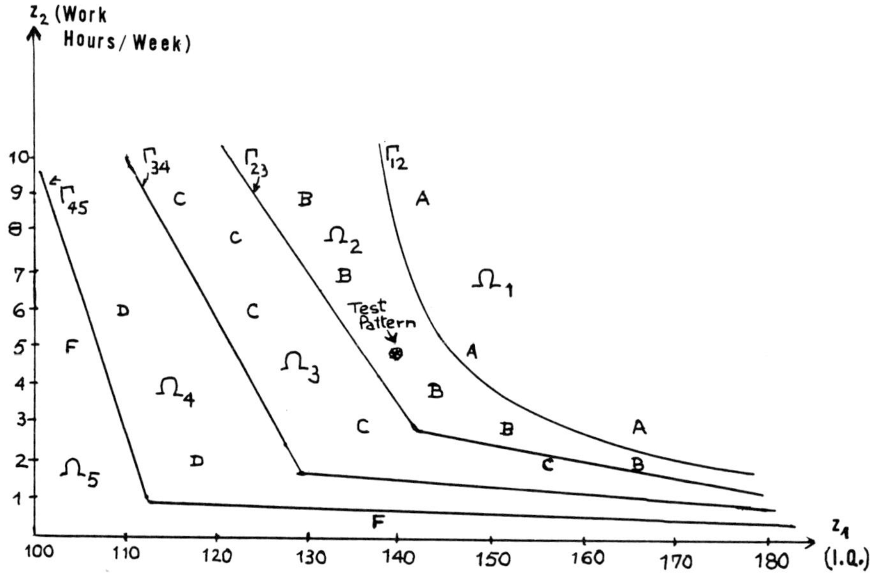

FIG. 4. Training feature vectors, decision regions, and decision region boundaries for example 1.

Actual implementation of the above classification by machine requires representing each decision region boundary by one or more equation of the form

$$d_{ij}(z_1,z_2) = 0, \tag{1}$$

where d_{ij} is a suitable real-valued function. In the case of Figure 4, we may elect to represent Γ_{12} by a quadratic polynomial and the remaining boundaries by piece-wise linear polynomials. Then the functions d_{ij} associated with each boundary are of the form:

$$\Gamma_{12}\colon d_{12}(z_1,z_2) = z_1^2 + 2\,\alpha_{12}z_1z_2 + \alpha_{22}z_2^2 + \alpha_1z_1 + \alpha_2z_2 + \alpha_3, \tag{2}$$

$$\Gamma_{23}\colon d_{23}^{(1)}(z_1,z_2) = z_1 + \beta_{12}z_2 + \beta_{11}, \tag{3a}$$

$$d_{23}^{(2)}(z_1,z_2) = z_1 + \beta_{22}z_2 + \beta_{21}, \tag{3b}$$

$$\Gamma_{34}\colon d_{34}^{(1)}(z_1,z_2) = z_1 + \gamma_{12}z_2 + \gamma_{11}, \tag{4a}$$

$$d_{34}^{(2)}(z_1,z_2) = z_1 + \gamma_{22}z_2 + \gamma_{21}, \tag{4b}$$

$$\Gamma_{45}\colon d_{45}^{(1)}(z_1,z_2) = z_1 + \delta_{12}z_2 + \delta_{11}, \tag{5a}$$

$$d_{45}^{(2)}(z_1,z_2) = z_1 + \delta_{22}z_2 + \delta_{21}, \tag{5b}$$

where the coefficients α_{12}, α_{22}, . . ., δ_{22}, δ_{21} are all obtained by approximation, or by some error correcting ("learning") algorithm in which the training samples are classified sequentially and a change in parameters is made whenever the classifier makes a wrong classification, the change correcting this error. Note also that each of the boundaries Γ_{23}, Γ_{34}, and Γ_{45} is represented by two equations corresponding to the two parts of the piece-wise linear curve which represents it.

Once functions (2) through (5b) are selected, then the decision region where any observed feature vector $\mathbf{z} = (z_1,z_2)$ lies may be determined by requiring that this point satisfy one or more inequalities involving some of those functions, as seen from Figure 4.

For example, for a point $\mathbf{z} = (z_1,z_2)$ [such as (140,5) mentioned previously] to belong to the decision region Ω_2, it must satisfy all of the following three inequalities:

$$d_{12}(z_1,z_2) < 0, \tag{6}$$

$$d_{23}^{(1)}(z_1,z_2) > 0, \tag{7}$$

$$d_{23}^{(2)}(z_1,z_2) > 0, \tag{8}$$

where the functions on the left side of expressions (6)–(8) have been defined in equations (2)–(3b).

A block diagram of a classifier which will implement the strategy indicated above is shown in Figure 5. The functions d_{ij} are called discriminant functions. In the system of Figure 5, the various discriminant functions are calculated for the observed feature vector, and their values are compared against the zero threshold. Depending upon which of these values are positive and which negative, a logic circuit is actuated to yield the label of the class to which the vector is assigned.

There are various ways of expressing discriminant functions and of thresholding them to perform the classification decision. The way selected depends heavily on the type of application under consideration and the taste of the designer.

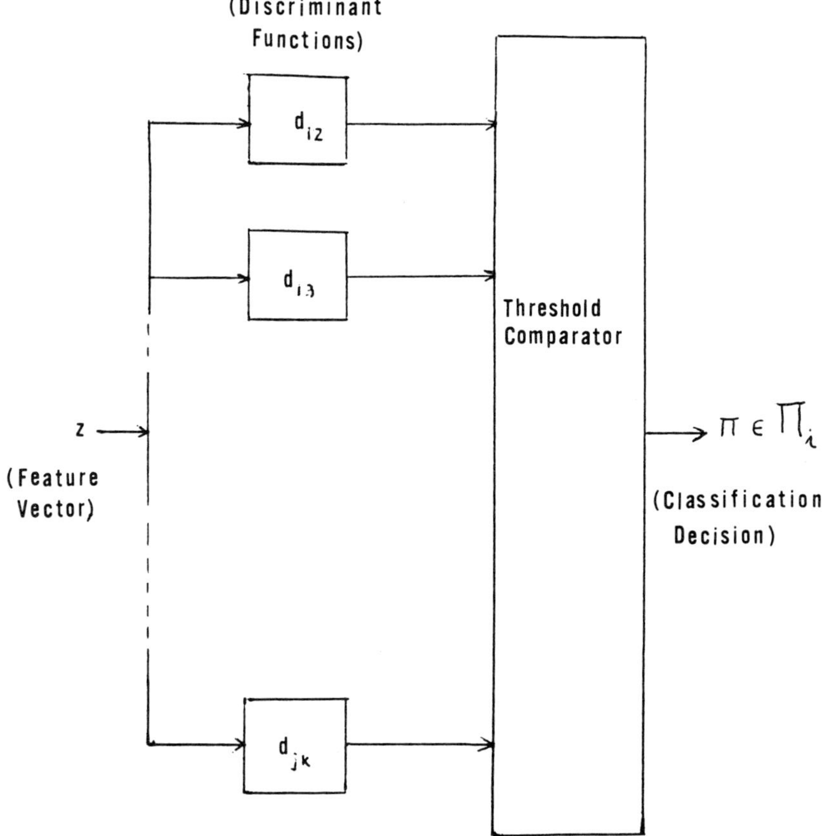

FIG. 5. Block diagram of a decision theoretic pattern classifier.

Example 2: Seismic trace event detection

Suppose that the observed pattern is the segment $y(t)$, $t_0 \leq t \leq t_1$ of a waveform, such as a segment of a trace, expressible in the form

$$y(t) = y_i(t) + w(t), \; t_0 \leq t \leq t_1 \tag{9}$$

where $w(t)$ is white Gaussian noise and $y_i(t)$ can only assume one of the two possible shapes $y_1(t)$ or $y_2(t)$. Depending upon which of these hypotheses is true, we create, respectively, the pattern classes Π_1 and Π_2 to which $y(t)$, $t_0 \leq t \leq t_1$ may belong. Figure 6 shows an example adapted from Dedman et al (1975) in which $y_1(t)$ and $y_2(t)$ are the shapes of a propagating wavelet, respectively, at shale/water-sand and gas-sand/water-sand interfaces.

The classification scheme, known to engineers as a ''correlation detector,'' which classifies $y(t)$, $t_0 \leq t \leq t_1$, with the least probability of error is depicted in Figure 7. Two features z_1 and z_2 are extracted from the observed waveform segment by correlating it, respectively, with each of its two possible shapes, i.e.,

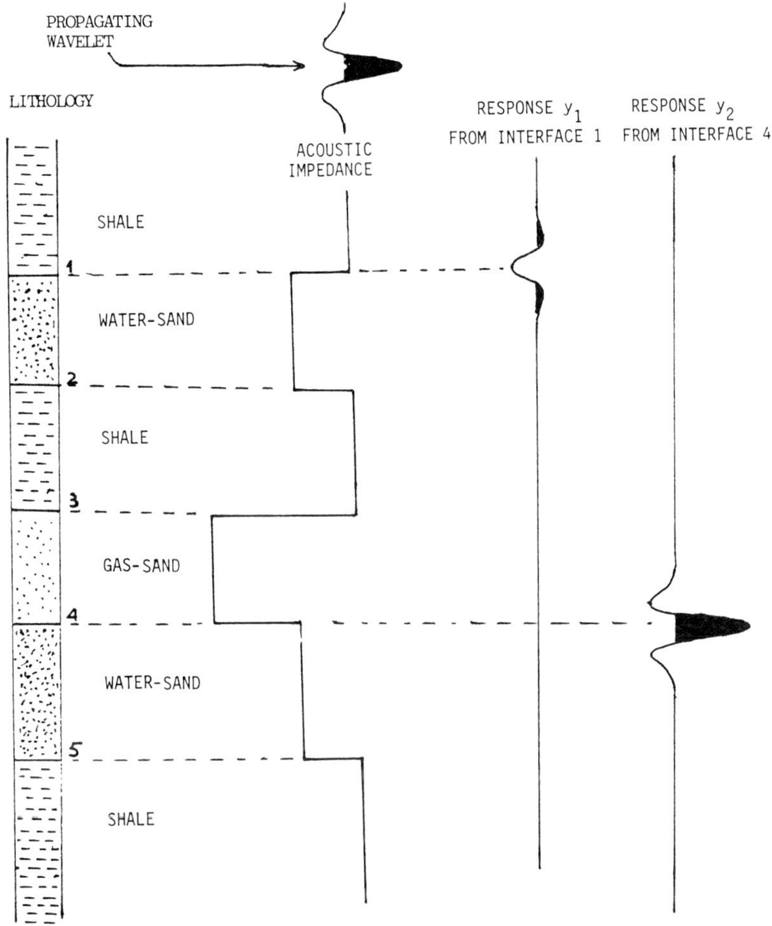

FIG. 6. Patterns in wavelet response at different types of lithological interfaces.

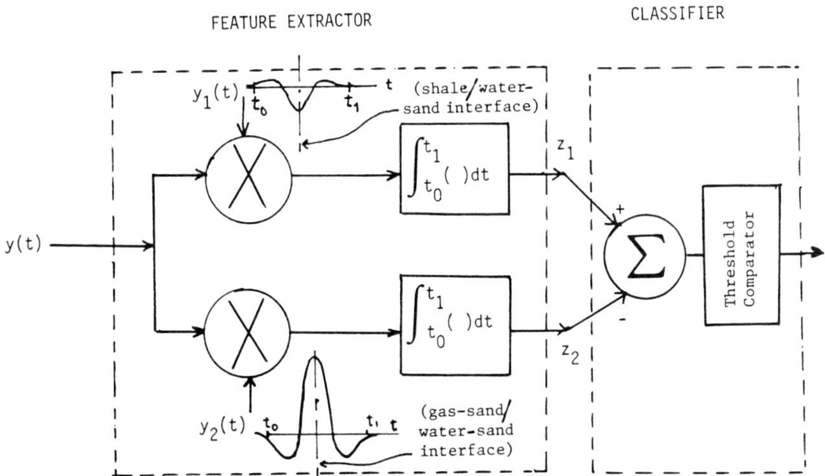

FIG. 7. Seismic trace event classifier.

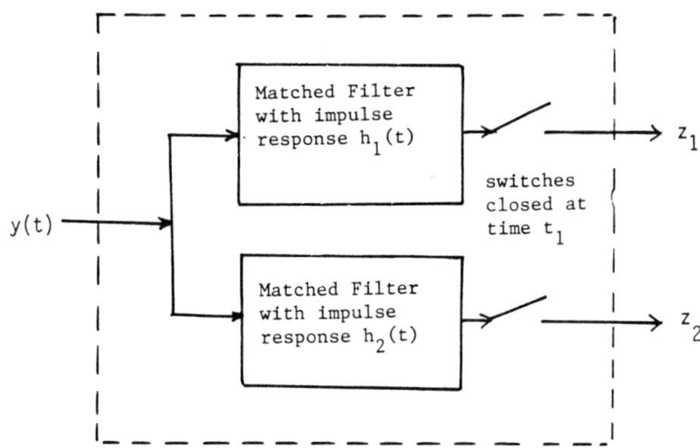

FIG. 8. Matched filter implementation of the feature extractor of Figure 7.

$$z_i = \int_{t_0}^{t_1} y_i(t)\, y(t)\, dt, \quad i = 1, 2. \tag{10}$$

Then the classifier compares the difference $z_1 - z_2$ with an appropriate threshold γ, classifying the observed waveform segment as belonging to the class Π_1 if

$$d_{12}(z_1,z_2) = z_1 - z_2 - \gamma > 0 \tag{11}$$

and to Π_2 otherwise.

The feature extractor of Figure 7 is commonly implemented using the matched filter scheme of Figure 8, where $h_i(t) = y_i(t_1 - t)$, $i = 1, 2$, are the impulse responses of the two filters shown.

The above approach is readily extendable to the M pattern class case.

In seismic interpretation, this approach may be used to address the general question: Is an event appearing in some trace also present somewhere else in the same trace or in some other trace?

Example 3: Well log interpretation

Some well log interpretation problems, such as gas detection, may be posed as decision-theoretic pattern recognition problems.

Consider the problem of detecting the presence of gas, oil, water, or none of these entities, from a set of open hole logs. Specifically, we use three porosity logs: neutron log $x_1(t)$, acoustic log $x_2(t)$, and density log $x_3(t)$, in addition to the gamma-ray log $x_4(t)$. Here, t denotes the depth at which the log readings are taken. We now have a four-class (Π_1, gas; Π_2, oil; Π_3, water; Π_4, none) pattern recognition problem which can be resolved using the decision-theoretic pattern recognition system of Figure 1.

For any fixed t, the four-dimensional vector $\mathbf{x}(t) = [x_1(t), x_2(t), x_3(t), x_4(t)]$ corresponds to the data vector $\mathbf{x}$ of Figure 1. The various corrections applied to the log data (see the earlier chapter by Jain) constitute the preprocessing done by the block P of Figure 1. Let $\mathbf{y}(t) = [y_1(t), y_2(t), y_3(t), y_4(t)]$ denote the output of the preprocessor. We can combine the functions of the blocks F and C and construct a set of four discriminant functions $d_i[\mathbf{y}(t)] = d_i[y_1(t), y_2(t), y_3(t), y_4(t)]$, $i = 1, 2, 3, 4$, which will classify the data as follows.

If

$$d_{1j}[\mathbf{y}(t)] \stackrel{\Delta}{=} d_1[\mathbf{y}(t)] - d_j[\mathbf{y}(t)] > 0, \quad j = 2, 3, 4, \tag{12}$$

classify the vector $\mathbf{y}(t)$ as arising from gas (pattern class Π_1). If

$$d_{2j}[\mathbf{y}(t)] \stackrel{\Delta}{=} d_2[\mathbf{y}(t)] - d_j[\mathbf{y}(t)] > 0, \quad j = 1, 3, 4, \tag{13}$$

classify $\mathbf{y}(t)$ as resulting from oil (pattern class Π_2). If

$$d_{3j}[\mathbf{y}(t)] \stackrel{\Delta}{=} d_3[\mathbf{y}(t) - d_j[\mathbf{y}(t)] > 0, \quad j = 1, 2, 4 \tag{14}$$

classify $\mathbf{y}(t)$ as arising from water (pattern class Π_3). If equations (12) through (14) do not hold, classify $\mathbf{y}(t)$ as resulting from some event other than the above three.

The classifier design problem now reduces to that of finding suitable expressions for the above discriminant functions. They can be linear, piece-wise linear, or nonlinear. The parameters in these expressions have to be determined by some curve-fitting or error-correcting algorithm as pointed out in example 1. For best results, the expressions used and the values of the parameters have to be conditioned on the type of formation (e.g., shale, uncompacted sand, compacted sand, shaly sand, . . .) prevailing at depth t.

Example 4: Porosity classification from seismic/well log data

As an example of an application of the pattern recognition approach to interpretation of actual exploration data, we briefly describe the contribution of Hagen (1981) to the classification of regions of a seismic section into porous and nonporous. More generally, as we explain a little further on, Hagen (1981) showed how the probability for each class (porous or nonporous) may be displayed by appropriate shading of the seismic section.

The seismic data used, consisting of a common-depth-point (CDP) stack from 1.4 to 1.8 sec, are shown in Figure 9. The particular objective was to partition a horizontal strip, in the depth interval I of 1.66 to 1.68 sec into porous and nonporous regions.

As shown in the figure, five wells penetrated the strip of interest, two of which were nonporous. The trace segments labeled $\mathbf{y}^1$ and $\mathbf{y}^2$ at shotpoints 214 and 230 were used as prototypes of porous and nonporous classes, labeled Π_1 and Π_2. The traces at well locations 153, 181, and 194 were used as test traces. The nearest-neighbor decision rule was used to classify these test trace segments as well as the remaining trace segments in the strip. The length of the sampling interval used was 2 msec; hence there were 10 samples associated with each trace segment in the strip.

Specifically, each trace segment $\mathbf{y}$ was represented as a ten-dimensional vector of instantaneous frequencies, i.e., as the set of values of instantaneous frequencies at the sampling points obtained from a Hilbert transform representation of the trace. By experimentation, the use of such a representation was found to enhance class discrimination. For m (a preselected positive integer less than 10), let $z_1, \ldots, z_m$

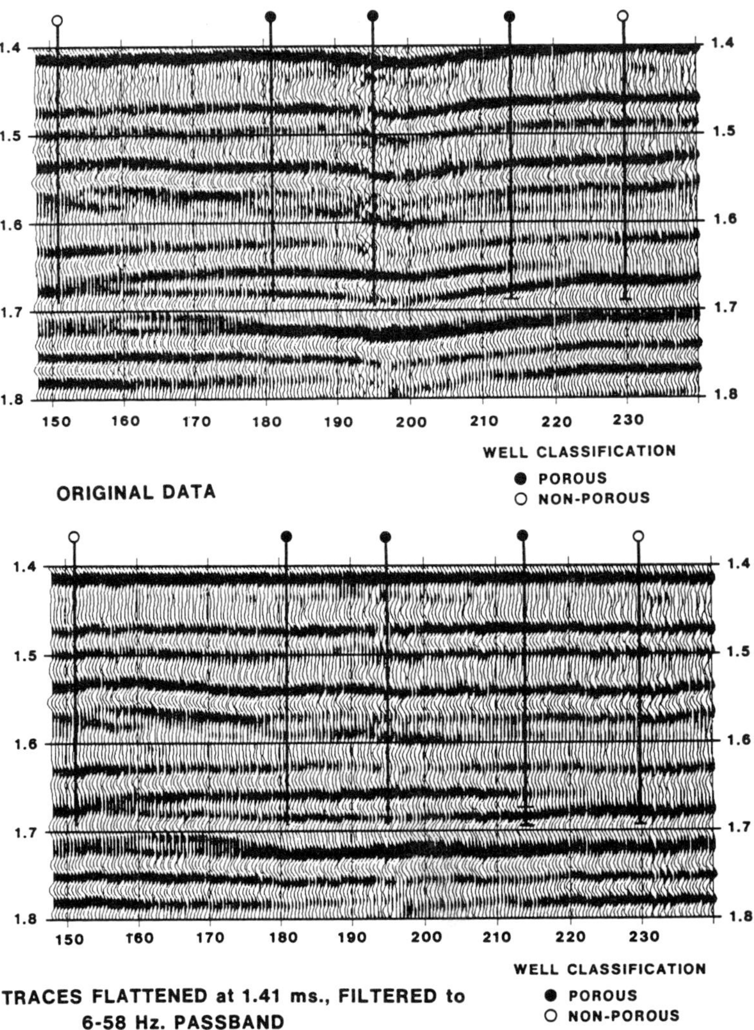

FIG. 9. Seismic data used by Hagen (1981).

denote the first m principal components[3] of a given vector $\mathbf{y}$. With these components the m-dimensional feature vector $\mathbf{z} = (z_1, \ldots, z_m)$ associated with $\mathbf{y}$ was formed. Let $\mathbf{z}^1 = (z_1^1, \ldots, z_m^1)$ and $\mathbf{z}^2 = (z_1^2, \ldots, z_m^2)$ denote the feature vectors associated with

[3]Let $\{\mathbf{y}\}$ be the set of vectors representing the trace segments in the entire strip under consideration. Their scatter is measured by their sample covariance matrix $\mathbf{K}$. Note that $\mathbf{K}$ is an $n \times n$ positive definite matrix, where n in this case is 10. Let $\lambda_1, \ldots, \lambda_n$ be the eigenvalues of $\mathbf{K}$ listed in order of decreasing values, and denote the corresponding normalized eigenvectors of $\mathbf{K}$ by $\mathbf{u}_1, \ldots, \mathbf{u}_n$. Then $\mathbf{u}_1$ is in the direction along which the scatter in the data is the highest; and $\mathbf{u}_2, \mathbf{u}_3, \ldots, \mathbf{u}_n$ are in the order of decreasing scatter in their respective directions. The component (projection) z_1 of any given $\mathbf{y}$ along $\mathbf{u}_1$ is called the first principal of $\mathbf{y}$. Similarly z_k, $k = 2, 3, \ldots, n$, the component of $\mathbf{y}$ along $\mathbf{u}_k$, is called the kth principal component of $\mathbf{y}$. It follows that the first m principal components $z_1, \ldots, z_m$ of $\mathbf{y}$ lie in an m-dimensional subspace of the n-dimensional data space, in which the variability of the data with the pattern classes is highest, compared to all other subspaces of dimension m. In this sense, the variables $z_1, \ldots, z_m$ thus chosen are considered to be the best features for the classification problem under discussion.

the prototype trace segments $\mathbf{y}^1$ and $\mathbf{y}^2$. A feature vector $\mathbf{z}$ (corresponding to a given $\mathbf{y}$) was classified as belonging to the class Π_1 if its distance to $\mathbf{z}^1$ was shorter than its distance to $\mathbf{z}^2$, and as belonging to the class Π_2 otherwise. In these calculations, the distance between any two feature vectors $\mathbf{z} = (z_1, \ldots, z_m)$ and $\tilde{\mathbf{z}} = (\tilde{z}_1, \ldots, \tilde{z}_m)$ was defined as the modified Euclidean distance

$$d(\mathbf{z}, \tilde{\mathbf{z}}) = \left[\sum_{k=1}^{m} \frac{1}{\lambda_k} |z_k - \tilde{z}_k|^2 \right]^{1/2}, \tag{15}$$

where λ_k is the eigenvalue, corresponding to the kth principal component, of the sample covariance matrix associated with the set of traces $\mathbf{y}$ in the entire strip.

Hagen (1981) calculated the probability of $\mathbf{z}$ belonging to class Π_1 (and similarly in the case of class Π_2) by the formula

$$P(\Pi_1/z) = \frac{\dfrac{1}{d(\mathbf{z},\mathbf{z}^1)}}{\dfrac{1}{d(\mathbf{z},\mathbf{z}^1)} + \dfrac{1}{d(\mathbf{z},\mathbf{z}^2)}}. \tag{16}$$

Figure 10 shows the result of one of the computer runs, for $m = 1, 2, 3$, and 5,

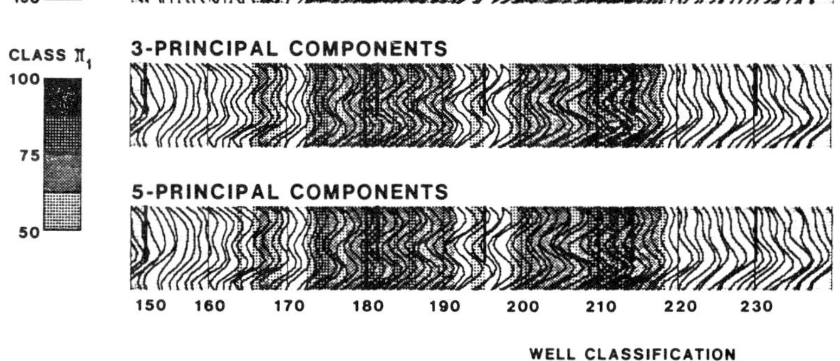

FIG. 10. Display of classification results of Hagen (1981).

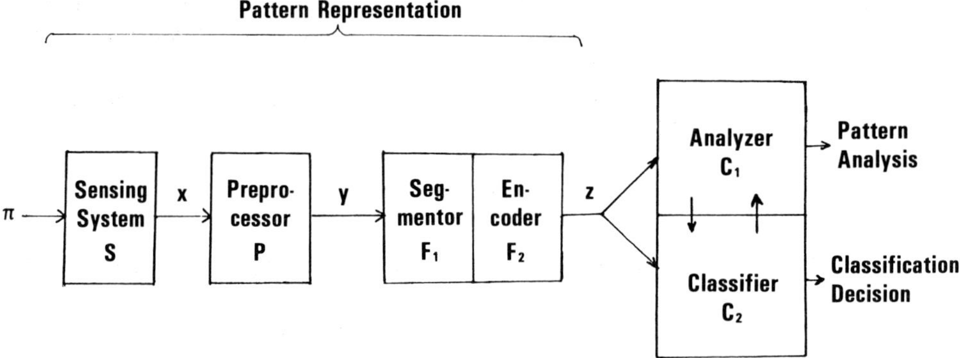

FIG. 11. Block diagram of a structural pattern recognition system.

where different values of the probabilities of class membership $P(\Pi_1/z)$ and $P(\Pi_2/z)$ are indicated by appropriate shading (varying from very dark for porous to very light for nonporous). The test trace segments at 151 and 181 were assigned to their correct classes with high probability. The classification of the test trace segment at 194 was less certain, presumably due to a recording anomaly in the area.

We have briefly introduced to the reader who may not be familiar with the subject the fundamental notions involved in the problem formulation and design of decision-theoretic pattern recognition systems and their relevance to problems of classification of exploration data. We now proceed to consideration of the second general type of pattern recognition systems.

Structural Pattern Recognition Systems

As indicated earlier, the simple vector description of patterns described thus far has been superseded during the past decade by the more elaborate pattern description methods from structural pattern recognition. These methods provide a framework for analysis and understanding of the structure, meaning, and relationships present in complex patterns. For this reason, they ought to be particularly appealing to the interpreter of complex patterns exhibited by signatures and by 2-D and 3-D configurations pertaining to exploration data.

The configuration of a structural pattern recognition system is depicted in Figure 11. The sensor and preprocessor blocks S and P are the same as before. The feature extractor stage of Figure 1 is now replaced by the system F consisting of the blocks F_1 and F_2. The block F_1 is a segmentor which breaks up the preprocessed data set $\mathbf{y}$ into elementary parts, called pattern primitives. The block F_2 is an encoder which represents these elementary parts as symbols (also called pattern primitives) from a finite alphabet, arranged as a set z possessing some structure. Thus z could be a string or an appropriate relational graph. z is called a *representation* of the pattern π, and is sometimes referred to, in abbreviated form, as the "pattern z."

The last stage C consists of an analyzer C_1 and a classifier C_2, which, respectively, analyze and classify z according to some appropriate set of rules, usually with some back-and-forth exchange of information between C_1 and C_2.

The three subcategories of structural pattern recognition systems mentioned previously will now be described.

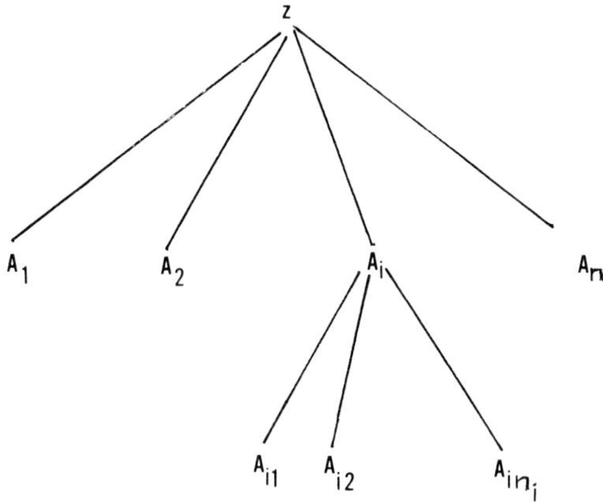

FIG. 12. Display of the hierarchical structural of a pattern z.

Syntactic/semantic systems

In the syntactic pattern recognition approach, a pattern z is viewed as possessing a hierarchical structure in the way indicated in Figure 12. In other words, z is regarded as an arrangement of subpatterns $A_1, \ldots, A_n$, these not being necessarily distinct. Any one or more of these subpatterns A_i may in turn consist of a set of lower level subpatterns $A_{i1}, \ldots, A_{in_i}$. This process continues until one reaches the subpatterns of the lowest level, which we denote by $a_1, \ldots, a_m$. These lowest level subpatterns $a_1, \ldots, a_m$ are what we earlier called pattern primitives. This hierarchical organization of the pattern z in the form of a tree with subpatterns at its nodes—in which sets of subpatterns at a given level are obtained from their immediate ancestors by means of appropriate substitution (production) rules—is the key idea behind syntactic description of patterns.

Formally, the above approach leads to modeling of the pattern z as a "sentence" from a formal language $L(G)$, where G is a phrase-structure grammar which generates the language $L(G)$. The sentence z is then a string $a_1a_2 \ldots, a_m$ of primitives, called "terminals." These terminals are grouped into sets, these sets being generated from intermediate variables, called "nonterminals," by means of "production rules" pertaining to the grammar G. Thus (referring again to Figure 12) in the linguistic approach, the structure of every pattern is characterized by some tree, which has a single node, denoted by the start symbol S as its root. It branches into intermediate nodes corresponding to nonterminal variables (subpatterns) and ends in a set of terminal nodes corresponding to the terminals constituting the string z.

There are a number of ways a pattern recognition problem can be formulated in the above setting.

One way is to model each pattern class Π_i, $i = 1, \ldots, M$, by a different grammar G_i, $i = 1, \ldots, M$. Then, given a pattern z, the pattern classification problem reduces to finding the language $L(G_i)$ to which z belongs.

Another way of approaching the problem is to characterize the pattern classes Π_i, $i = 1, \ldots, M$, by prototype sentences $z^1, \ldots, z^M$ from a language $L(G)$ to which z belongs, and classify z by a nearest-neighbor algorithm on the basis of those prototypes. In such a case, an appropriate distance function must be selected in the "space" of sentences.

Yet another way of solving the problem is to model z as a sentence from a stochastic language and use a probabilistic decision scheme for the classification of z.

The merit of the formal language approach is that it not only permits *classification* but also provides an *analysis* or *interpretation* (description of the hierarchical configuration of the pattern in terms of subpatterns) of the pattern being observed. The analysis portion of the recognition process illuminates one's understanding of the way information is organized within the pattern. For this reason, the linguistic approach can be of special value in the recognition and understanding of patterns that are very complex such as those in geologic and geophysical prospects in petroleum exploration. However, in many instances the amount of processing this approach entails is very large; hence much of the current research in the field is directed toward increasing the efficiency of processing.

Note that in some problems, the decision-theoretic approach described earlier may be combined with partial linguistic analysis simply to eliminate the options that one would have to consider if the linguistic analysis were not carried out.

Finally, the power of the formal language approach described above can be enhanced by introducing semantics into the framework by the use of attributed grammars in the linguistic models.

One of the first successful applications of syntactic pattern recognition was made by Ledley (1964) in the classification of chromosomes into "submedian" and "telocentric" from microscope photographs. Ledley developed a context-free grammar to classify each type based on the shape of the contour delineating the boundary of the chromosome. Figure 13 shows the primitives (arcs) of which the boundaries of both types are made, as well as how, in Ledley's formal language, the boundary of each chromosome is represented as a juxtaposition of appropriate primitives.

It is of considerable interest to use a similar formal language approach to recognize and interpret contours present in seismic sections. In this connection, we note that a seismic section may be regarded as an image whose gray levels consist of trace values or values of some trace transform variable such as the instantaneous frequency. If a subsurface hydrocarbon trap, such as an anticline, fault, salt dome, pinchout, reef, or stratigraphic trap is present, its boundaries may be extracted from such a representation of the section in the form of a set of contours, by appropriate edge detection and thinning algorithms. This and segmentation of the contours into primitive arcs can be realized in much the same way as that mentioned in connection with the chromosome pictures. The question of how best to execute edge detection and primitive extraction, under the specific imaging conditions prevailing in this

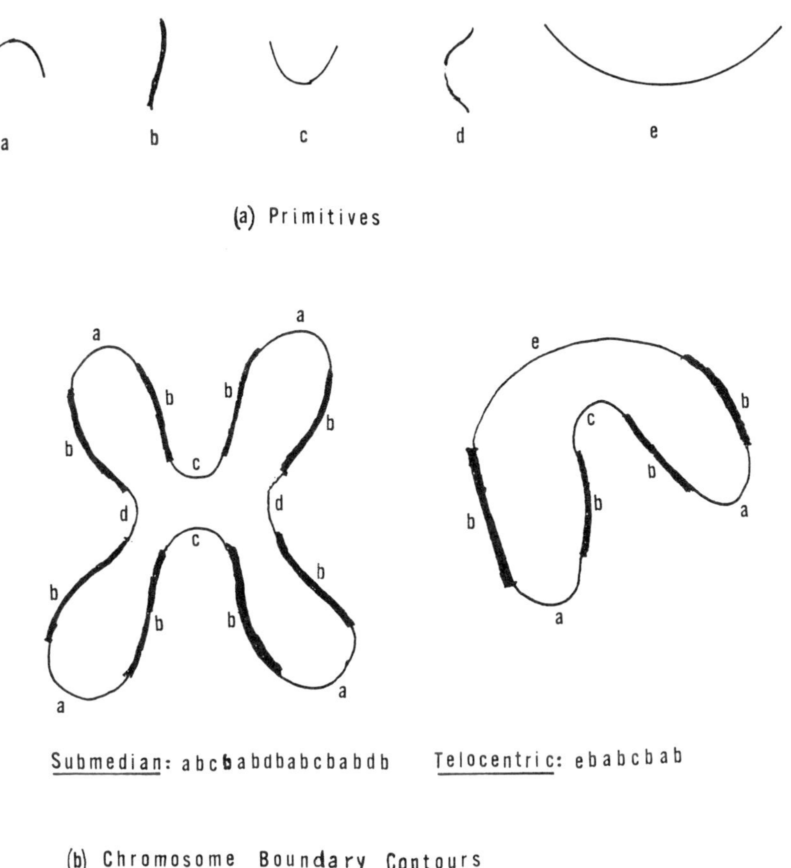

FIG. 13. Primitives and sentences pertaining to the submedian and telocentric chromosomes of Ledley's (1964) grammar.

problem, is under active investigation in seismic research laboratories. For simple examples of how formal languages, and in particular Shaw's (1970) picture description language, may be applied to such problems, the reader is referred to deFigueiredo (1982).

Systems based on relational structures

While the preceding pattern recognition systems have a number of attractive features, even greater flexibility and descriptive power can be achieved using the relational structure approach.

In most cases, a relational structure may be represented by a graph, with the nodes of the graph representing the various objects present in a complex pattern and the arcs representing the relationships existing between pairs of objects. In addition, with each node one can associate a set of properties (real variables) describing the properties of the object represented by that node.

Relational structures are used to describe and interpret complex patterns present in 2-D and 3-D scenes, such as how objects of interest are configured in the scene and how these configurations relate to other known configurations. There is considerable research activity in the pattern recognition area directed toward the analysis and use of relational structures. Its influence ought to be felt in the future on the interpretation of 2-D and 3-D seismic displays by computers.

Artificial intelligence approach

While the techniques from the field of artificial intelligence (AI) cover a wide range, a specific approach that appears promising is the one applied to the analysis of mineral exploration problems by members of the Stanford Research Institute (SRI) AI group (Duda et al, 1976, 1978; Hart et al, 1978). One of the products of this approach has been the program for computer-based consultation for mineral exploration called Prospector (Hart, et al, 1978). According to its creators, work on this program was partly motivated by the success achieved by the then-existing computer-based consultation programs for medical diagnosis such as Mycin (Shortliffe, 1976). In fact, these systems have revealed levels of competence in their performance rivaling those of expert physicians.

It is interesting to consider the application of approaches similar to that in the Prospector to oil and gas exploration problems. In such an application, knowledge of various types of formations, collected and organized by experts, would be encoded and stored in a computer in the form of models; then these models would be used by an interactive man-machine system to interpret the field data pertaining to a current prospect.

In short, one may state that AI approaches such as the one outlined above provide a methodology for (1) organizing and using very large data bases pertaining to exploration, (2) incorporating the experience and opinion of experts in the construction of models of pertinent knowledge, (3) updating these models on the basis of new evidence, and (4) implementing the interpretation strategy by an interactive computer program. For these reasons, AI approaches are likely to play an increasingly significant role in the interpretation of exploration data by computer-based systems.

Conclusion

Some of the fundamental concepts from pattern recognition theory have been presented with the view of their application to petroleum exploration problems. It is believed that they can provide a framework for the transfer, and in fact for the enhancement, of many of the interpretation functions presently carried out by humans to machines and man-machine systems. It is also presumed that these developments will receive a great incentive from the astonishing strides currently being made by computer technology.

REFERENCES

Dedman, E. V., Lindsey, J. P., and Schramm, M. W., Jr., 1975, Stratigraphic modeling: A step beyond bright spot: World Oil, May.

de Figueiredo, R. J. P., 1982, Pattern recognition and interpretation in oil and gas exploration: Rice Univ. Tech. Rep., EE82-8201.

Duda, R. O., and Hart, P. E., 1973, Pattern classification and scene analysis: New York, Wiley-Interscience.

Duda, R. O., Hart, P. E., and Nillson, N. J., 1976, Subjective Bayesian methods for rule-based inference systems: AFIPS Conf. Proc., v. 45, p. 1075–1082.

Duda, R. O., Hart, P. E., Nillson, N. J., and Sutherland, G. L., 1978, Semantic network representations in rule-based inference systems, *in* Pattern-directed inference systems: D. A. Waterman and F. Hayes-Roth, Eds., New York, Academic Press, p. 203–221.

Hagen, D. C., 1981, The application of principal component analysis to seismic data sets: Proc. 2nd Intl. Symp. on computer aided seismic analysis and discrimination, C. H. Chen, Ed., IEEE no. 81CH1687-3, p. 98–109.

Hart, P. E., Duda, R. O., and Einaudi, M. T., 1978, PROSPECTOR—A computer-based consultation system for mineral exploration: Math. Geol. v. 10, p. 589–610.

ICPR-V, 1980, Proc. of the 5th Intl. Conf. on Pattern Recognition (2 vol.): IEEE no. 80CH1499-3, December 1–4.

Ledley, R. S., 1964, High-speed automatic analysis of biomedical pictures: Science, v. 146, p. 216–223.

Shaw, A. C., 1970, Parsing of graph-representable pictures: J. ACM, v. 3, p. 453–481.

Shortliffe, E. H., 1976, Computer-based medical consultations—MYCIN: New York, American Elsevier.

REFERENCES FOR GENERAL READING

Bois, P., 1972, Analyse sequentielle: Geophys. Prosp., v. 20, p. 497–513.

——— 1976, Reconnaissance des horizons sismiques par analyse factorielle discriminante: Geophys. Prosp., v. 24, p. 696–718.

——— 1980, Autoregressive pattern recognition applied to the delimitation of oil and gas reservoirs: Geophys. Prosp., v. 28, p. 572–591.

Brown, A. R., 1979, 3-D seismic survey gives better data: Oil and Gas J., November 5.

Fu, K. S., 1974, Syntactic methods in pattern recognition: New York, Academic Press.

Hanson, A., and Riseman, E., Eds., 1979, Computer vision systems: New York, Academic Press.

Hartigan, J. A., 1975, Clustering algorithms: New York, John Wiley.

Hemon, Ch., and Mace, D., 1978, Essai d-une application de la transformation de Karhunen-Loeve aut traitement sismique: Geophys. Prosp., v. 26, p. 600–626.

Khattri, K., Sinvhal, A., and Awasthi, A. K., 1979, Seismic discriminants of stratigraphy derived from Monte Carlo simulation of sedimentary formation: Geophys. Prosp., v. 27, p. 168–195.

Knuth, D. E., 1968, Semantics of context-free languages: J. Math. Syst. Theory, v. 2, p. 127–146.

Mathieu, P. G., and Rice, G. W., 1969, Multivariate analysis used in the detection of stratigraphic anomalies from seismic data: Geophysics, v. 34, p. 507–515.

McQuillin, R., Bacon, M., and Barclay, W., 1979, An introduction to seismic interpretation: Houston, Gulf Publishing Co.

Payne, C. E., Ed., 1977, Seismic stratigraphy—Applications to hydrocarbon exploration: AAPG memoir 26, Tulsa.

Pratt, W. K., 1978, Digital image processing: New York, Wiley-Interscience.

Robinson, E. A., and Treitel, S., 1980, Geophysical signal analysis: Englewood Cliffs, N.J., Prentice-Hall.

Schlumberger, 1972, Log interpretation—v. I: Principles: New York, Schlumberger Ltd.

Tou, J. T., and Gonzalez, R. C., 1974, Pattern recognition principles: Reading, MA, Addison-Wesley.

Tucker, P., and Yorston, H., 1973, Pitfalls in seismic interpretation: SEG, monograph no. 2, Tulsa.

Waters, K. H., and Rice, G. W., 1975, Some statistical and probabilistic techniques to optimize the search for stratigraphic traps on seismic data: Proc. of the World Petroleum Congress, no. 9, Tokyo.

INDEX